THE ESSENTIALS
OF FINITE ELEMENT
MODELING AND ADAPTIVE
REFINEMENT

THE ESSENTIALS OF FINITE ELEMENT MODELING AND ADAPTIVE REFINEMENT

FOR BEGINNING ANALYSTS TO ADVANCED RESEARCHERS IN SOLID MECHANICS

JOHN O. DOW

UNIVERSITY OF COLORADO AT BOULDER

MP MOMENTUM PRESS

MOMENTUM PRESS, LLC, NEW YORK

The Essentials of Finite Element Modeling and Adaptive Refinement: For Beginning Analysts to Advanced Researchers in Solid Mechanics
Copyright © Momentum Press®, LLC, 2012.

First published by Momentum Press®, LLC
222 East 46th Street, New York, NY 10017
www.momentumpress.net

ISBN-13: 978-1-60650-332-4 (hard back, case bound)
ISBN-10: 1-60650-332-4 (hard back, case bound)
ISBN-13: 978-1-60650-334-8 (e-book)
ISBN-10: 1-60650-334-0 (e-book)

DOI: 10.5643/9781606503348

Cover design by Jonathan Pennell
Interior design by Exeter Premedia Services Private Ltd.,
Chennai, India

10 9 8 7 6 5 4 3 2 1

Printed in Canada

Contents

PREFACE

As the title declares, this book is largely concerned with finite element modeling and the improvement of these models with adaptive refinement. The intended audience for this book consists of readers who are either early in their technical careers or mature users and researchers in computational mechanics.

The width of the intended audience for this book makes it unique and presents an interesting challenge to the author. There is only one way a technical book (versus a popularization of a technical topic) can address such a wide audience. The book must present developments **that simplify, improve, and extend the discipline.**

This book contains material that accomplishes these ends. The finite element method is simplified, improved, and extended. Furthermore, the finite element and finite difference methods are unified as a result of these developments. These capabilities are used here to simplify, improve, and extend the adaptive refinement process.

These capabilities derived from earlier work done by the author, his colleagues, and graduate students in which the continuum properties are extracted from frames and trusses. This was done so it could be determined, with a minimum of analysis, whether the torsion and flexure was uncoupled in lattices for the space station in order to reduce the effect from the docking of the Space Shuttle. In other words, this process inverted the finite element method.

Later, the author recognized that the process for extracting continuum properties for skeletal structures could be inverted to produce an approach that both simplified and improved the formulation of finite element stiffness matrices. In fact, this new approach for forming finite element stiffness matrices renders the isoparametric formulation procedure obsolete as is discussed in Chapter 3.

The key to this new formulation procedure is the recognition that the displacement interpolation functions are actually truncated Taylor series expansions whose coefficients could be expressed in terms of physically interpretable quantities. Later still, it was recognized that finite difference templates or operators were embedded in the new element formulation process. These observations provide the basis for this book and make its contents accessible for the breadth of the intended audience.

Recently, this work was extended to create new types of error estimators and refinement guides. In addition to being easy to compute because they are point-wise quantities, the new error estimators report the error in terms of quantities that are basic to continuum mechanics.

One of the new error estimators quantifies the errors in terms of strains. The other new error estimator reports the errors with a metric related to the applied loads. The evaluation of an element in terms of quantities of direct interest in continuum mechanics contrasts with the strain energy metric used in the well-known error estimator developed by Zienkiewicz and Zhu (Z/Z), which is of secondary interest in analysis.

The new *in situ* refinement guides differ from previous refinement guides because they are based on physical principles instead of correlations with error estimates. In the new approach, the modeling deficiencies in individual elements are identified by comparing the modeling capabilities of the elements to an estimate of the exact solution over the domain of the element using physically interpretable notation.

This book integrates these recent developments to adaptive refinement with earlier improvements to the finite element and finite difference methods. One-dimensional problems are used to illustrate the underlying concepts. This is done so the ideas are not submerged in the large amount of data produced in multidimensional problems. As a result, the presentation is compact and the basic ideas are easy to understand.

Most of the background material presented here for the one-dimensional case is available in higher dimensions in my previous book entitled, *A Unified Approach to the Finite Element Method and Error Analysis Procedures*. Only the recent improvements to error estimators and refinement guides are not included in the previous book. Extensions of these topics are identified as possible research avenues in the final section of the concluding chapter of this book.

This book is accessible to someone new to finite element analysis because it contains the necessary background information from the following topics:

1. Taylor series expansions from undergraduate calculus.
2. The idea of basis functions from linear algebra.
3. Definition of rigid body motions from linear elasticity.
4. Definition of strains from mechanics of materials.
5. Integration over an area using Green's theorem.
6. Solution of simultaneous equations.

This book is useful to advanced researchers because it presents the following topics from a fresh point of view:

1. An element formulation process that makes the isoparametric approach obsolete.
2. Procedures for evaluating the modeling capabilities of individual elements during the formulation procedure.
3. Improvements and extensions to the finite difference method.
4. New types of error estimators.
5. A new *a posteriori* approach for identifying the level of refinement needed to rapidly improve finite element models.
6. The identification of Gauss points in terms of Chebyshev polynomials.

A synopsis of each of the 12 Chapters that comprise this book follows.

CHAPTER 1:   INTRODUCTION

Chapter 1 provides an overview of the book and major topics contained in the book, namely, error analysis and adaptive refinement. The two problems facing an analyst using an approximate solution technique are identified in the first section as: (1) the need to assess the accuracy of the approximate solution and (2) the identification of modifications to the model that will produce results of the desired accuracy. The overall objectives and the specific tasks that are required to produce satisfactory results with the finite element method are presented. The deficiencies in a finite element model are identified as the inability of the simple polynomial basis functions of the individual finite elements to capture the complexity of the exact solution being sought. The errors are quantified by identifying metrics that measure the amount by which the approximations produced by the individual elements fail to capture the exact solution. Then, procedures that use physically interpretable interpolation polynomials to identify the changes needed to rapidly improve the model are presented. These techniques are demonstrated and compared to earlier error analysis and adaptive refinement procedures in the following chapters.

CHAPTER 2:   AN OVERVIEW OF FINITE ELEMENT MODELING CHARACTERISTICS

Chapter 2 provides an intuitive introduction to the errors that exist in finite element results as a consequence of replacing the continuum with a discrete number of finite elements. These **discretization errors** are produced when the exact solution is too complex to be represented by the interpolation functions of the individual elements. The discretization errors are literally **seen** in the finite element results as interelement jumps in the strain distributions. The interelement jumps provide an ideal metric for an error estimator and a termination criterion. On one hand, the interelement jumps are shown in Chapter 4 to be aggregated residual quantities that measure the failure of the finite element solution to satisfy the governing differential equation being approximated. On the other hand, the interelement jumps in the strain express the errors in terms of quantities that are being sought in computational mechanics, namely, strain. This means, for example, that if the material being analyzed fails at 1000 units, the magnitude of the error can be interpreted in terms of the failure criterion.

CHAPTER 3:   STRAIN MODELING CAPABILITIES OF INDIVIDUAL FINITE ELEMENTS

Chapter 3 presents a procedure for identifying the strain modeling capabilities of individual finite elements during the formulation of the element stiffness matrix. This capability is made possible by the recognition that finite element interpolation polynomials can be interpreted as Taylor series expansions whose coefficients can be expressed in terms of quantities that produce displacements in the continuum, namely, rigid body motions, constant strains, and derivatives of the strain components. This physically interpretable notation reduces the level of mathematics required for the development of the finite element method, error analysis, and adaptive refinement processes to that which is learned in undergraduate calculus. That is to say, functional analysis is not needed. The advances that this notation brings to computational mechanics are discussed at length in the concluding chapter.

CHAPTER 4: THE SOURCE AND QUANTIFICATION OF DISCRETIZATION ERRORS

Chapter 4 identifies the source of the interelement jumps in the nodal strains that quantify the discretization errors in finite element solutions. The interelement jumps consist of the nodal equivalent values of the point wise residuals on the domain of an element. The residuals quantify the failure of the finite element solution to satisfy the governing differential equation being solved. This understanding of the source of the discretization errors allows error estimators and refinement guides to be based on first principles instead of on a correlation with secondary quantities, such as strain energy. The refinement guides that utilize this knowledge and the physically interpretable notation presented in Chapter 3 are developed in Chapter 11. As discussed in the concluding chapter, the availability of refinement guides based on first principles changes the structure of the adaptive refinement process.

CHAPTER 5: MODELING INEFFICIENCIES IN IRREGULAR ISOPARAMETRIC ELEMENTS

Chapter 5 introduces a better and more efficient way to form stiffness matrices than the isoparametric approach. The stiffness matrices formed using the physically interpretable strain gradient notation are simpler to compute because significantly fewer integrals must be evaluated and isoparametric mappings are not required. Furthermore, the way in which the isoparametric mapping causes modeling errors in distorted isoparametric elements is demonstrated. The inefficiencies that exist in finite element models formed with distorted isoparametric elements are demonstrated with examples. In addition to identifying a source of errors in finite element models, this chapter further demonstrates the efficacy of strain gradient notation.

CHAPTER 6: INTRODUCTION TO ADAPTIVE REFINEMENT

Chapter 6 introduces and applies the adaptive refinement process. The presentation is designed to identify the three necessary components of adaptive refinement: (1) an error estimator, (2) a termination criterion, and (3) a refinement strategy. In this demonstration of adaptive refinement, the jumps in the interelement nodal strain are used as the error estimator and to define the termination criteria. A simple refinement strategy that divides an element that fails the termination criteria into two elements is used in this demonstration. A refinement guide based on first principles is developed and presented in Chapter 12. In one demonstration of adaptive refinement presented in this chapter, the inaccuracies in the strain representation for distorted isoparametric elements that have been identified in the previous chapter are shown to produce inefficient finite element models.

CHAPTER 7: STRAIN ENERGY-BASED ERROR ESTIMATORS

Chapter 7 introduces and demonstrates the first practical error estimator. At the heart of this procedure, which was developed by Zienkiewicz and Zhu, is the formation of a smoothed strain representation that is assumed to be closer to the exact solution than the discontinuous finite element result. The error is estimated by computing the difference in the strain energy between the smoothed solution and the finite element solution. This error estimator is put on a solid theoretical foundation in this chapter by showing that the smoothed solution is, indeed, closer to the exact result than the finite element solution. The idea of an improved solution that is closer to the exact solution than the discontinuous finite element solution is extended in Chapter 11 to form a refinement guide based on first principles that significantly improves the adaptive refinement process.

CHAPTER 8: A HIGH RESOLUTION, POINT-WISE RESIDUAL ERROR ESTIMATOR

Chapter 8 develops and applies a high resolution, point-wise error estimator that forms a residual quantity using finite difference operators. This error estimator identifies the magnitude of the failure of the finite element solution to satisfy the finite difference approximation of the governing differential equation at individual points. The rationale behind this error estimator is the idea that both the finite element and the finite difference methods attempt to solve the same problem with different approaches. Since both solutions will approach the exact solution in the limit, the residual will approach zero as the solution gets close to the exact solution. However, the primary importance of this development is the demonstration that the finite element and the finite difference methods share the same Taylor series basis. This means that finite difference templates can be applied to any point on any finite element model. In addition to possibly infusing the finite difference method with new life as it pertains to solving solid mechanics problems, the use of finite difference templates provides an alternative, highly effective way to form a smoothed solution for use in the refinement guide that is developed in Chapter 11.

CHAPTER 9: MODELING CHARACTERISTICS AND EFFICIENCIES OF HIGHER ORDER ELEMENTS

Chapter 9 demonstrates that higher order elements are more efficient on a node-for-node basis than lower order elements when they **represent complex strain distributions with stringent termination**

criteria. This behavior is identified by replacing single five-node bar elements with increasing numbers of four-node elements in a problem with a complex strain distribution. The lower order elements compensate for their inability to represent the additional strain states that higher order elements can represent with a finite representation of the higher order strain representations. That is to say, the curvature in the four-node elements changes in order to represent the change of curvature that a single five-node element can represent. This result is exploited to produce refinement guides based on first principles that are developed in Chapter 11. As we will see, this new approach to forming refinement guides allows the role of the error estimator to be simplified. It need only serve as a termination criterion. It need not serve as the basis for the degree of refinement needed to improve a finite element model.

CHAPTER 10: FORMULATION OF A 10-NODE QUADRATIC STRAIN ELEMENT

Chapter 10 presents the formulation of a 10-node finite element using the physically interpretable strain gradient notation for two primary reasons. The first is to make this higher order element available because of the efficiencies that were demonstrated for higher order elements in the previous chapter. The second is to demonstrate the advantages of using the strain gradient approach over the isoparametric approach. These advantages include: (1) the visual identification of the strain modeling capabilities of the 10-node element, (2) the clarification of the role and use of the compatibility equation, (3) the need to evaluate significantly fewer integrals than are required to form an isoparametric element (15 versus 210), and (4) the fact that the required integrals have a simple form and can be integrated exactly with little difficulty. Two Appendices are associated with this chapter. The first provides a numerical example for a 10-node element. Its purpose is to provide a sample so that the implementation can be checked. The second Appendix is comprised of a heavily annotated set of Matlab m-files for forming the stiffness matrix presented in the first Appendix. It is designed to provide the details for forming strain gradient based finite elements stiffness matrices.

CHAPTER 11: PERFORMANCE-BASED REFINEMENT GUIDES

Chapter 11 develops a new type of refinement guide that is based on first principles instead of a correlation with the error estimator. This approach to refinement changes the structure of the adaptive refinement process because the error estimator now needs only to function as a termination criterion since it is not directly involved with identifying the level of refinement needed. This means that an error estimator can be chosen that is directly related to the problem being solved. For example, the use of the interelement jumps in strain as the error estimator provides a metric, that is, strains, that can be directly related to a failure criterion.

   This approach to model refinement, first, decomposes an improved strain distribution through the use of strain gradient notation to estimate the participation of the higher order strain representations that the elements in the finite element model cannot represent. These modeling deficiencies are then compared to the modeling characteristics of the elements that make up the finite element model in order to identify the level of refinement needed to satisfy the termination criterion by using the knowledge developed in Chapter 9 concerning the behavior of higher order elements.

CHAPTER 12:   SUMMARY AND RESEARCH OPPORTUNITIES

Chapter 12 provides a list of research opportunities associated with multidimensional problems that are extensions of the developments presented here. These opportunities are given in such a way that they constitute a summary of the developments contained in this book.

In closing, I wish to thank the approximately 25 graduate students that worked on these related topics and acknowledge their hard work and innumerable contributions to these advances in the finite element method, the finite difference method, and adaptive refinement. Without their contribution this book could not have been written.

CHAPTER 1

INTRODUCTION

1.1 PROBLEM DEFINITION

The finite element method is capable of producing accurate approximate solutions for a wide variety of differential equations. The domain of the problem is broken into a finite number of geometrically simple subdivisions. These subdivisions, known as finite elements, are shown in Fig. 1.1. The exact solution is approximated on an individual element by low-order polynomial interpolation functions that attempt to represent the actual displacements that exist on the domain of the finite element.

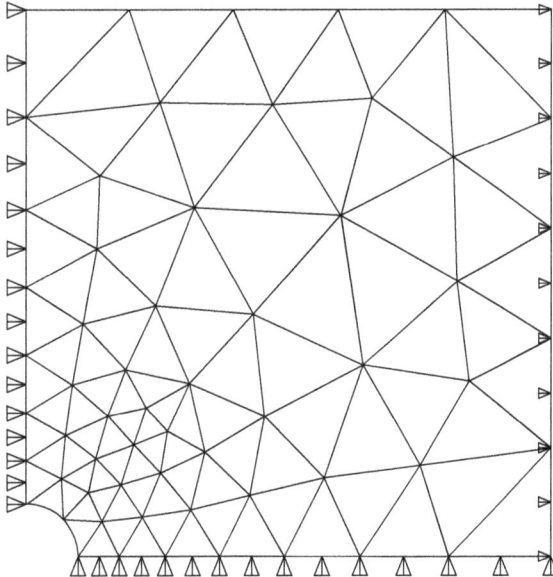

Figure 1.1. A finite element mesh.

1

Errors are produced by the finite element model when the low-order interpolation functions are incapable of representing the complexity of the exact solution on the domain of the individual elements. These modeling deficiencies appear as discontinuities in the strains on the boundaries between the elements.

Examples of such errors are shown in Fig. 1.2 for a one-dimensional longitudinal bar problem that is modeled with five three-node finite elements. This figure consists of the discontinuous strain representation that is produced by this finite element model when it is superimposed on the exact answer to the problem.

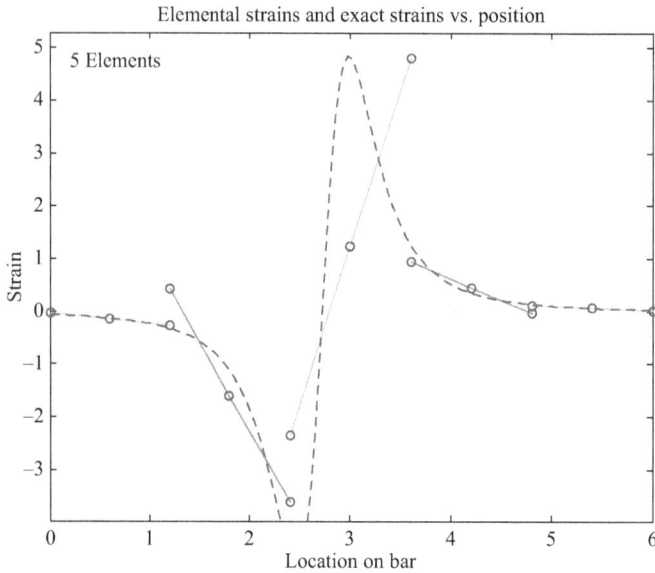

Figure 1.2. Five element strain results.

As can be seen by comparing the two strain representations, this five-element model does not produce a good approximation of this exact solution. More specifically, the linear strain representations available from the individual three-node elements approximate the complexity of the exact solution with varying degrees of success. The sizes of the interelement jumps in the finite element strain representation are related to the inaccuracy in the finite element result.

For example, the interelement jump in the strain representation between the two elements on the right side of the bar is relatively small. In contrast, the interelement jump in the strain between the two elements on the left side of the bar is relatively large. This difference exists because the two elements on the left end of the bar do not represent the exact strain distribution as well as the two elements on the right end of the bar do.

In practice, we cannot assess the accuracy of a finite element approximation by directly comparing it with the exact result because exact answers do not exist for most practical problems.

As a consequence, **we are faced with two difficult problems in order to produce approximate results with an acceptable level of accuracy**. The first problem requires that we assess the accuracy of the finite element approximation. The second problem requires that we modify the existing finite element model so that it efficiently produces a result that is sufficiently accurate.

1.2   OVERALL OBJECTIVES

The overall objectives of this work are the following:

1. To develop a procedure based on basic or first principles of continuum mechanics for identifying refinements to the finite element model that will improve the accuracy of the resulting solution.
2. To present procedures for estimating the errors in a finite element result in terms of quantities that are meaningful to the analyst; for example, the errors are given as a percentage of a critical strain value.
3. To give an overview of different error analysis procedures.
4. To demonstrate the improved modeling efficiency that results from using higher order finite elements to represent complex strain distributions.
5. To present an improved and more intuitive element formulation procedure.

The first two objectives apply directly to the components of the adaptive refinement process that are developed here. The last three objectives provide the necessary background and developments that serve as the basis for the error estimators and refinement guides presented in this work.

1.3   SPECIFIC TASKS

Since we cannot identify the errors in a finite element result by a direct comparison with an exact result, we must assess the accuracy of the finite element approximation and improve the model indirectly. The indirect assessment and refinement of the existing model is accomplished with information that is available to us, namely, the modeling characteristics of the individual finite elements and the results produced by the finite element model.

The overall objectives of this work are accomplished with the following seven ambitious tasks:

1. Identify the two sources of errors in finite element models, namely, the modeling errors that may exist in individual elements and the inability of individual elements to capture the complexity of the solution being sought (Chapters 2, 3, 4, 5, 10, and 11).
2. Relate the modeling errors and the interelement jumps in the finite element result (Chapters 4 and 9).

3. Present the two primary approaches to error analysis, namely, the residual and the recovery approaches (Chapters 4, 6, and 7).

4. Develop a new simplified approach to error analysis that integrates the residual and recovery approaches to error analysis (Chapters 7 and 8).

5. Develop a new approach for forming refinement guides to improve finite element models that is based on first principles and not on a correlation based on the magnitudes of the elemental error estimates, that is, the errors are identified by comparing an estimate of the emerging exact solution with the modeling capability of an individual element (Chapter 11).

6. Demonstrate that the finite element and the finite difference methods are directly related by showing that finite element interpolation functions and finite difference derivative approximations can be formed from the same truncated Taylor series expansions (Chapters 3, 5, and 10).

7. Present an improved procedure for forming finite element stiffness matrices by specializing the displacement interpolation functions for solid mechanics problems by using physically interpretable coefficients expressed in terms of the quantities that produce displacements in the continuum, namely, rigid body motions and strains (Chapters 3 and 5).

1.4   THE CENTRAL ROLE OF THE INTERPOLATION FUNCTIONS

As pointed out in Section 1.1, the finite element method attempts to represent portions of the exact solution on the domains of the individual elements with low-order polynomial interpolation functions. If the exact solution is too complex over the domain of an element for the element's interpolation function to represent, errors will occur in the finite element solution.

Figure 1.3 presents an example of the improvement to a finite element representation that occurs when individual elements containing errors are subdivided. The subdivision of an element can be interpreted as reducing the complexity of the portion of the exact solution that an individual element is attempting to represent.

In Fig. 1.3a, the region of maximum strain for the exact strain distribution shown in Fig. 1.2 is represented by two elements. In Fig. 1.3b, the two elements shown in Fig. 1.3a are refined by

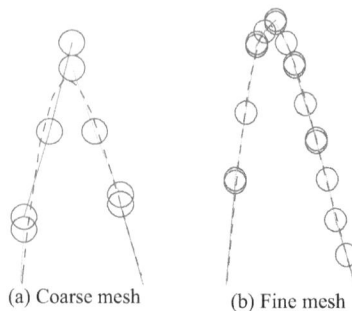

(a) Coarse mesh   (b) Fine mesh

Figure 1.3. Two meshes at critical points.

subdividing each of them into four elements. This means that each element in Fig. 1.3b is being asked to represent a smaller, less complex portion of the exact solution than are the individual elements in Fig. 1.3a.

When the two finite element representations of the points of maximum strain shown in Fig. 1.3 are compared with the exact solution, the refined model produces a better result than does the initial model. As a consequence of the improvement to the model, the interelement jumps in the strain are smaller for this better representation of the exact solution. It should be noted that the Weierstrass approximation theorem gives the finite element method a solid theoretical foundation[1] and guarantees that the refinement process will improve the finite element model.

1.5   A CLOSER LOOK AT THE INTERPOLATION FUNCTIONS

The overview of the way a finite element model forms an approximate solution presented in the previous section identifies that **the heart of the finite element method is contained in the interpolation functions**. The tasks listed in Section 1.3 are achieved by first recognizing that the interpolation polynomials are truncated Taylor series expansions. This recognition performs two functions. First, it identifies that the finite element and the finite difference methods have a common basis.[2] The identification of a common basis for these two powerful approximation techniques provides a starting point for the development of the point-wise error estimator presented in Chapter 8 and for the refinement guides based on first principles that are presented in Chapter 10.

Second, this recognition allows the coefficients of the interpolation functions to be expressed in terms of rigid body motions and strain quantities. As a result, the displacements produced by the interpolation functions are expressed in terms of the quantities that produce displacements in the continuum. This physical interpretation of the Taylor series coefficients that comprise the interpolation functions is possible because the rigid body rotations and the strain quantities are defined in terms of derivatives of the displacements in the continuum. As we will see, the transparency produced by this physically interpretable notation provides insights into the modeling capabilities of both individual elements and of overall finite element models that were not previously possible.

Next, we will show three forms of the same interpolation function as it progresses from a standard form to the physically interpretable form. This physically interpretable notation, known as strain gradient notation, is developed in detail in Chapter 3 and in Reference [3]. The interpolation

---

[1] The Weierstrass approximation theorem states that any continuous function can be uniformly approximated on an interval by polynomials to any degree of accuracy. Note that this theorem does not mean that the function can be represented exactly. The theorem implies that the exact solution can be approximated as closely as desired [1, 2].

[2] The common basis for the finite element and the finite difference methods exists because the process for formulating the interpolation functions for a finite element from the truncated Taylor series expansion is identical to the process for forming the derivative approximations in the finite difference method. This duality is discussed in detail in Reference [3].

functions for displacements in the x direction for a four-node planar element are used in this example. The three forms of the interpolation functions in this progression are the following:

$$u(x, y) = a_1 + a_2 x + a_3 y + a_4 xy \tag{Eq. 1.1a}$$

$$u(x, y) = (u)_0 + (\partial u/\partial x)_0 x + (\partial u/\partial y)_0 y + (\partial^2 u/\partial x \partial y)_0 xy \tag{Eq. 1.1b}$$

$$u(x, y) = (u_{rb})_0 + (\varepsilon_x)_0 x + (\gamma_{xy}/2 - r_{rb})_0 y + (\varepsilon_{x,y})_0 xy \tag{Eq. 1.1c}$$

The standard form of the interpolation function used in most developments of the finite element method is given in Eq. 1.1a. The physical meaning of the coefficients in this form of the displacement interpolation polynomials, denoted by the a's, cannot be seen by inspection.

However, when the interpolation functions are recognized as truncated Taylor series expansions as shown in Eq. 1.1b, the coefficients can be interpreted in terms of rigid body motions and strain quantities. This is the case because the Taylor series coefficients are expressed in terms of the displacements and the derivatives, or gradients, of the displacements that are evaluated at the origin of the local coordinate system.

Finally, the Taylor series form of the interpolation function is specialized for solid mechanics problems in Eq. 1.1c. The constant coefficient, a quantity that identifies the displacement of the local origin, is interpreted as the rigid body displacement of the finite element in the x direction. The next term, the normal strain term in the x direction, is simply the definition of the normal strain from linear elasticity, $\varepsilon_x = \partial u/\partial x$. The remaining terms are associated with the shear strain, the rotation around the z-axis, and gradient of the normal strain component, ε_x, in the y direction.

The coefficient of the xy term in Eq. 1.1c has a y following a comma in the subscript. This symbol and its location indicate that a derivative of ε_x with respect to y has been taken. This term, $\varepsilon_{x,y}$, indicates the rate of change in the y direction of normal strain in the x direction. As a result of interpreting the rigid body displacements as zeroth order gradients, this notation is designated as **strain gradient notation**.

The significance of expressing the interpolation functions in terms of quantities that have meaning in solid mechanics problems allows the equations produced from these interpolation functions to be evaluated by inspection. The transparency produced by the inclusion of physical meaning into the interpolation functions allows: (1) modeling errors in individual elements to be identified (see Chapter 3); (2) a computationally simpler element formulation procedure to be developed (see Chapters 3 and 10 and Reference [3]); (3) *in situ* errors in finite element results to be identified (see Chapter 11); and (4) the development of refinement guides based on first principles (see Chapter 11).

This physically interpretable notation is an example of self-referential notation. Self-reference means that cause and effect are related symbolically in the equations. For example, Eq. 1.1c indicates that the displacements in the x direction are produced by the quantities on the right-hand side of the equation. As we will see in the next section and later in the text, if the cause and effect relationship is not true, the equation will contain an error that is discernable by inspection.

1.6   PHYSICALLY INTERPRETABLE INTERPOLATION FUNCTIONS IN ACTION

An example of the insights provided by strain gradient notation is presented in this section. This is accomplished by identifying one of the modeling errors that exists in the four-node quadrilateral element. The detailed development that identifies this error and other modeling deficiencies in the four-node element is presented in Chapter 3 and in Reference [3].

When the standard form of the interpolation functions for a four-node quadrilateral (see Eq. 1.1a) is introduced into the linear elasticity definition of shear strain, we get the following:

$$\gamma_{xy}(x, y) = \frac{\partial u}{\partial y} + \frac{\partial v}{\partial x} = (a_3 + b_2) + b_4 x + a_4 y \qquad \text{(Eq. 1.2)}$$

As can be seen, this equation contains a constant term and linear terms in x and y. Because of the arbitrary nature of the coefficients in Eq. 1.2, the shear modeling characteristics of the four-node element are not obvious. In fact, one might be tempted to assume that this equation provides a complete linear representation of the shear strain. As we will see when this shear strain expression is expressed in strain gradient notation, such an assumption is wrong.

In fact, the two linear terms contained in Eq. 1.2 are modeling errors. As we will see in Chapter 3, the existence of these errors and others in the four-node element reduces the effectiveness of a four-node element to that of a three-node triangle. As a result, the efficiency of the overall finite element model is reduced.

When the shear strain representation for a four-node element is expressed in the physically interpretable strain gradient notation (see Eq. 1.1c), the modeling errors introduced by these linear terms can be identified by visual inspection. The shear strain expression that is formed with the physically interpretable notation follows:

$$\gamma_{xy}(x, y) = \frac{\partial u}{\partial y} + \frac{\partial v}{\partial x} = (\gamma_{xy})_0 + (\varepsilon_{x,y})_0 \, x + (\varepsilon_{y,x})_0 \, y \qquad \text{(Eq. 1.3)}$$

As was the case for the shear strain expression given by Eq. 1.2, Eq. 1.3 contains a constant term and linear terms in x and y. However, in this case the physical meaning of the coefficients contained in Eq. 1.3 is clearly visible. As a result, the shear strain modeling characteristics of a four-node element can be identified by inspection.

The constant term, γ_{xy}, represents a state of constant shear strain on the full domain of the element. The coefficient of the x term, $\varepsilon_{x,y}$, represents the gradient in the y direction of the normal strain in the x direction. Similarly, the coefficient of the y term, $\varepsilon_{y,x}$, represents the gradient in the x direction of the normal strain in the y direction. As will be reiterated throughout the text, the values of the strain gradient coefficients apply the local origins of the individual elements.

As a result of being able to identify the physical meaning of the coefficients in Eq. 1.3, the errors introduced by $\varepsilon_{x,y}$ and $\varepsilon_{y,x}$ can be identified by visual inspection. On one hand, if these two terms were correct, they would have to be gradients of the shear strain because of the Taylor series nature of the representation. That is to say, these two terms would be expressed as the following shear strain terms, namely, $\gamma_{xy,x}$ and $\gamma_{xy,y}$. On the other hand, the two linear terms in Eq. 1.3 cannot be correct because the normal strains and the shear strains are not coupled in the theory of linear elasticity.

The fact that the linear terms in the shear strain expression are incorrect was known long before the advent of the physically interpretable strain gradient notation. These terms, known as parasitic shear terms, are typically removed from a four-node quadrilateral element by the application of a procedure called **reduced-order Gauss quadrature**.

Even if these erroneous terms are removed, a four-node element is no more effective than a three-node constant strain triangle. The element with the least number of nodes that can represent the strain components with a linear model is a six-node triangular element. It should be noted that if care is not taken and a reduced-order Gauss quadrature is applied to higher order elements, other errors can be introduced [3].

1.7 THE OVERALL SIGNIFICANCE OF THE PHYSICALLY INTERPRETABLE NOTATION

The use of this physically interpretable notation for the interpolation functions can be likened to the simplification to computing that occurred with the introduction of graphical user interfaces (GUIs). With the introduction of GUIs, the power of computing was extended to nonexperts because knowledge of the commands for an operating system was not required. In many cases, computing became a matter of pointing and clicking. As a result, a four-year-old could accomplish things with computers that had been out of reach for a majority of the population before the introduction of GUIs.

As we will see, the advent of the physically interpretable strain gradient notation allows nonexperts to: (1) clearly see the modeling capabilities of individual finite elements, (2) understand the metric (measures, quantities) that is being used to evaluate the accuracy of a finite element model in terms of quantities that are significant to computational mechanics, and (3) relate the refinement guide directly to the modeling deficiencies contained in the finite element model. That is to say, this development makes the finite element method more easily understandable to a nonspecialist.

As a result of making the details of finite element modeling easily understandable, computational mechanics now more closely approximates the situation where it can be viewed as a "utility." As a further consequence, nonspecialists are more likely to produce results that accurately represent the exact solution to the problem being solved.

1.8 EXAMPLES OF MODEL REFINEMENT AND THE NEED FOR ADAPTIVE REFINEMENT

As discussed earlier, the errors in the solutions produced by finite element models occur when the interpolation functions on the individual elements are incapable of capturing the complexity of the exact solution. The obvious approach for improving a finite element model is to subdivide each of the individual elements. This process is called **uniform refinement**.

In this section, we will demonstrate the improvements to finite element solutions that are produced by uniform refinement as well as address its major deficiency. As we will see, uniform refinement produces overly large finite element models because it introduces elements into regions that accurately represent the exact solution with the existing number of elements.

The introduction of unnecessary elements identifies the need for the capability to evaluate the errors in a finite element result so that the model can be improved only in regions where the error exceeds a predetermined limit. That is to say, the deficiency in uniform refinement identifies the need for adaptive refinement, which will be discussed in the next section.

The deficiency that exists with uniform refinement will now be demonstrated by uniformly refining an example problem several times. The result of dividing each of the five individual elements contained in the initial model (see Fig. 1.2) into two elements is presented in Fig. 1.4a. When the strain distribution produced by the ten-element model is compared with the exact result, this uniformly refined model shows an improvement on the five-element model, but it still does not accurately represent the maximum and minimum points of the exact solution. We will quantify these errors later when we develop error estimators. Note, however, that the interelement jumps at the critical points have been reduced in the refined model.

When the ten-element model is uniformly refined again, the model improves. As shown in Fig. 1.4b, the regions of the strain distribution with rapidly changing curvature before the minimum point and after the maximum point better represent the exact solution. Although the representations of the maximum and minimum points are improved, the finite element approximation of the strain distribution still does not give an accurate picture of the strain distribution in these regions. Note that a significant interelement jump in the strain has appeared in the center of the bar where the strain is essentially zero.

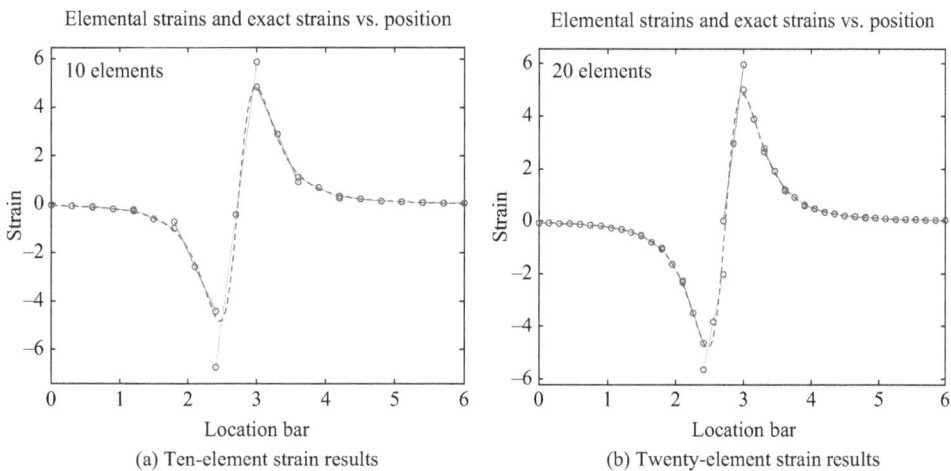

Figure 1.4. Strain results for two successive uniform refinements of Fig. 1.2.

These two examples identify the difficulties inherent in uniform refinement. On one hand, elements are added to regions that adequately represent the exact solution. On the other hand, not enough elements are added in regions of high error. These deficiencies are further highlighted when the 20-element model is uniformly refined two more times as shown in Fig. 1.5.

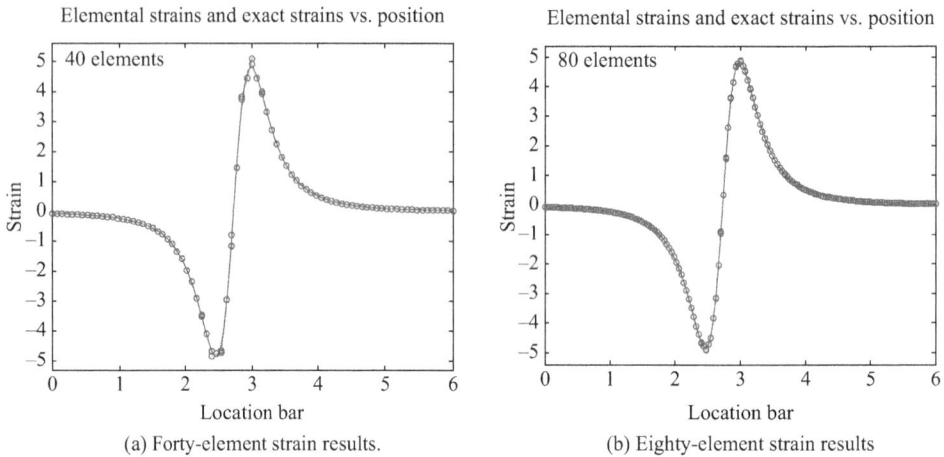

(a) Forty-element strain results.   (b) Eighty-element strain results

Figure 1.5. Strain results for two successive uniform refinements of Fig. 1.4b.

As can be seen in Fig. 1.5a, the strain representation produced by this 40-element model is better than the approximation produced by the 20-element model. However, the region of maximum strain still does not match the exact solution well and the interelement jumps in the strain are clearly visible. When the finite element model is uniformly refined again as shown in Fig. 1.5b, the contour of the region of maximum error is captured reasonably well by the 80-element model. Note that the interelement jumps have been further reduced. They are primarily seen as a thickening of the circles that represent the interelement nodes.

The foregoing examples of uniform refinements highlight the problem with uniform refinement. Too many elements are placed in regions where the actual strain distributions are adequately represented and not enough elements are placed in the critical regions with high rates of change in the strain distributions. An alternative approach that eliminates this deficiency is outlined and demonstrated in the next section.

1.9 EXAMPLES OF ADAPTIVE REFINEMENT AND ERROR ANALYSIS

The initial finite element model of a problem rarely provides a solution that is accurate enough for use in the design process. The obvious strategy for improving the model is to repeatedly subdivide every element in the model until the change in two successive results is acceptably small. However, as was seen in the previous section, this brute force approach for reaching an acceptable solution leads to finite element models that are unmanageably large because it needlessly introduces elements into regions that contain little or no error.

The excessive growth produced by uniformly refining a finite element model can be eliminated by selectively improving the model only in the regions that contain high levels of error. A procedure for identifying regions of unacceptable error and improving the model in these regions is shown schematically in Fig. 1.6.

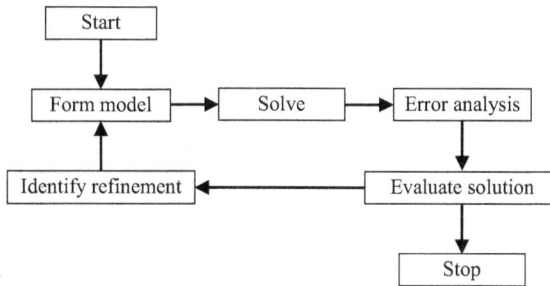

Figure 1.6. Adaptive refinement schematic.

This procedure, known as **adaptive refinement**, begins by forming an initial finite element model. The errors in the solution of the initial finite model are then estimated. If the specified level of accuracy is not achieved, the model is improved by refining the mesh in regions of unacceptable error. The process is repeated, starting with the improved model, until an acceptable solution is obtained.

Figure 1.7 illustrates an example of adaptive refinement of a shear panel with an internal circular hole. This finite element model, an approximation of the Kirsch problem [3], contains a stress concentration at the upper-most point on the one-quarter circle in this doubly symmetric problem.

The initial mesh that consists of six-node triangles is shown in Fig.1.7a. This mesh contains 430 degrees of freedom. The model is loaded with a uniform load on the right end of the panel. When this problem is adaptively refined so that the estimated error in the strain energy content of each element is less than 5%, the final adaptively refined mesh contains 11,454 degrees of freedom. This result is shown in Fig. 1.7b [3]. Note that the elements on the boundary are not subdivided. This means that these elements represent the exact solution with an adequate degree of accuracy.

For the sake of comparison, it is estimated that if the initial model is uniformly refined until the same level of error is achieved at the stress concentration, the model would contain over 10^6 degrees of freedom. This means that the nodal density for the uniformly refined mesh would be as dense on the whole domain of the problem as it is in the densest portion of the adaptively refined model. That is to say, the final figure would look entirely black. This comparison highlights the fact that adaptive refinement is necessary if accurate approximate solutions are to be achieved with efficiency.

In the previous section, the initial mesh shown in Fig. 1.2 was uniformly refined until it contained 80 elements. The resulting strain distribution was first shown in Fig. 1.5b and is repeated here as Fig. 1.8a so that it can be seen adjacent to the error analysis performed on it.

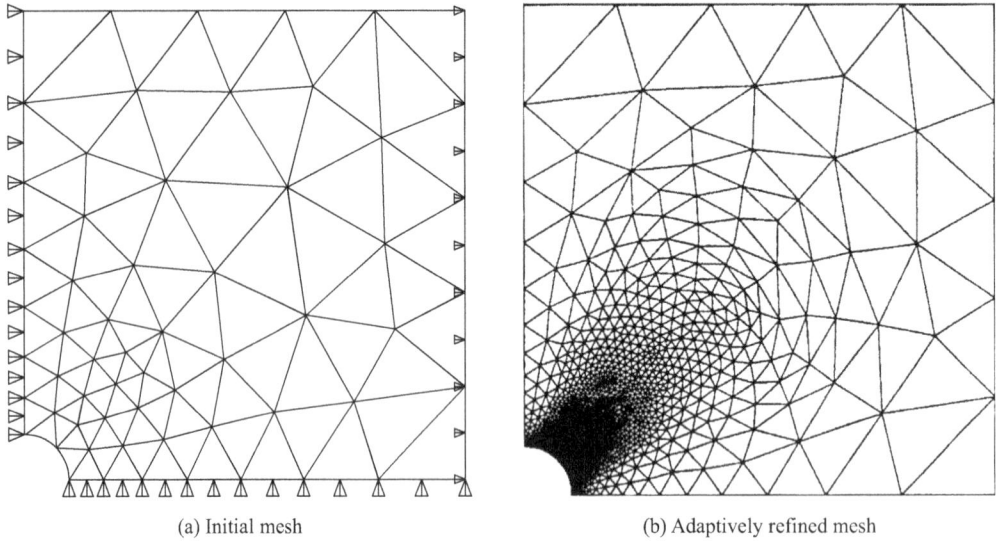

(a) Initial mesh   (b) Adaptively refined mesh

Figure 1.7. Adaptively refined stress concentration.

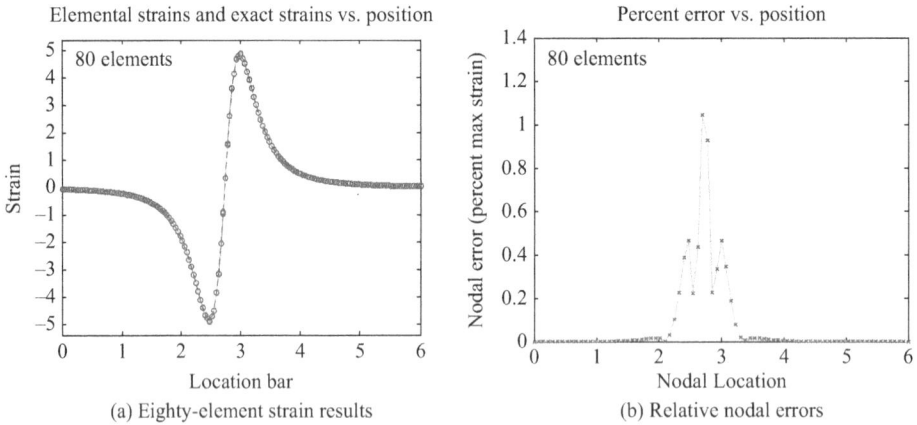

(a) Eighty-element strain results   (b) Relative nodal errors

Figure 1.8. Fourth uniformly refined finite element model.

The results of an error analysis performed on the 80-element model are shown in Fig. 1.8b. In this analysis, the errors are quantified in terms of the interelement jumps in strain. The errors in Fig. 1.8b are given as a percentage of the maximum absolute strain in the model. In this case, the maximum error is approximately 1.1%. The maximum acceptable error for this case was defined as 4.0%. The initial model and the three subsequent uniform refinements with 10, 20, and 40 elements all had maximum errors in excess of 4.0%.

An example of adaptive refinement will now be presented. The initial model presented in Fig. 1.2 is adaptively refined by subdividing any element that had an error in excess of the 4.0% termination criterion. The result after four cycles of adaptive refinement is presented in Fig. 1.9. When the finite element result is compared with the exact strain distribution in Fig. 1.9a, the exact solution is well represented by the finite element approximation. The errors for this result are presented in Fig. 1.9b. As can be seen, the largest error is less than 3.5%, so the termination criterion is satisfied.

The efficacy of the adaptive refinement process has been demonstrated with the above example. The adaptively refined model satisfied the termination criterion of 4.0% error with only 19 elements. The fact that the uniformly refined model required 80 elements to satisfy the termination criterion demonstrates the advantages of adaptive refinement over uniform refinement.

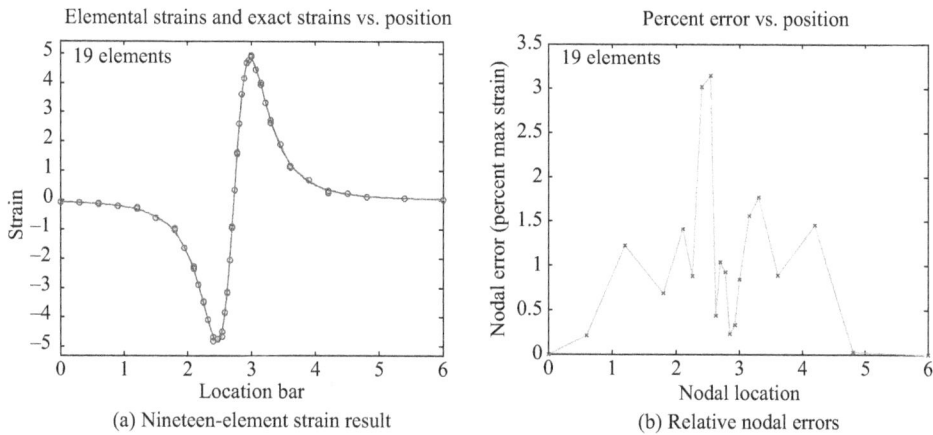

(a) Nineteen-element strain result   (b) Relative nodal errors

Figure 1.9. Fourth adaptively refined finite element model.

1.10 SUMMARY

This chapter has identified the two problems that surface when an approximate solution technique is used: (1) the existence of errors in the approximate solution and (2) the need to improve the approximate model in order to achieve the desired level of accuracy. These problems are addressed for the finite element method by presenting extensive examples and developments that provide detailed insights into the modeling characteristics of individual finite elements and of finite element models.

The insights into these modeling characteristics are used to develop new approaches for the two major components of the adaptive refinement process, namely, error estimation procedures and model refinement strategies. In addition to presenting these new approaches, an overview and examples of the existing error estimation and adaptive refinement procedures are presented.

The errors in finite element results are due to the inability of the interpolation polynomials of the individual elements to capture the exact solution that they are being asked to represent. The new point-wise error estimators presented here quantify the errors in terms of quantities that have direct physical meaning to an analyst, for example, the percentage of a significant strain quantity.

The new approaches to mesh refinement are based on the recognition that the finite element interpolation polynomials are actually truncated Taylor series expansions. This, in turn, allows the coefficients of the Taylor series expansions to be interpreted in terms of rigid body motions and strain quantities. The use of this physically interpretable, strain gradient notation allows the modeling capacities of the individual elements to be compared with an estimation of the exact solution that is emerging from the finite element solution to identify model refinements that lead to the rapid improvement of the finite element model.

1.11 REFERENCES

1. Scheid, F., *Numerical Analysis*, 2nd ed., McGraw-Hill, Inc, Schaum's Outline Series, New York: McGraw-Hill Book Company, 1989, p. 267.
   This presentation of the Weierstrass approximation theorem is readily accessible to the non-mathematician.
2. Madson, J. C. and Handscomb, D. C., *Chebyshev Polynomials*, Boca Raton, FL: Chapman and Hall/CRC Press, 2003, p. 45.
   This is the first new book on Chebyshev polynomials to be written in several decades. Its primary thesis is that much of numerical analysis can be based on Chebyshev polynomials.
3. Dow, J. O., *A Unified Approach to the Finite Element Method and Error Analysis Procedures*, New York: Academic Press, 1999.
   This book puts the finite element method, the finite difference method, and error analysis on a common Taylor series basis. It develops and applies the physically based notation introduced in Chapter 3.

CHAPTER 2

AN OVERVIEW OF FINITE ELEMENT MODELING CHARACTERISTICS

2.1 INTRODUCTION

The introductory chapter includes an overview of the errors that can exist in finite element results as a consequence of replacing the continuum with a discrete representation. Figure 1.3, which is reproduced here as Fig. 2.1, identifies two salient characteristics of the errors that can result from this discretization of a continuous problem.

As can be seen in Fig. 2.1, the first characteristic of these errors is the existence of interelement jumps in the finite element strain representation. In Chapter 4, we show that the interelement jumps are directly related to the failure of the finite element solution to satisfy the governing differential equation being solved on the domains of the individual elements. That is to say, the magnitudes of the interelement jumps quantify the failure of the finite element representation to capture the exact solution. Since these errors are due to deficiencies in the discrete representation of the continuum, these errors are known as **discretization errors**.

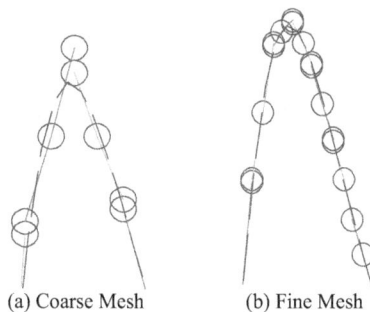

(a) Coarse Mesh   (b) Fine Mesh

Figure 2.1. Two meshes at critical points.

The second characteristic of these errors is the reduction in the interelement jumps that occurs when the elements that exhibit errors are subdivided. An example of this reduction is seen when Fig. 2.1a is compared to Fig. 2.1b. When the two elements in Fig. 2.1a are subdivided to produce the model shown in Fig. 2.1b, the interelement jumps in the strain produced by the refined model are smaller than those in the initial model. This reduction in the interelement jumps is a consequence of modeling badly represented portions of the exact solution with an additional number of smaller elements. The smaller elements are required to represent a smaller and, hence, less complex portion of the exact solution. As a result, the interpolation functions of the individual elements produce a better approximation of the exact solution.

The practical approach for reducing the discretization errors in a finite element model to an acceptable level was identified in the introductory chapter to be the adaptive refinement process that is shown schematically in Fig. 2.2. The goal of adaptive refinement is to iteratively identify regions of unacceptable error and to improve the finite element model until an acceptable result is produced.

As shown in Fig. 2.2, the adaptive refinement process consists of three distinct components. The first component is an error analysis that estimates the magnitude of the errors in the individual elements. The solution is then evaluated by comparing the elemental errors to a termination criterion. If the error estimate for each element satisfies the termination criterion, the result is considered satisfactory and the process is terminated. If any element possesses an unacceptable level of error, the mesh is refined in order to improve the accuracy of the finite element model and the process is repeated.

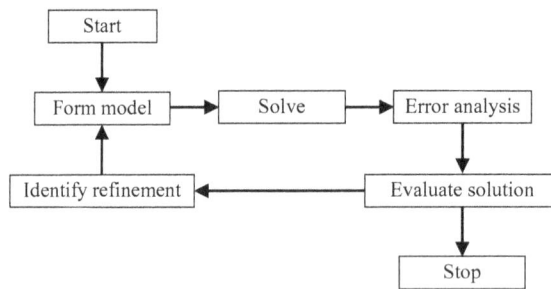

Figure 2.2. Adaptive refinement schematic.

The further development and/or confident application of each of these three components of the adaptive refinement process depend on an understanding of the following three characteristics of a finite element solution, namely: 1) the source of the discretization errors, 2) the effect of these errors on the approximate solution and 3) the requirements for improving the finite element model to reduce the errors.

This chapter is designed to provide an intuitive understanding of the errors in finite element models and of their characteristics. This overview of finite element modeling characteristics is designed to put the detailed developments presented in the following chapters in perspective.

The specific objectives of this chapter are the following:

1. To demonstrate the effect of discretization errors on finite element solutions.
2. To provide visual demonstrations of the causes of discretization errors that exist in finite element solutions.
3. To demonstrate approaches for improving the finite element representation of the exact solution.

These objectives will be achieved with a series of increasingly complex example problems that present various aspects of finite element results and the errors that exist in these results. The theoretical bases that explain these results are presented in later chapters.

In the next section, two examples of finite element models that produce exact results are presented. We will see that these finite element solutions contain no interelement jumps in the strain representations. That is to say, these solutions are continuous, which is the signature of an exact solution. These examples are shown to emphasize the fact that the goal of the adaptive refinement process is to produce finite element results in which the interelement jumps are below a threshold value.

After the finite element models that produce exact results are presented, results from various finite element models that are incapable of representing the exact solutions are presented. As expected, the inability to represent the exact solution is seen by the presence of interelement jumps in the strain representations. Then these models will be improved in different ways in order to demonstrate that the interelement jumps in the strains are reduced as the model is improved so that it better represents the exact solution.

Since this book attempts to present the major concepts of finite element modeling in a clear and compact way, one-dimensional problems are used almost exclusively in this study of discretization errors in finite element models. However, two sources of strain modeling errors do not surface when one-dimensional elements are considered.

One source of error that exists in multidimensional elements that does not exist in one-dimensional problems is known as aspect ratio stiffening. This error occurs when an element is too long and thin. This source of error and its control is discussed in Appendix 2A of this chapter.

A second source of error that exists in multidimensional elements that does not exist in one-dimensional elements is the incorrect representation of the strain components in individual elements. This source of error and its identification were introduced in Section 1.6 of Chapter 1. In Section 1.6, it was shown that the errors in the higher dimensional strain representations occur when the application of the definitions of strain to the displacement interpolation functions do not produce the correct strain representations. These errors are identified with the aid of displacement interpolation polynomials that are expressed in terms of physically interpretable notation.

Because of the importance of strain modeling errors in multidimensional finite elements, the strain modeling characteristics in individual two-dimensional elements are discussed in the following locations:

1. In Chapter 3, the strain modeling errors in a four-node quadrilateral element are identified.
2. In Chapter 5, the modeling deficiencies in isoparametric elements are discussed.
3. In Chapter 10, the stiffness matrix for a 10-node triangle is developed using the physically interpretable strain gradient notation.

2.2   CHARACTERISTICS OF EXACT FINITE ELEMENT RESULTS

This section demonstrates the characteristics of a finite element solution when it is identical to the exact solution. As we will see, an exact finite element solution exhibits no discontinuities in the strain representation. In other words, a finite solution is exact if it contains no interelement jumps in the strain representation. This characteristic of an exact solution will be put on a solid theoretical foundation in Chapter 4.

The continuous nature of an exact finite element solution is demonstrated with the two example problems shown in Fig. 2.3. The finite element model consists of five four-node bar elements of equal length fixed at both ends as shown in Fig. 2.3. Only the interelement nodes are shown in this figure. The model is loaded with the two distributed loads that are illustrated in Figs. 2.3a and b, respectively.

In Fig. 2.3a, the finite element model is loaded with a constant axial load distribution. In Fig. 2.3b, the model is loaded with a linearly varying axial load. It should be noted that the loads shown are actually in the horizontal direction so they produce tension and compression in the bar.

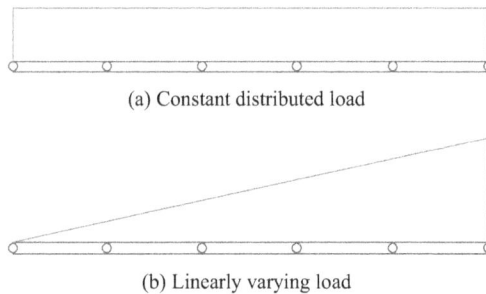

(a) Constant distributed load

(b) Linearly varying load

Figure 2.3. Two finite element models with distributed loads.

The strain distributions produced by the finite element model for these two loading conditions are shown in Figs. 2.4a and b, respectively. In these figures, the interelement nodes are shown as "*o*'s" and the interior nodes are shown as "*x*'s." Both of the finite element strain distributions in Fig. 2.4 are continuous. If the finite element results were not continuous, jumps would exist in the strain representation at the interelement nodes that are identified with the "*o*'s." The fact that there are no discontinuities in the strain representations means that both finite element results shown in Fig. 2.4 match the exact solutions presented in Appendix 2B.

In Chapter 3, we will investigate the modeling capabilities of individual four-node bar elements. We will see that this element is capable of representing strain distributions that are comprised of a linear combination of constant strain, linearly varying strain, and quadratically varying strain. When we look at the behavior of the individual elements in Fig. 2.4a, we see that each element represents the same linear variation of strain and that no interelement jumps exist in the strain representation. It should be noted that the displacement representation also contains a rigid body motion component.

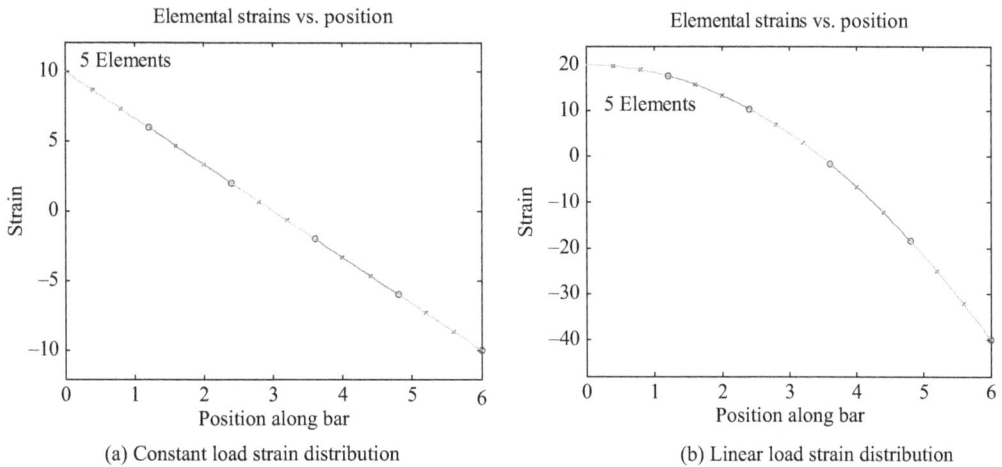

Figure 2.4. Exact finite element strain distributions.

In Fig. 2.4b, the individual elements each represent a parabolic strain distribution and there are no interelement jumps in the strain representation. Since each of the four-node elements is capable of representing the strain distribution of the exact result on its domain, there are no discretization errors and, hence, no interelement jumps in the strains. This means that the finite element result is identical to the exact result.

The two foregoing examples identify the two primary characteristics of finite element representations that are at the heart of adaptive refinement:

1. A finite element solution contains no errors if the interpolation polynomial of each element is able to capture the exact strain distribution in the portion of the solution that it represents.
2. No interelement jumps occur in the strains if the finite element exactly reproduces the actual solution.

Since an exact finite element result does not contain any interelement jumps in its strain representations, the goal of adaptive refinement is to modify a finite element model so that the interelement jumps are reduced to a level that produces acceptable stress and strain results. In some examples, we will use the magnitude of the interelement jumps as a termination criterion and as the basis for identifying the level of refinement to be given an element.

2.3   MORE DEMANDING LOADING CONDITIONS

The next objective of this chapter is to demonstrate the existence of a direct connection between the errors in a finite element result and the size of the discontinuities in the interelement strains. We will accomplish this goal by applying two loading conditions that produce strain distributions that are too complex for finite element models that are constructed with four-node bar elements to represent.

These two loading conditions are presented in this section. As we will see in the next section, these loading conditions produce strain distributions that are difficult for finite element models to represent. We will see that the initial five-element model contains significant interelement jumps in the strain. That is to say, the individual elements are incapable of representing the complexity of the exact strain distribution that exists on the domains of the individual elements that are produced by these two loading conditions.

The first loading condition is centered on the bar as shown in Fig. 2.5a. It produces the exact strain distribution shown in Fig. 2.5b. This strain distribution contains the following types of variation in the strains: 1) regions that are close to varying linearly, 2) regions that approximate parabolic variations, 3) a maximum extreme point, 4) a minimum extreme point, and 5) three inflection points that are indicated on the figure by x's.

The second loading condition shown in Fig. 2.6a is similar to the loading condition in Fig. 2.5a, except that the point of symmetry is shifted slightly to the left of center. This loading condition produces the exact strain distribution shown in Fig. 2.6b. This nonsymmetric strain distribution is nearly identical to the one for the symmetric case, but as we will see in later sections, it surfaces modeling characteristics that are submerged in the symmetric case when evenly spaced finite elements are used to represent the problem.

The two loading conditions shown in Figs. 2.5a and 2.6a are not random choices. They are used because they produce displacements in the continuous bar that are Runge functions. Runge functions are often used to test interpolation procedures because they are difficult for a polynomial interpolation function to reproduce [1–3]. The displacement of the bar produced by the symmetric loading condition is shown in Fig. 2.7.

The Runge functions that represent the displacements can be integrated twice to produce the loading conditions applied to the bars. The loading condition shown in Fig. 2.4.a is derived from the following Runge function: $f(x) = 300/(x + 30.0/2.0)^2$. In the interest of completeness, the off-center

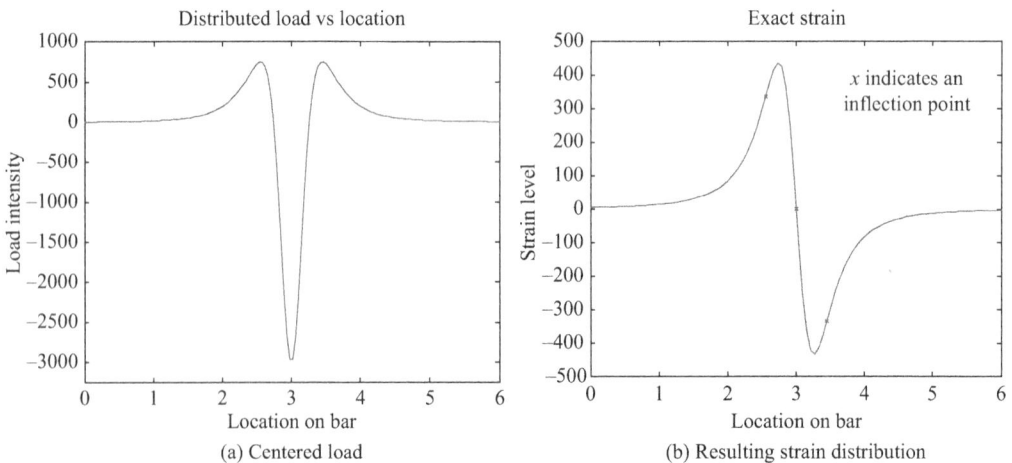

Figure 2.5. A "High-Demand" loading condition and resulting strain distribution.

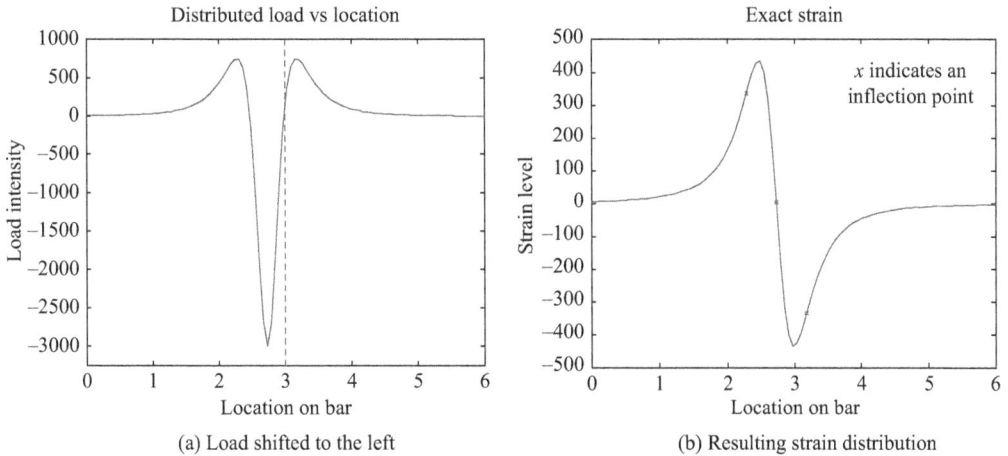

(a) Load shifted to the left   (b) Resulting strain distribution

Figure 2.6. A "High-Demand" loading condition and resulting strain distribution.

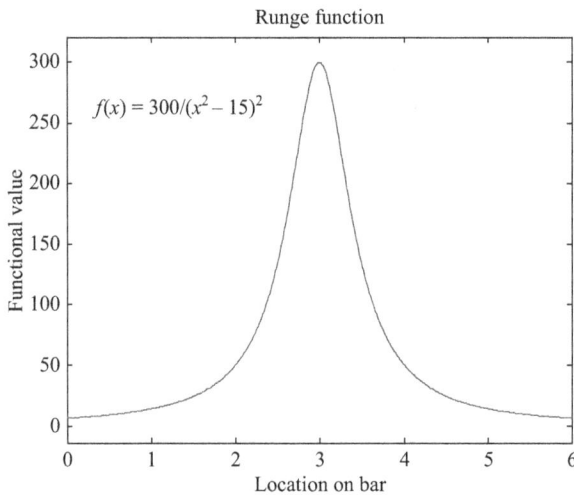

Figure 2.7. A Runge function.

distributed load shown in Fig. 2.5a is found by integrating the following displacement function twice, $f(x) = 300/(x + 30.0/2.2)^2$. It should be noted that the exact strains contained in the example problems shown in Figs. 2.5b and 2.6b are the first derivative of the displacement functions.

2.4   DISCRETIZATION ERRORS IN AN INITIAL MODEL

When the symmetric distributed load shown in Fig. 2.5a is applied to a bar problem modeled with five four-node elements and fixed ends, the strain distribution produced by the finite element model is overlaid on the exact result in Fig. 2.8a. The difference between the two curves identifies the

actual error in the finite element solution. In order to clearly identify the boundaries of the individual elements, the end nodes of the individual elements are designated with "*o*'s."

In Fig. 2.8b, only the finite element approximation of the strain distribution is shown. The interior nodes of the individual elements are designated as "*x*'s" in this figure. The finite element solution is shown separately in this figure because in later refinements it can be difficult to visually separate the exact and the approximate solutions since the interelement jumps in the strain representations are so small that they are not always clearly visible.

As can be seen in Fig. 2.8a, the finite element solution differs markedly from the exact solution in the region containing the critical points that correspond to the maximum and minimum strains. However, the finite element solution is close to the exact solution in the portions of the bar that start from each of the fixed boundaries. This figure demonstrates the two primary objectives of this chapter, which are to show that: (1) discretization errors are produced when the individual finite elements cannot represent the exact strain distribution and (2) the interelement jumps in the strain identify the location and magnitude of the discretization errors.

When we focus on the modeling characteristics of the individual elements in this particular problem, we see that the magnitudes of the errors in the finite element result are inversely proportional to the ability of the individual finite elements to represent the exact result. The nearly linear strains represented by the first and last elements in Fig. 2.8 are very close to the exact result. Similarly, the strains represented by the second and next to last element are nearly parabolic and are quite close to the exact result. As we saw in the previous section, four-node finite elements can represent linear and quadratic distributions exactly. Since each of these elements is very close to representing the exact strain distribution, the contributions made by these elements to the interelement jumps are small.

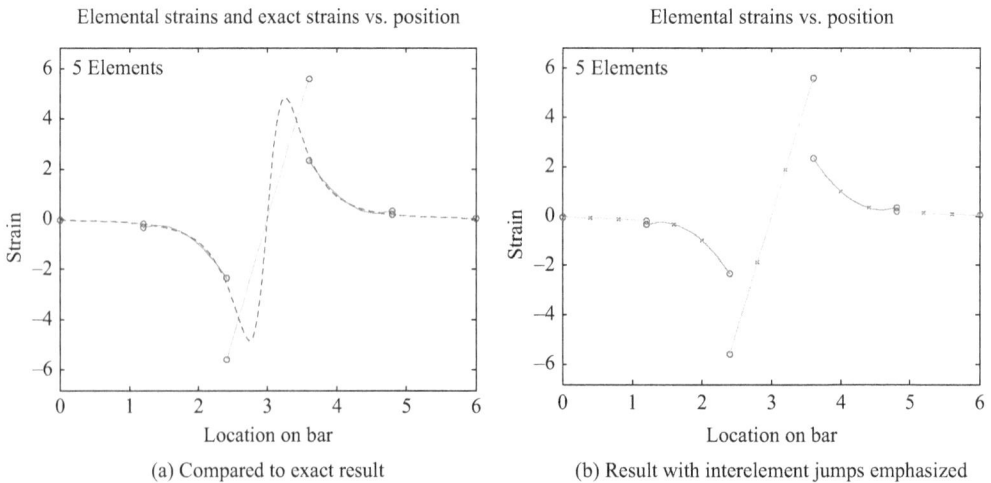

(a) Compared to exact result   (b) Result with interelement jumps emphasized

Figure 2.8. An inaccurate finite element result.

In contrast to the accurate representation of the exact strain distribution at the ends of the bar, the strain distribution produced by the center element is not even close to the exact result. The exact result has an "S" shape that contains five extreme points, the maximum point, the minimum point, and three inflection points. A single four-node element does not possess the capacity to represent this complex strain distribution. As a result, the interelement jumps contributed by this element are large.

Figure 2.8a has demonstrated the correlation between the failure of the finite element model to represent the exact strain and the size of the interelement jumps. The interelement jumps in the finite element strain representation are small at the two end portions of the bar where the finite element result is close to the exact result. Conversely, the interelement jumps are large in the center of the bar where the approximate solution and the exact result differ widely. Thus, we can deduce that the discretization errors are quantified by the size of the interelement jumps. As mentioned earlier, we will give this observation a solid theoretical basis in later chapters.

This section has demonstrated two important characteristics of finite element results. First, the sizes of the interelement jumps are directly related to the difference between the exact result and the finite element representation. Second, the differences between the exact solution and the exact result depend on the ability of the interpolation function in the finite element to capture the exact strain distribution.

2.5   ERROR REDUCTION AND UNIFORM REFINEMENT

We will now refine the problem that was just solved by subdividing each element into two equal length elements. The model is uniformly refined in order to: (1) examine the behavior of individual elements as the finite element model is improved and (2) identify the need for the adaptive refinement process. Two uniform refinements will be applied to the five-element model that was solved in the previous section to produce models with 10 and 20 elements, respectively. The result for the first application of uniform refinement is shown in Fig. 2.9.

When Figs. 2.9a and b are compared to Figs. 2.8a and b, we see that the 10-element approximation is significantly closer to the exact result than the five-element representation. Replacing the single center element with two elements allows the finite element model to capture the general "S" shape of the exact result because of the ability of the individual elements to represent quadratic strain distributions. Since the individual finite elements provide a better representation of the exact solution everywhere in the bar than was the case for the five-element representation, the interelement jumps are reduced. This reduction in the errors is expected since increasing the number of elements in the model should better represent the exact solution.

The interelement jumps associated with the two elements in the center of the 10-element model are still relatively large compared to the errors in the remainder of the problem. These errors exist because the exact strain distribution in the region of these two elements is still too complex for the two four-node elements to capture. Specifically, the modeling capabilities of the two elements in the center of the bar are incapable of capturing the complex shape of the exact result that contains both the extreme values and the inflection points that were identified in Fig. 2.6b.

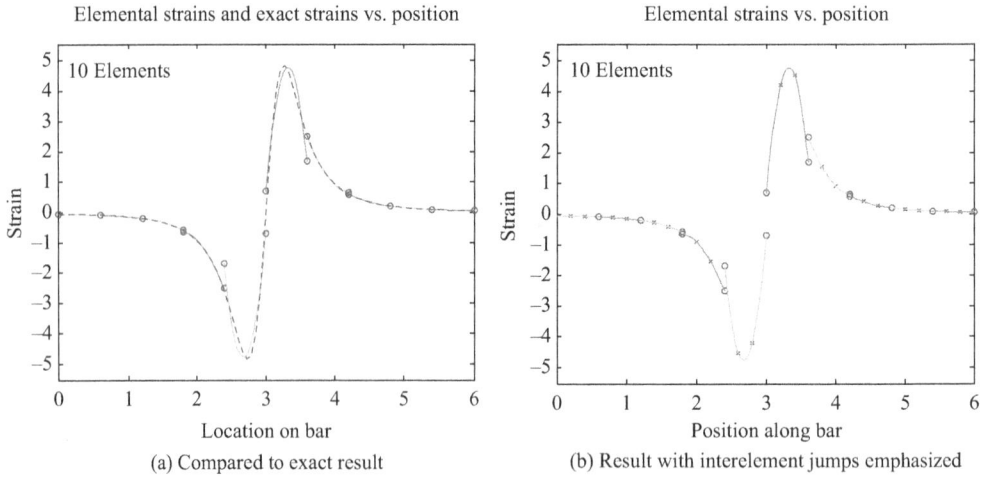

Figure 2.9. A uniformly refined finite element result.

The strain distribution produced by the 20 elements contained in the second uniform refinement is shown in Fig. 2.10. As can be seen, the four elements representing the center portion of the bar are better able to capture the "S" shape than were the two elements in the previous model. Due to the improvement in the finite element strain model, the interelement jumps are reduced, as would be expected.

The primary flaw in uniform refinement can be seen in this sequence of refinements. These examples show that uniform refinement introduces too many elements into regions of the model with little or no error. This can be seen at the two ends of the bar.

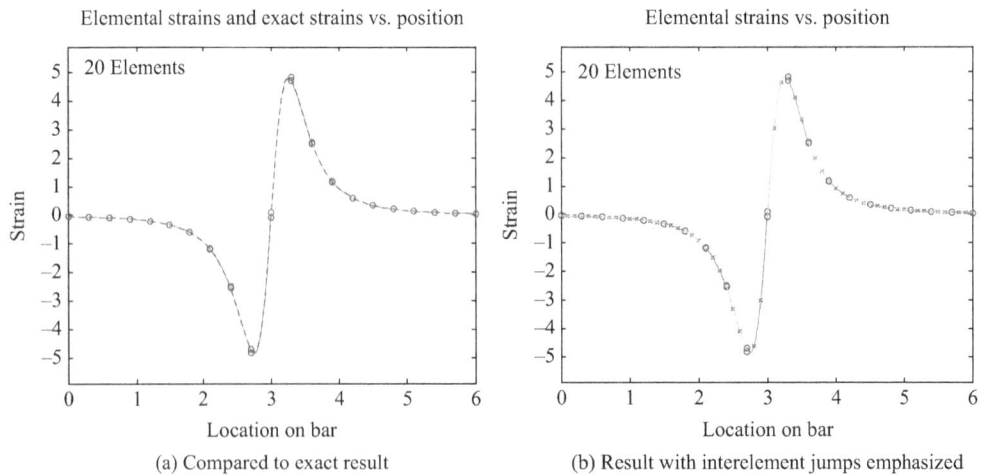

Figure 2.10. A uniformly refined finite element result.

Little, if any, error exists in these regions after the first uniform refinement has been applied. Hence, little improvement is possible in these two regions as a result of a second uniform refinement. Thus, little, if any, improvement occurs in the finite element representation at the two ends of the bar when the number of elements is doubled in these two regions. It can be concluded that this use of additional modeling capacity in regions of low error is wasted.

To summarize, this section has identified the need for adaptive refinement by demonstrating that uniform refinement introduces unnecessary modeling capability in regions that are already well represented. Inversely, a comparison of Figs. 2.9 and 2.10 demonstrated that the finite element model was significantly improved when additional elements were introduced into regions of high error. This improvement is produced because the subdivision of the exact result simplifies the strain distribution that an individual element must represent.

2.6   ERROR REDUCTION AND ADAPTIVE REFINEMENT

This section demonstrates the effectiveness of the adaptive refinement process by subdividing only the elements with the highest levels of interelement jumps and comparing the results to those produced by uniform refinement. We will see that the selective refinement of the finite element model produces a more efficient use of the modeling capabilities of a finite element model.

In this demonstration, the elements to be subdivided are identified by visual inspection. In later chapters, refinement guides are developed that improve the model by using a wide variety of criteria and strategies. In its simplest form, a refinement guide subdivides individual elements by quantifying the discretization errors using metrics based on the estimates of the error. A later chapter develops refinement guides based on basic principles of finite element modeling, not correlations.

As can be seen in Fig. 2.8, the largest error in the five-element mesh is contained in the center element. In this first example of adaptive refinement, only the center element is subdivided. It is replaced by two equal-length elements. The finite element solution for the resulting six-element model is presented in Fig. 2.11 along with the 10-element uniformly refined model.

When the result from the six-element adaptively refined model shown in Fig. 2.11b is compared to the result for the 10-element uniformly refined model shown in Fig. 2.11a, the results are seen to be quite similar. Both approximations capture the two extreme values and the largest interelement jumps are similar in size and location. Minor differences occur in the interelement jumps at points with low levels of strain. We will quantify the differences between the two solutions in later chapters after we have developed error estimators.

The two elements in the center of the model in Fig. 2.11b are easily identified as containing the highest level of errors because they are associated with the largest jumps in the interelement strains. To further demonstrate the advantages of adaptive refinement, we will subdivide these two higher-error elements.

As mentioned earlier, when the two results are compared, we see that slightly more error is present in the two elements flanking the center elements in the adaptively refined model than in the elements of the corresponding region of the 10-element uniformly refined model, even though the strain is nearly zero in this region, that is, this is not a region of critical strain. These two elements

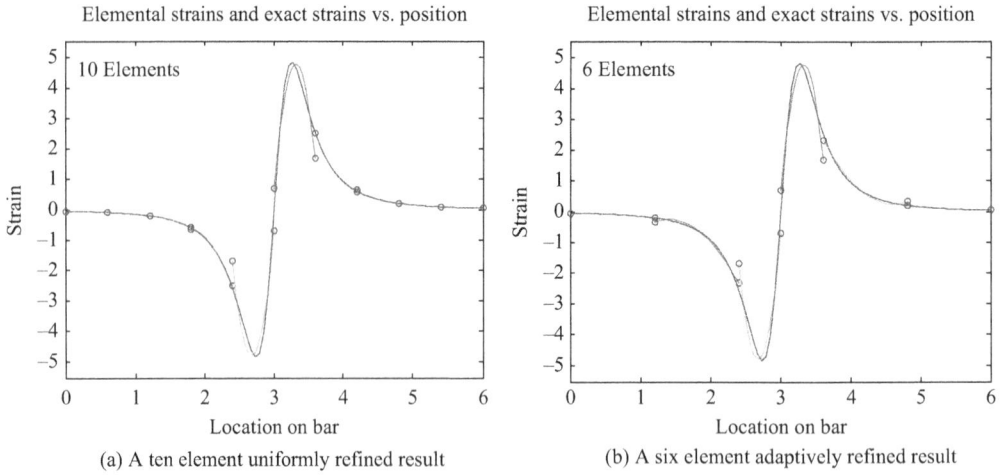

(a) A ten element uniformly refined result

(b) A six element adaptively refined result

Figure 2.11. A uniformly refined result and an adaptively refined finite element result.

will be adaptively refined because the error is slightly higher in this region than in the corresponding region of the uniformly refined model. When the six-element model is adaptively refined, the result is a 10-element model.

The strain distribution for the adaptively refined 10-element model is shown side-by-side with the result from the uniformly refined model with 20 elements in Fig. 2.12. When the results are compared, we see that the two approximations are nearly identical even though the adaptively refined model contains only 10 elements. This example further demonstrates that judicious refinement of a model produces an efficient use of modeling capacity.

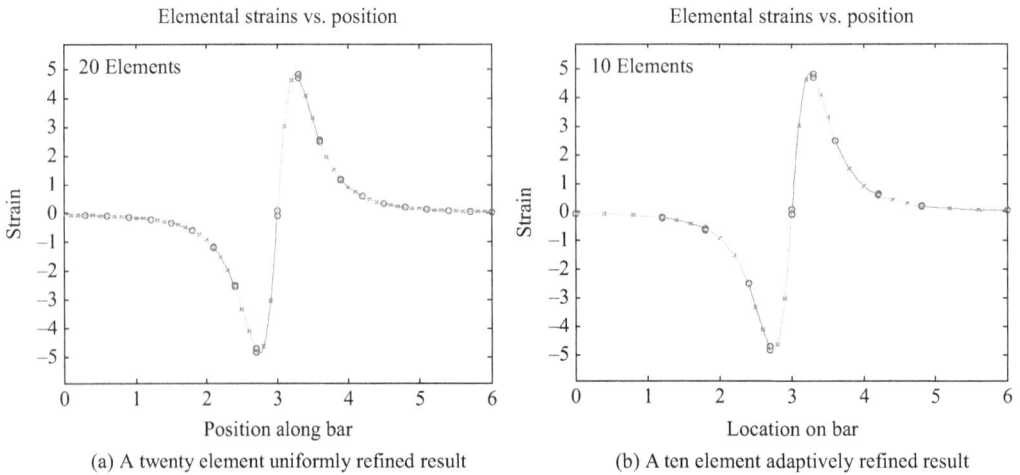

(a) A twenty element uniformly refined result

(b) A ten element adaptively refined result

Figure 2.12. A uniformly refined result and an adaptively refined finite element result.

To summarize, this section has shown the modeling efficiency that results from applying adaptive refinement. However, there is still work to do. We must replace the visual identification of elements that can be profitably subdivided with procedures that can be applied automatically. This will be demonstrated in later chapters.

2.7 THE EFFECT OF ELEMENT MODELING CAPABILITY ON DISCRETIZATION ERRORS

The objective of this section is to further highlight the direct relationship between the discretization errors and the modeling characteristics of individual finite elements. We will accomplish this by comparing and contrasting the results for a problem that is modeled using both four-node and three-node elements. The two models being compared have different numbers of elements but both models contain the same number of degrees of freedom. We will solve an initial model of 23 degrees of freedom as well as a uniformly refined model of 47 degrees of freedom. This problem is loaded with the nonsymmetric loading condition shown in Fig. 2.6.

The exact solution and the approximate strain distributions for the two initial models are shown in Fig. 2.13. The primary purpose of this example is to compare and contrast the way the two different elements attempt to capture the extreme points in the stress distribution. Both models overshoot the maximum strain. Neither element can capture both extreme points because the exact strain distribution is more complex than either element can represent.

The most significant qualitative difference between the modeling capabilities of the four- and the three-node elements is revealed in their attempts to represent the minimum peak. Because the strain representation of the four-node element is capable of representing a quadratic curve, it closely

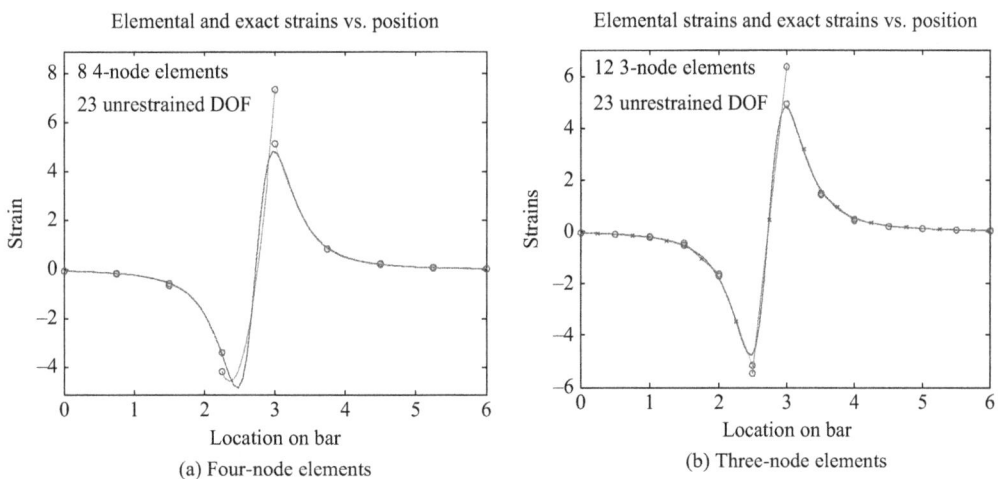

Figure 2.13. Twenty-three degrees of freedom models.

captures the actual shape near the minimum extreme point. The three-node model cannot capture the extreme point on the domain of an element because a three-node element can only represent linear strain.

In Figs. 2.13a and b, the magnitudes of the interelement jumps for both of the finite element models are not acceptable because they are too large with respect to the absolute value of the maximum strains. We will now uniformly refine the two models in order to improve the representations of the strains. The results of this improvement to the models are presented in Fig. 2.14.

As can be seen in Fig. 2.14a, the four-node model closely captures both extreme values of the strain distribution. In the case of the three-node model shown in Fig. 2.14b, significant interelement jumps occur in the regions of the extreme values as well as at the node in the center of the model.

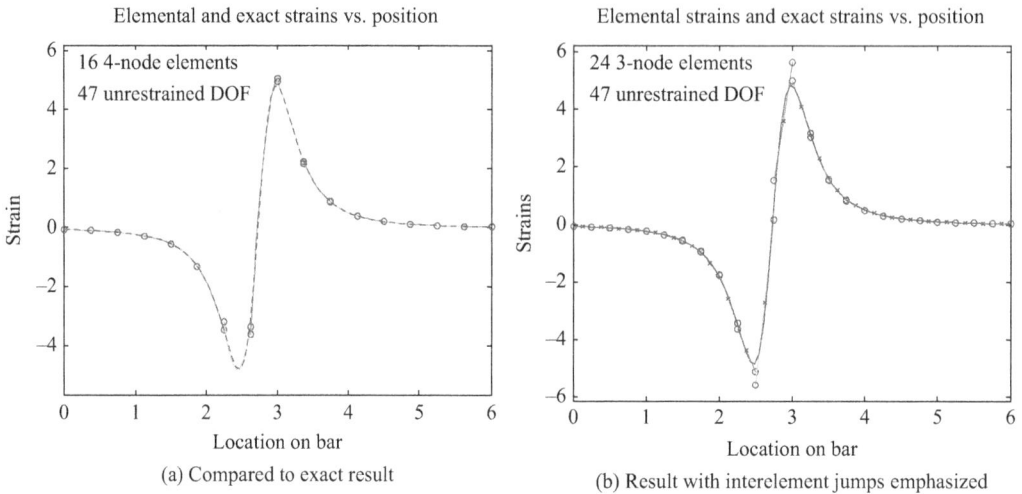

(a) Compared to exact result

(b) Result with interelement jumps emphasized

Figure 2.14. An unconverged finite element result.

The contrasting behavior of the two models at the minimum extreme point is highlighted in Fig. 2.15. In this figure, the regions containing the minimum extreme values for the two refinements of both the four-node and the three-node models are magnified as if examined under a microscope. In both four-node representations shown in Figs. 2.15a and c, the extreme value is captured on the domain of a single element. When these two figures are compared, we see that the location and the magnitude of the representation of the minimum strain are improved in the second model so that it almost matches the exact result.

When the two representations produced by the two models formed with three-node elements shown in Figs. 2.15b and d are examined, it can be seen that further refinement is needed if the three-node model is going to more closely capture the location of the extreme value. Note that the interelement jump is larger in the region of the minimum point for the refined model. This behavior will be explained in Chapter 4.

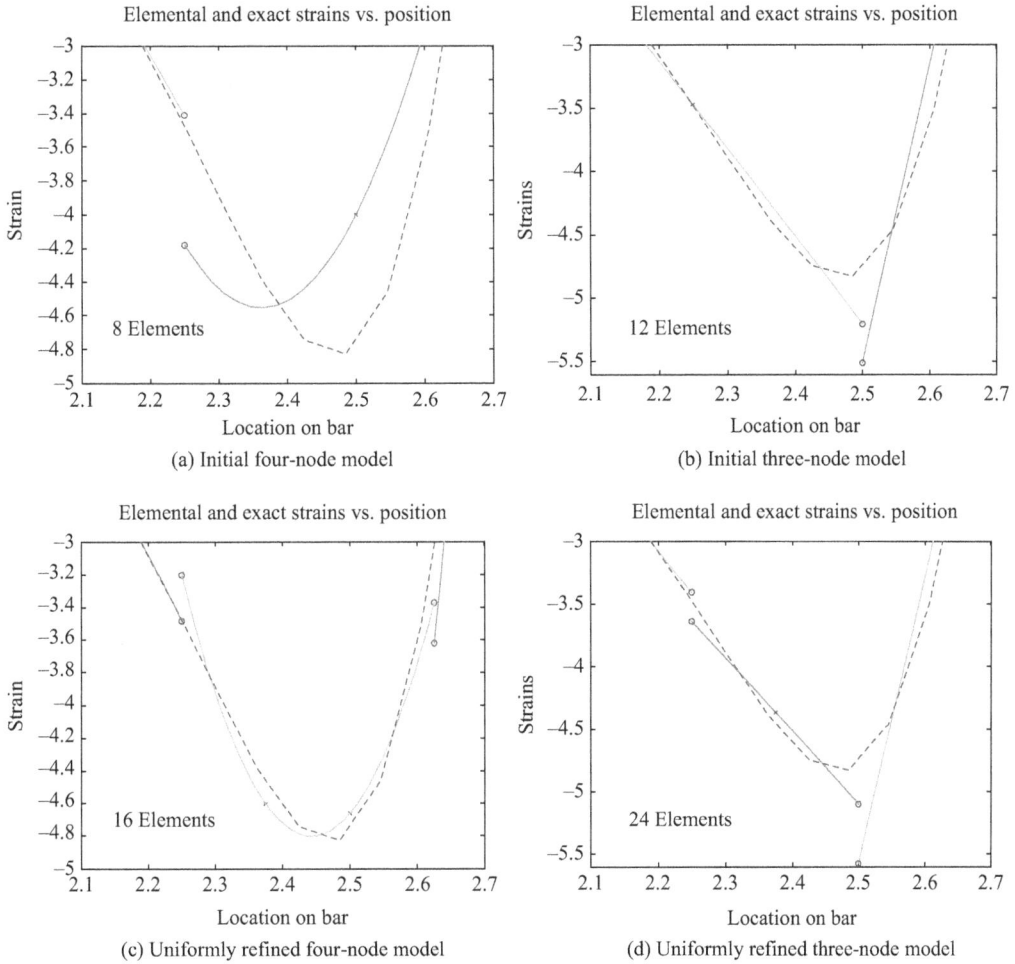

Figure 2.15. Magnification of the minimum points.

In summary, this section has shown that the modeling capabilities of the individual elements are related to the accuracy of the solution. In Chapter 9, we will see that higher order elements are significantly more efficient than lower order elements when the modeling capabilities of the elements are fully utilized.

It should be made clear that no single best element exists for every application. For example, if the maximum stress or strain concentration occurs on the boundary, an element capable of representing linear strain distributions is adequate. This can be seen in Figs. 2.13 and 2.14 by the accuracy of the results as they approach the extreme points.

If the stress concentration takes the form of an extreme value, elements that are capable of representing a quadratic strain distribution have the advantage of being able to capture the extreme value on the domain of a single element. This is one of the lessons that has been shown clearly in this section.

2.8 SUMMARY AND FUTURE APPLICATIONS

This chapter has accomplished the three objectives identified in the Introduction of this chapter, namely:

1. We have shown that modeling errors appear in finite element models as interelement jumps in the strains and that the size of the jump is related to the severity of the modeling error.
2. We have shown that the modeling errors are directly connected to the ability of the individual elements to represent the complexity of the exact solution to the problem, that is, the errors are related to the modeling characteristics of the individual finite elements.
3. We have shown that models can be improved with uniform refinement, adaptive refinement and with the use of higher order finite elements.

The applications of uniform and adaptive refinement have shown that the practical application of the finite element method requires the use of adaptive refinement. If uniform refinement is used instead of adaptive refinement, inefficient models with too many degrees of freedom are generated. The use of higher order elements is studied in detail in Chapter 9.

As has been described in this chapter, the implementation of adaptive refinement requires the following three capabilities:

1. The ability to identify the location and to estimate the magnitude of discretization errors in the finite element model.
2. The ability to identify when to terminate the adaptive refinement process, that is, the ability to identify when the solution is as accurate as desired.
3. The ability to identify how to effectively refine the finite element model in regions of excessive error, that is, develop refinement guides.

We will use the knowledge gained in this chapter as the basis for developing the three capabilities just identified for the implementation of the adaptive refinement process. That is to say, in later chapters, we will develop (1) error estimators, (2) termination criteria, and (3) refinement guides.

The demonstrations presented in this chapter are given a solid theoretical foundation in Chapters 3 and 4. Chapter 3 presents and applies procedures for formally identifying the strain modeling capabilities of individual finite elements. This knowledge will be used to define the local refinement of the finite element model that will improve the finite element model. Chapter 4 contains the proof that the jumps in the interelement strains are directly related to the inability of an individual element to represent the strain distribution in the exact solution.

Chapter 11 develops a refinement guide that is based on an estimation of the exact solution that is emerging from the finite element result. This estimate is compared to the modeling capabilities of the elements used in the finite element model to identify the number of subdivisions that are given to an element. These two capabilities make use of the physically interpretable strain gradient notation introduced in Chapter 1 and that is used in several of the chapters that follow.

2.9 REFERENCES

1. Scheid, F. *Numerical Analysis*, 2nd Ed., McGraw-Hill, Inc., New York: Schaum's Outline Series, 1989, p. 267.
2. Madson, J. C. and Handscomb, D. C. *Chebyshev Polynomials*, Boca Raton, FL: Chapman and Hall/CRC, 2003, p. 45.
3. Van Loan, C. F. *Introduction to Scientific Computing*, 2nd Ed., Upper Saddle River, NJ: Prentice-Hall, 2000, pp. 90–1.

Elements of Two-Dimensional Modeling

2A.1 INTRODUCTION

The focus on the adaptive refinement of one-dimensional problems leaves one significant constraint on finite element models untouched, namely, the requirement that the aspect ratio of the elements must be controlled. If an element is too long and narrow, significant elemental modeling errors are introduced into the model. Elements with acceptable and unacceptable aspect ratios are shown in Fig 2A.1.

When the aspect ratio of an element is too high, the element will contain modeling errors because it will be overly stiff for some of the deformation patterns [A1]. Mathematically, this modeling deficiency will produce an overly wide range of eigenvalues in the element's stiffness matrix. That is to say, if the aspect ratio in an element is not controlled, the element contains modeling errors and the resulting global stiffness matrix can be ill-conditioned.

In order to control the aspect ratio during adaptive refinement, it is often necessary to subdivide elements that accurately represent the strain distribution on their domain. The first objective of this Appendix is to demonstrate the need to subdivide elements that accurately represent the exact solution in order to maintain the proper aspect ratio of elements containing excessive error. The second

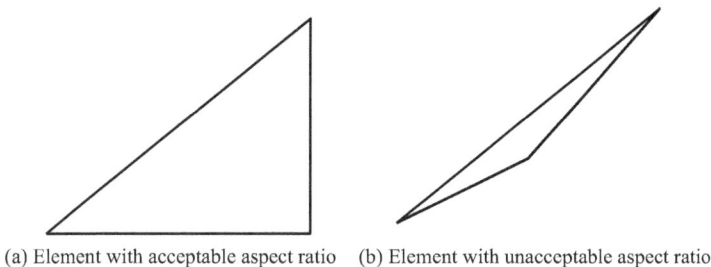

(a) Element with acceptable aspect ratio   (b) Element with unacceptable aspect ratio

Figure 2A.1. Elements with acceptable and unacceptable aspect ratios.

objective is to demonstrate the advantage of using error estimates based on point-wise strain quantities instead of on the widely used strain energy metrics.

The strain-based error measures are better than the strain energy metrics for two reasons. First, the strain-based error estimators are quantified in terms of quantities that are of primary interest in continuum mechanics. This contrasts with strain energy, which is a secondary or derivative quantity in continuum mechanics problems. Secondly, the strain energy error estimators aggregate the errors over the domain of an element instead of estimating the error with a point-wise quantity. This means that integration is required over the domain of each element. The well-known strain energy-based error estimators developed by Zienkiewicz and Zhu are presented in detail in Chapter 7 of this book.

2A.2   SUBMODELING REFINEMENT STRATEGY

If a large problem contains several stress concentrations, the application of adaptive refinement to the whole problem can lead to problems that are too large to be solved with the available computing capacity. In this demonstration of mesh refinement in a two-dimensional problem, a submodeling approach that takes advantage of St. Venant's principle is presented [A2, A3].

In this approach, a region containing high levels of error is separated from the overall problem at a boundary where the exact solution is accurately represented. The relationship between an overall problem and a submodel containing a stress concentration that is not accurately modeled is shown in Fig. 2A.2. The domain of the overall problem is shown with the applicable boundary conditions.

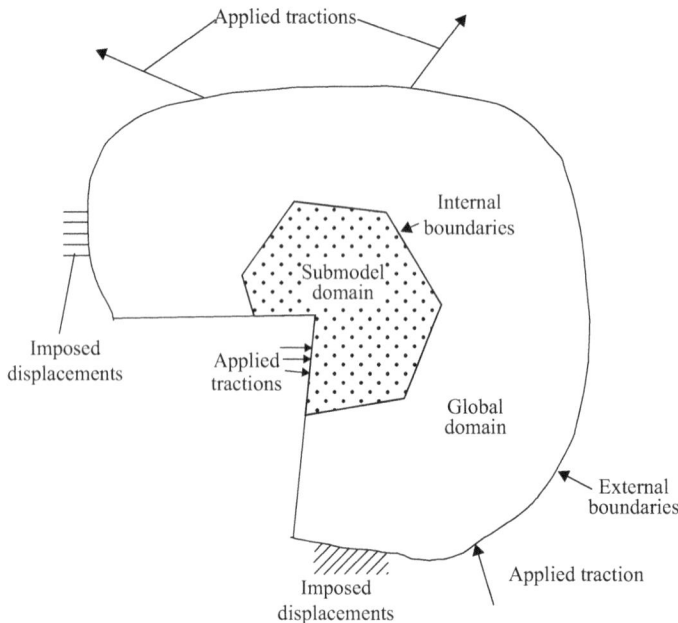

Figure 2A.2. Submodel definition.

The internal boundary that separates the subdomain from the larger problem passes through elements that adequately represent the exact solution. A finite element model of the subdomain is formed and the subdomain is treated as a separate problem. The displacements on the internal boundary found by the finite element model are taken as boundary conditions for the new problem. Any boundary conditions from the original problem that appear in the subdomain are also applied as boundary conditions to the reduced problem.

In this Appendix, the initial model of the subdomain is then adaptively refined. In another version of this approach presented in Reference [A1], the physical size of the region containing the stress concentration is sequentially reduced as the model is improved because the outer elements in the region are now accurately representing the exact solution.

2A.3   INITIAL MODEL

The initial model from which the subdomain is extracted in this example is shown in Fig. 2A.3. This model represents one-quarter of the doubly symmetric Kirsch problem of a circular hole with a loading that is equivalent to 1000 psi on the right end of the panel. Young's modulus is taken as 3000 ksi and a Poisson's ratio of 0.3 is used. This initial model contains 430 degrees of freedom. In the standard form of the Kirsch problem, the panel has an infinite length and the stress concentration factor is 3.0, so the maximum stress would be 3000 psi [A4]. For a panel of this reduced size, the maximum stress in 3047 psi [A5].

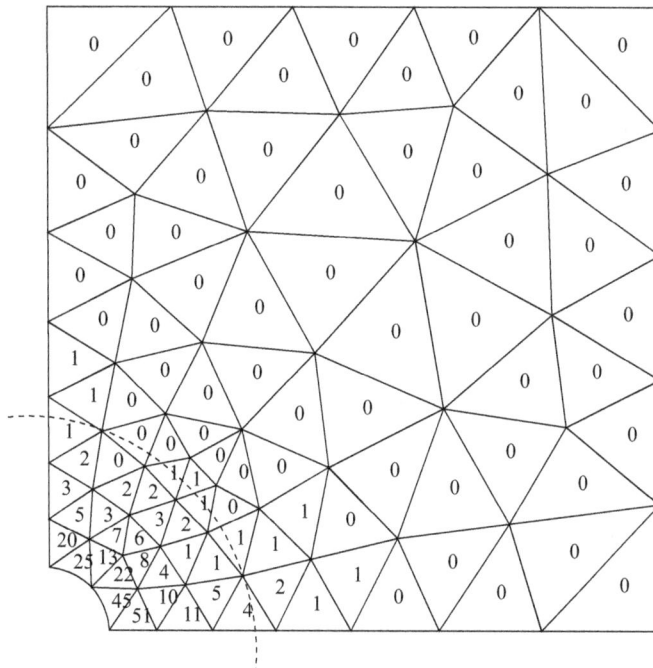

Figure 2A.3. Original model with Z/Z error estimates.

The strain energy errors in the individual elements are identified on the individual elements in Fig. 2A.3. The errors range from 51% at the point of maximum stress contained in the left-most element in the bottom row of elements to 0% for the elements that make up the outer boundary of the subdomain.

2A.4   ADAPTIVE REFINEMENT RESULTS

The subdomain that is adaptively refined is separated from the overall problem by the dashed circular arc shown in Fig. 2A.3. A termination criterion of 4.0% error in the elemental strain energy is used to identify the elements that must be subdivided. The results of the adaptive refinement process are presented in Table 2A.1.

Two significant results are contained in Table 2A.1. On one hand, the maximum stress at the critical point, the point of the stress concentration, converges more rapidly than the error in the elemental strain energy content. This conclusion can be reached by comparing the rates of change of the contents of Columns 2 and 3 with Column 4. This difference in convergence exists because the stress result is a point-wise measure instead of an aggregated measure like the strain energy.

In addition to this functional advantage, the meaning of the error in the stress result is more useful than the error in the elemental strain energy. The stress metric that quantifies the error estimate is directly related to the primary reason for performing the analysis in the first place, which is the identification of the location of the maximum stress or strain level in the problem.

The fact that the need to control the aspect ratio leads to the sub-division of elements that accurately represent the continuum can be seen in Fig. 2A.4. This figure contains the meshes for the refinements containing 942 and 4466 degrees of freedom, respectively. These meshes are chosen for display because they clearly show the subdivision of elements that accurately represent exact solution. As can be seen in Fig. 2A.4a, the outer edge of the subdomain is represented with eight elements that have a low level of estimated error.

In the next refinement shown in Fig. 2A.4b, the eight elements on the outer edge have been replaced by 12 elements. That is to say, elements with acceptable levels of error have been subdivided to maintain aspect ratio control. Note that there are no long and thin elements in these models so the

Table 2A.1. Subdomain adaptive refinement summary

Iteration number	Maximum stress	Stress concentration error	Maximum elemental error	Degrees of freedom
1	2984	2.07	25.0	282
2	2977	2.29	20.0	442
3	3005	1.38	12.9	942
4	3017	0.98	7.2	4466
5	3018	0.95	5.1	9174
6	3018	0.95	4.0	11,394

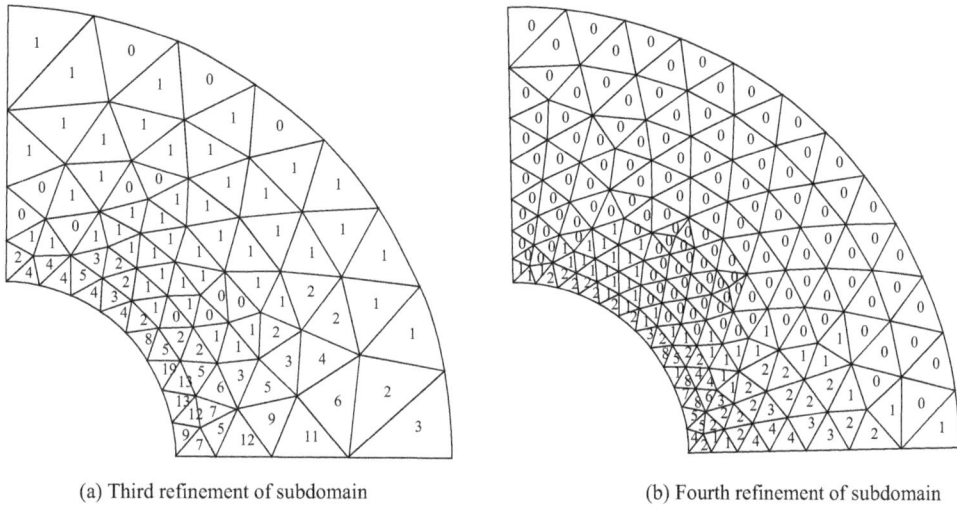

(a) Third refinement of subdomain

(b) Fourth refinement of subdomain

Figure 2A.4. Successive adaptive refinements of the subregion.

aspect ratio has been controlled. An inspection of these figures identifies other elements that accurately represent the continuum that have been subdivided.

2A.5 SUMMARY

This abstraction of results from References [A2, A3] has demonstrated one aspect of finite element modeling for multidimensional problems that is not encountered in the study of one-dimensional problems. The need to control the aspect ratio of the individual elements can lead to the refinement of elements that accurately represent the exact solution.

In the process of demonstrating this characteristic of multidimensional problems, the advantage of using point-wise estimates of the strain as an error estimator has been seen. The metric has direct meaning in terms of the reason an analysis is performed, namely, will the material exceed the stress or strain limit? Furthermore, the higher resolution of the point-wise metric leads to faster convergence than does the error estimator based on the aggregation of the elemental strain energy. Further examples of the efficacy of point-wise error estimators are presented in Reference [A6].

Since no mention has been made of how to accomplish aspect ratio control, a very brief introduction to the topic will be given. A widely used approach to mesh generation and refinement is Delaunay triangulation. A simplified version of this approach to aspect ratio control is given in Reference [A7].

In this simplified approach, an element with a high level of error is subdivided by putting a node in the center of the longest side of the element. This strategy insures that the primary goal of mesh refinement is accomplished, namely, the reduction of h, the radius of the circle that circumscribes the element. This reduction in size means that each of the subdivisions will be representing a smaller, less complex portion of the exact solution.

In addition, this procedure for subdividing elements ensures that the aspect ratio is controlled. However, if the long side of the element is also an edge of an element that accurately represents the exact solution, this element must also be subdivided. It is this situation that causes the refinement of elements that exhibit adequately low error levels.

2A.6 REFERENCES

A1. Dow, J. O. *A Unified Approach to the Finite Element Method and Error Analysis Procedures*, New York: Academic Press, 1999.

A2. Sandor, M. J. *Sub-Model Boundary Identification for Use in a Global/Local Adaptive Refinement Technique, Master's Degree Thesis*, University of Colorado, 1993.

A3. Dow, J. O. and Sandor, M. J. "Submodeling Approach to Adaptive Refinement," *AIAA Journal* 33 (1995): 1550–4. DOI: http://dx.doi.org/10.2514/3.12589.

A4. Budynas, R. G. *Advanced Strength and Applied Stress Analysis*, 2nd Ed., New York: WCB/McGraw-Hill, 1999, pp. 235–238.

A5. Young, W. C. *Roark's Formulas for Stress and Strain*, 6th Ed., New York: McGraw-Hill, 1989, p. 725.

A6. Anderson, R. C. *Point-wise Extensions to the Zienkiewicz/Zhu Adaptive Refinement Procedure, Master's Degree Thesis*, University of Colorado, 2010.

A7. Nambiar, R. V., Valera, R., Lawrence, K. L., Morgan, R. B. and Amil, D. "An Algorithm for Adaptive Refinement of Triangular Finite Element Meshes," *International Journal for Numerical Methods in Engineering* 36 (1993): 499–509. DOI: http://dx.doi.org/10.1002/nme.1620360308.

Exact Solutions for Two Longitudinal Bar Problems

2B.1 INTRODUCTION

This Appendix presents the exact solution for a bar fixed at both ends and deformed by two relatively simple distributed loads: (1) a constant load and (2) a linearly varying load. These loads are shown in Fig. 2.3 in the main body of this chapter. They are reproduced here for the convenience of the reader as Fig. 2B.1.

The form of the governing differential equation presented here differs slightly from the form shown in some presentations. In this case, the positive load is on the left-hand side of Eq. 2B.1. This sign on the applied loading occurs when the governing equation is derived using equilibrium or the Calculus of Variations. In some presentations, the distributed load on the right-hand side of the equation has a positive sign. This difference can be explained by Newton's third law concerning equal and opposite reactions.

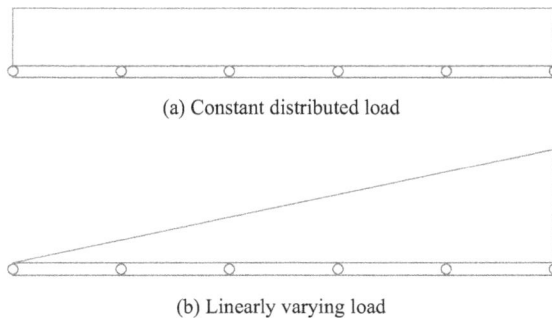

(a) Constant distributed load

(b) Linearly varying load

Figure 2B.1. Two finite element models with distributed loads.

2B.2 GENERAL SOLUTION OF THE GOVERNING DIFFERENTIAL EQUATION

The governing differential equation and boundary conditions for finding the axial displacement, u, for a uniform bar with fixed boundary conditions are the following:

$$AE \frac{d^2u}{dx^2} + f(x) = 0 \qquad \text{(Eq. 2B.1)}$$

$$u(0) = 0.0 \quad u(L) = 0.0$$

where A is the area of the bar, E is Young's modulus, $f(x)$ is the applied load, and L is the length of the bar.

This problem is solved using two applications of separation of variables. After the first application of separation of variables to Eq. 2B.1, we have the following result:

$$AE \int d\left(\frac{du}{dx}\right) = -\int f(x)\, dx \qquad \text{(Eq. 2B.2)}$$

$$AE \frac{du}{dx} = -\int f(x)\, dx + C_1$$

If we had a free boundary, we could evaluate the constant C_1 at this point in the development by imposing the constraint that the boundary load at one end of the bar is equal to zero. If this were the case, the stress and the strain at the free end of the bar would be equal to zero.

2B.3 APPLICATION OF A FREE BOUNDARY CONDITION

A free-end boundary condition would be imposed on the problem as follows:

$$P(\text{free end}) = A\sigma_x = AE\varepsilon_x = AE\left(\frac{du}{dx}\right)_{end} \qquad \text{(Eq. 2B.3)}$$

The boundary condition given by Eq. 2B.3 identifies the load at a free end of the bar in terms of the applied load. The applied load is equal to the area of the bar multiplied by the applied stress applied to the bar. In Eq. 2B.3, the stress is expressed as a function of the strain.

When the free end is unloaded, the boundary condition becomes the following:

$$AE\left(\frac{du}{dx}\right)_{end} = 0$$

$$\left(\frac{du}{dx}\right)_{end} = 0 \qquad \text{(Eq. 2B.4)}$$

$$(\varepsilon_x)_{end} = 0$$

As a result, C_1 in Eq. 2B.2 would be equal to zero. However for the cases being solved here, we do not have a free end for this bar. This treatment of a free boundary is presented for completeness.

2B.4   SECOND APPLICATION OF SEPARATION OF VARIABLES

When we apply separation of variables the second time to Eq. 2B.2, we get the following:

$$AE\int du = \int \left(\int -f(x)\,dx + C_1 \right) dx$$
$$AE\,u(x) = \int\int -f(x)\,dx + C_1 x + C_2$$

(Eq. 2B.5)

We are now in a position to evaluate the arbitrary constants for the case of fixed boundary conditions.

2B.5   SOLUTION FOR A CONSTANT DISTRIBUTED LOAD

We will now find the solution for a bar fixed at both ends when it is loaded with a distributed load that is constant, that is, $f(x) = C$. In this case, the boundary conditions are: $u(0) = 0$ and $u(L) = 0$. When the boundary conditions are substituted into Eq. 2B.5, we have the following:

$$AE\,u(x) = -C\frac{x^2}{2} + C_1 x + C_2$$
$$0 = 0 + 0 + C_2; \qquad x = 0$$
$$0 = -C\frac{L^2}{2} + C_1 L; \quad x = L$$

(Eq. 2B.6)

When these results are rearranged, the arbitrary constants become:

$$C_1 = C\frac{L}{2}$$
$$C_2 = 0$$

(Eq. 2B.7)

When these constants are substituted into Eq. 2B.6, we have the following result for the displacement of the bar:

$$u(x) = \frac{1}{AE}\left(-C\frac{x^2}{2} + CL\frac{x}{2} \right)$$
$$= \frac{C}{2EA}\left(Lx - x^2 \right)$$

(Eq. 2B.8)

As we can see, the displacement function is a quadratic polynomial.

However, in the main body of this chapter, most of the plots are of the strain distributions. The strain distribution for the bar loaded with a constant load is formed from Eq. 2B.8 to give the following:

$$\varepsilon_x(x) = \frac{du}{dx}$$

$$= \frac{C}{2EA}(L - 2x)$$

(Eq. 2B.9)

The strain distribution for a constant distributed load was shown in the main body of the text as Fig. 2.3a. This figure is reproduced here for convenience as Fig. 2B.2. As can be seen, the strain distribution is a linear function as defined in Eq. 2B.9.

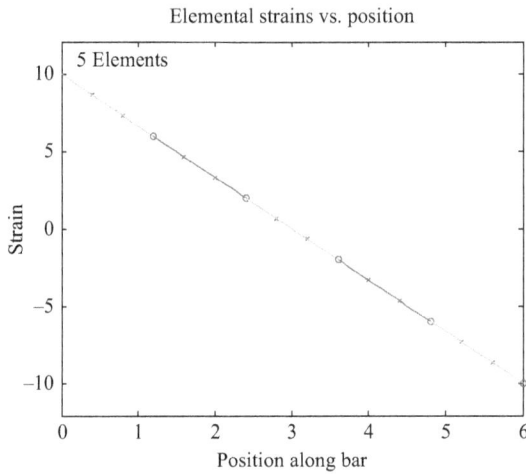

Figure 2B.2. Constant load strain distribution.

2B.6 SOLUTION FOR A LINEARLY VARYING DISTRIBUTED LOAD

When a linear distributed load, $f(x) = Cx$, is applied to the bar and the boundary conditions are applied as was done for the constant load in Eq. 2B.6, we get the following:

$$AE\,u(x) = -C\frac{x^3}{6} + C_1 x + C_2$$

$$0 = 0 + 0 + C_2 \qquad\qquad x = 0$$

$$0 = -C\frac{L^3}{6} + C_1 L \qquad x = L$$

(Eq. 2B.10)

When these results are rearranged, the arbitrary constants become:

$$C_1 = C\frac{L^2}{6}$$ (Eq. 2B.11)

$$C_2 = 0$$

When these constants are substituted into Eq. 2B.10, we have the following result for the displacement of the bar:

$$u(x) = \frac{1}{AE}\left(-C\frac{x^3}{6} + CL^2\frac{x}{6}\right)$$ (Eq. 2B.12)

$$= \frac{C}{6EA}\left(L^2 x - x^3\right)$$

As we can see, the displacement function is a cubic polynomial.

The strain distribution for the bar loaded with a linear load distribution is formed from Eq. 2B.12 to give the following:

$$\varepsilon_x(x) = \frac{du}{dx}$$ (Eq. 2B.13)

$$= \frac{C}{6EA}\left(L^2 - 3x^2\right)$$

The strain distribution for a linear load distribution was given in the main body of the text as Fig. 2.3b. This figure is reproduced here for convenience as Fig 2B.3. As can be seen, the strain distribution is a quadratic function as defined in Eq. 2B.13.

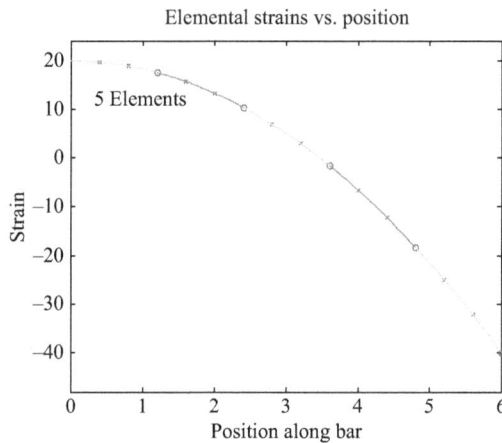

Figure 2B.3. Linear load strain distribution.

2B.7   SUMMARY

These two loading conditions are applied to a finite element model in the main body of this chapter to demonstrate that the interelement jumps in the stresses and/or strains are equal to zero when the finite element solution can represent the exact solution without error.

It should be noted that if there is a change in material or a change in section at some point along the length of the bar, there will be jumps in either the stress or the strain.

CHAPTER 3

IDENTIFICATION OF FINITE ELEMENT STRAIN MODELING CAPABILITIES

3.1 INTRODUCTION

In the previous chapter, the errors in finite element solutions were seen to appear as jumps in the interelement strains. In the next chapter, these jumps are shown to be residual quantities that are created when the finite element solution does not satisfy the governing differential equations being solved. These errors are produced when the interpolation polynomials are unable to capture the complexity of the exact strain distribution that exists on the domains of the individual elements. In other words, the errors result from the limited modeling capacity of the individual elements.

The relationship between the modeling errors and the modeling capability of the interpolation polynomials is illustrated in Fig. 3.1. In this figure, the strain representations of a minimum point produced by models formed with three- and four-node elements and with the **same number of degrees of freedom** are superimposed on the exact solution. When Fig. 3.1a and Fig. 3.1b are compared, we see that the four-node elements better represent the complexity of the exact solution than do the three-node elements.

Four-node elements are better able to represent the minimum point than three-node elements because they can represent curvature in the strain distribution. In contrast, a three-node element can only represent linear strain distributions. This difference in modeling capability exists because the interpolation polynomial of a four-node element contains one more term than does the interpolation function for a three-node element. As a consequence, the four-node strain representation is better able to represent the exact solution so the interelement jumps are smaller for models formed with four-node elements instead of three-node elements.

Chapter 2 and Fig. 3.1 have implicitly identified the inability of the interpolation functions to capture the exact solution as the "root cause" of the errors in finite element models. The **overall**

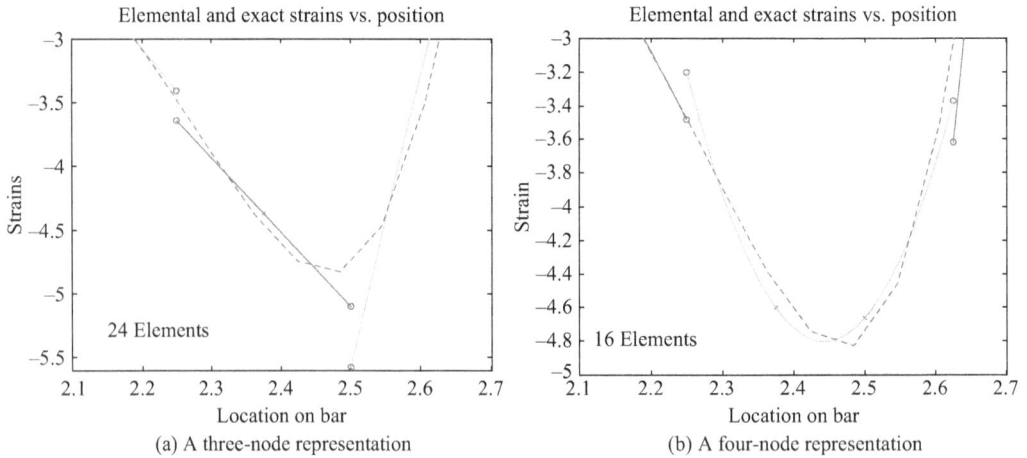

Figure 3.1. Representations of a minimum point.

objective of this chapter is to provide an approach that clearly identifies the modeling capabilities of the interpolation polynomials in individual finite elements. The essence of this procedure is to embed knowledge of solid mechanics into the notation of the finite element displacement interpolation functions.

The **specific objectives of this Chapter** are threefold: **(1)** To express the interpolation functions in terms of strain quantities that allow the modeling capabilities of individual finite elements to be evaluated by visual inspection, **(2)** To visually evaluate the modeling capabilities of selected elements, and **(3)** To extend the application of this transparent, physically interpretable notation to identify the *in situ* strain representations that exist in individual elements.

The first objective is accomplished with a two-step process. The displacement interpolation functions are first recognized as truncated Taylor series expansions. This allows the coefficients of the interpolation functions to be transformed into rigid body motions and strain quantities that produce displacements and deformations in the continuum. A by-product of this development that is important to later developments is the recognition that the finite element and the finite difference methods have a common basis.

The second objective is accomplished in this and subsequent chapters by evaluating the modeling capabilities of selected elements. It should be noted that this capability allows modeling errors in the strain representations of individual elements to be identified if they exist and, in some cases, to be eliminated.

The third objective is accomplished by extracting the strain representations that actually exist in individual elements from the displacements of a solution. In Chapter 11, this capability is extended to form a new type of refinement guide.

3.2   IDENTIFICATION OF THE STRAIN MODELING CAPABILITIES OF A THREE-NODE BAR ELEMENT

The strain modeling capabilities of a three-node element are identified qualitatively in terms of nodal displacements in this section. This is done in preparation for embedding this knowledge into the coefficients of the interpolation functions in the next two sections.

The nodal displacements in a three-node bar are a combination of three linearly independent displacement patterns. Each of these displacement patterns can be chosen so that it corresponds to a clearly defined strain state as shown in Fig. 3.2. The three physically interpretable strain states consist of a rigid body displacement, a deformation due to a constant strain, and a deformation due to a linearly varying strain. Positive axial displacements along the length of the bar are presented as positive vertical displacements in the figures to enhance visual clarity.

The displacement pattern for a rigid body displacement of a three-node element is shown in Fig. 3.2a. As the name denotes, an element experiencing a rigid body displacement does not deform. Every point, including the three nodes, has the same displacement. Since there is no strain in the element, a rigid body displacement is denoted as a zeroth order strain state.

When a three-node bar is deformed with a state of constant strain, the nodal displacements vary linearly in the element as shown in Fig. 3.2b. The right-hand node moves to the right and the left-hand node moves to the left. The nodal displacements are such that both sections of the bar have the same level of tension or compression. The displacements shown in the figure produce tension in the bar. The center node is chosen as the local origin so it does not displace. Again, positive axial displacements in the bar are depicted as upward vertical displacements in the figure. As a consequence of this sign convention, negative axial deformations are shown as downward vertical displacements in the figure.

The third linearly independent displacement pattern that corresponds to a well-defined strain state is

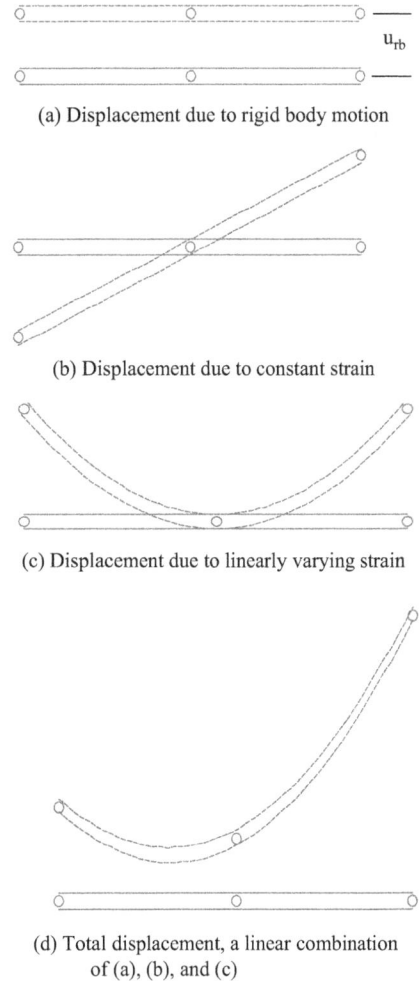

(a) Displacement due to rigid body motion

(b) Displacement due to constant strain

(c) Displacement due to linearly varying strain

(d) Total displacement, a linear combination of (a), (b), and (c)

Figure 3.2. Linearly independent deformation patterns for a three-node bar and their sum.

shown in Fig. 3.2c. This parabolic displacement pattern produces a linearly varying strain distribution. In this case, both end nodes move to the right and the center node is stationary because it is the local origin of the Taylor series expansion. As a result, the left-hand section of the bar is put in compression and the right-hand section is put in tension. The normal strain varies along the length of the bar in the x direction, that is, the quantity $d\varepsilon_x/dx$ is not equal to zero.

Every displacement of the three-node bar is a linear combination of these three linearly independent modeling capabilities. An example of such a linear combination is shown in Fig. 3.2d. This deformation is an example of the most complex shape that a three-node element can represent because it is composed of a linear combination of the three deformation patterns shown in Figs. 3.3a–3.3c.

3.3 AN INTRODUCTION TO PHYSICALLY INTERPRETABLE INTERPOLATION POLYNOMIALS

An overview of the motivation for and an outline of the derivation of the physically interpretable coefficients of the displacement interpolation polynomial is presented in this section. This presentation gives an analytic perspective to the modeling capacities illustrated in the previous section. The contents of this section are condensed in Fig. 3.3. The detailed derivation presented in the next section puts this physically interpretable notation on a solid theoretical foundation.

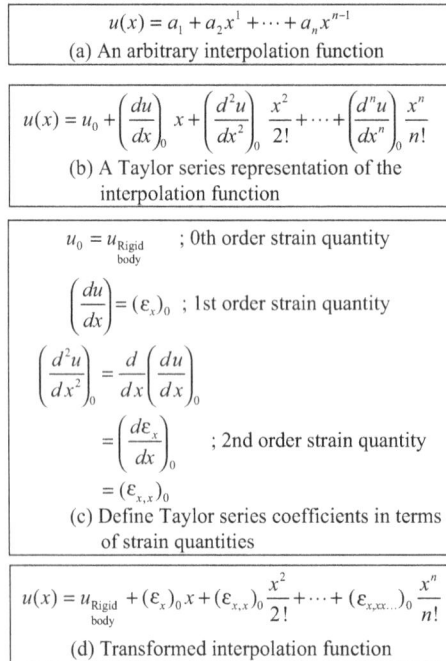

$$u(x) = a_1 + a_2 x^1 + \cdots + a_n x^{n-1}$$
(a) An arbitrary interpolation function

$$u(x) = u_0 + \left(\frac{du}{dx}\right)_0 x + \left(\frac{d^2u}{dx^2}\right)_0 \frac{x^2}{2!} + \cdots + \left(\frac{d^nu}{dx^n}\right)_0 \frac{x^n}{n!}$$
(b) A Taylor series representation of the interpolation function

$$u_0 = u_{\text{Rigid body}} \quad ; \text{0th order strain quantity}$$

$$\left(\frac{du}{dx}\right) = (\varepsilon_x)_0 \quad ; \text{1st order strain quantity}$$

$$\left(\frac{d^2u}{dx^2}\right)_0 = \frac{d}{dx}\left(\frac{du}{dx}\right)_0$$

$$= \left(\frac{d\varepsilon_x}{dx}\right)_0 \quad ; \text{2nd order strain quantity}$$

$$= (\varepsilon_{x,x})_0$$
(c) Define Taylor series coefficients in terms of strain quantities

$$u(x) = u_{\text{Rigid body}} + (\varepsilon_x)_0 x + (\varepsilon_{x,x})_0 \frac{x^2}{2!} + \cdots + (\varepsilon_{x,xx\cdots})_0 \frac{x^n}{n!}$$
(d) Transformed interpolation function

Figure 3.3. The formulation of physically interpretable interpolation polynomials.

Before outlining the derivation, let us identify the limitations of the standard form of the displacement interpolation polynomial shown in Fig. 3.3a. The arbitrary coefficients, the a's, have no clearly identified physical meaning. More specifically, these coefficients do not relate the displacements to the theory of continuum mechanics. As a result, we cannot evaluate the strain modeling capabilities of an individual element by visually inspecting the strain representations formed from the displacement interpolation polynomial presented in Fig. 3.3a.

We can eliminate this deficiency by expressing the interpolation polynomial in terms of strain quantities since these quantities produce the deformations in the continuum. As a result, the strain modeling capabilities of individual elements can be identified by visually inspecting the strain expressions formed from the interpolation functions containing the physically interpretable coefficients. This capability will be demonstrated in this and in later chapters.

The derivation of the physically interpretable interpolation polynomial consists of a two-step process. In the first step of the derivation, shown in Fig. 3.3b, the standard form of the interpolation function is interpreted as a truncated Taylor series expansion of the displacements with respect to a local origin. This step replaces the content-free a's with coefficients expressed in terms of the displacement and its derivatives. The constant term is interpreted as a rigid body displacement of the local origin.

In the next step, shown in Fig. 3.3c, the derivatives of the displacements in the Taylor series expansion are expressed in terms of strain and the derivatives of strain. This is accomplished by introducing the strain–displacement relation, $\varepsilon_x = du/dx$. The coefficients of the displacement interpolation function are now expressed in terms of quantities that are directly related to the displacements and deformation of the continuum.

Finally, the coefficients expressed in terms of the strain quantities contained in Fig. 3.3c are substituted into the Taylor series presented in Fig. 3.3b. The final form of the physically interpretable interpolation polynomial is shown in Fig. 3.3d.

The physically interpretable coefficients that produce displacements for a three-node bar element are the following: $u_{Rigid\ Body}$, $(\varepsilon_x)_0$ and $(\varepsilon_{x,x})_0$. These coefficients indicate that a three-node bar element can represent a rigid body displacement, a constant strain state, and a linearly varying strain distribution. The physical meaning of these three terms is discussed in detail in Section 3.2 and illustrated in Fig. 3.2.

Two features of the notation need to be emphasized. The subscripts written after the comma indicate a derivative and the subscript zero indicates that the quantity is evaluated at the local origin of the finite element. As mentioned in the previous section, we will identify a rigid body motion as a zero-th order strain term since a rigid body motion does not cause the continuum to deform.

3.4 IDENTIFICATION OF THE PHYSICALLY INTERPRETABLE COEFFICIENTS

In the previous section, the interpolation functions for one-dimensional problems expressed in terms of strain quantities were introduced. In this section, the process for forming these physically interpretable coefficients is presented. This is done to show that this notation is not created by an arcane or mysterious process. The coefficients are formed with a straightforward procedure that is easily extended to two and three dimensions. The details of the derivations of the multidimensional cases are presented in Reference [1].

The standard interpolation polynomial for the three-node element is the following:

$$u(x) = a_1 + a_2\,x + a_3\,x^2 \qquad \text{(Eq. 3.1)}$$

In Eq. 3.1, the coefficients of the interpolation function do not have any specified physical meaning. For example, the physical meaning of the coefficient a_3 cannot be identified by visual inspection.

The first step toward expressing the arbitrary coefficients as strain quantities is to interpret Eq. 3.1 as a truncated Taylor series expansion. When Eq. 3.1 is recognized as a three-term Taylor series expansion, we have:

$$u(x) = u_0 + \left(\frac{du}{dx}\right)_0 x + \left(\frac{d^2u}{dx^2}\right)_0 \frac{x^2}{2} \qquad \text{(Eq. 3.2)}$$

where the subscript zero refers to the local origin which is located at the center of the element in this analysis.

The recognition of the interpolation polynomial as a Taylor series expansion changes the nature of interpolation polynomials. This interpretation introduces physical meaning into the coefficients of the polynomial. The coefficients of the interpolation function given by Eq. 3.2 are expressed in terms of the displacement of the continuous bar being analyzed.

In order to evaluate the strain modeling capabilities of an element by visual inspection, the interpolation polynomials must be expressed in terms of strain quantities. This means that we must express the coefficients of Eq. 3.2 in terms of strain quantities. The necessary relationships between the displacements and the strains are shown qualitatively in Fig. 3.2. We will now form the analytic relationships for these relationships by transforming the dependent variables in the coefficients of Eq. 3.2 from displacement quantities to strain quantities.

When Eq. 3.2 is evaluated at $x = 0$, Eq. 3.2 reduces to $u(0)=u_0$. That is to say, the displacement at the origin of the local coordinate system is equal to the constant term in the Taylor series expansion. We will interpret this term as a rigid body displacement and express it as:

$$u_0 = (u_{\text{rb}})_0 \qquad \text{(Eq. 3.3)}$$

where rb designates a rigid body displacement in the x direction.

Since this coefficient is implicitly multiplied by x to the zeroth power, that is, the number 1, the rigid body displacement will be considered as a zeroth order strain term.

We will now transform the coefficients of the linear and quadratic terms of Eq. 3.2 to quantities expressed in terms of strain instead of displacement. This is a straightforward process once we recognize that the coefficient of the linear term is identical to the linear elasticity definition of the normal strain in the x direction, that is, $\varepsilon_x = du/dx$.

When the definition of normal strain in the x direction is applied to Eq. 3.2 and the result is evaluated at the origin of the local coordinate system, the result is the following:

$$\varepsilon_x(x) = \frac{du(x)}{dx} = \left(\frac{du}{dx}\right)_0 + \left(\frac{d^2u}{dx^2}\right)_0 x$$

$$\varepsilon_x(0) = (\varepsilon_x)_0$$

(Eq. 3.4)

Again, the subscript zero is included in order to make it clear that the $(\varepsilon_x)_0$ term is a Taylor series coefficient evaluated at the local origin. Thus, the coefficient of the linear term in Eq. 3.2 can be interpreted as the magnitude of the constant normal strain that exists at the local origin of the finite element.

The coefficient of the quadratic term is expressed as a strain quantity by taking two derivatives of Eq. 3.2 with respect to x. When this coefficient is expressed in terms of the derivatives of the normal strain, we have the following:

$$\left(\frac{d^2u}{dx^2}\right)_0 = \frac{d}{dx}\left(\frac{du}{dx}\right)_0 = \frac{d}{dx}(\varepsilon_x)_0 = \left(\frac{d\varepsilon_x}{dx}\right)_0$$

$$= (\varepsilon_{x,x})_0$$

(Eq. 3.5)

where a subscript after the comma, for example, "x,x" indicates a derivative with respect to x.

Equation 3.5 indicates that the coefficient of the linear term in Eq. 3.2 can be interpreted as the rate of change of the normal strain in the x direction at the local origin. This term is a gradient of the normal strain. As a result, this physically interpretable form of the interpolation polynomial is designated as strain gradient notation.

When the coefficients of Eq. 3.2 are expressed in terms of the physically interpretable quantities as presented in Eqs. 3.3–3.5, the result is the following:

$$u(x) = (u_{rb})_0 + (\varepsilon_x)_0 x + (\varepsilon_{x,x})_0 \frac{x^2}{2}$$

(Eq. 3.6)

where the subscript 0 refers to the local origin and the subscript following the comma indicates differentiation with respect to x.

Equation 3.6 indicates that the displacements in a three-node bar element are produced by a rigid body displacement, a constant strain, and a linearly varying strain. As mentioned earlier, the displacements produced by these three independent components are illustrated in Fig. 3.2.

3.5 THE DECOMPOSITION OF ELEMENT DISPLACEMENTS INTO STRAIN COMPONENTS

In later applications, we will want to identify the participation levels of the individual strain states that exist in the individual elements in a finite element result. That is to say, we want to extract the magnitudes of the coefficients of the physically interpretable interpolation functions from the nodal displacements of a finite element result. In Chapter 11, this capacity is extended to provide the basis for a new approach to creating refinement guides.

The participation factors of the strain states due to the displacements of an element are extracted from a set of equations formed from Eq. 3.6. The nodal locations and displacements are introduced into Eq. 3.6, one at a time, to form the set of equations. This is equivalent to treating the displacement at each of the nodes on an individual element as a boundary condition. The nodal locations for the example presented here are shown in Fig. 3.4. The local origin for this element is located in the center of the element at node 2.

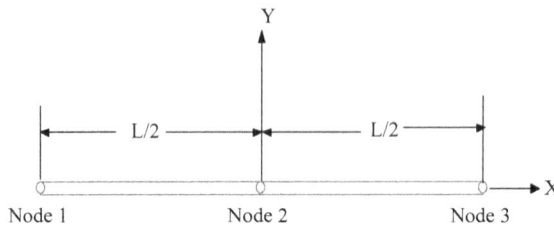

Figure 3.4. Nodal locations for a three-node element.

When we evaluate Eq. 3.6 at node 1, the displacement is equal to the nodal displacement, u_1, and the nodal location is given as $x_1 = -L/2$. When these conditions are introduced into Eq. 3.6, we get the following:

$$u_1 = u(x = x_1)$$

$$= (u_{rb})_0 + (\varepsilon_x)_0 \, x_1 + (\varepsilon_{x,x})_0 \frac{x_1^2}{2} \qquad (\text{Eq. 3.7})$$

$$= (u_{rb})_0 + (\varepsilon_x)_0 \, (-L/2) + (\varepsilon_{x,x})_0 \frac{1}{2}(-L/2)^2$$

When Eq. 3.6 is evaluated at each of the three nodal locations of the finite element as shown in Fig. 3.4, the following set of simultaneous linear equations is produced:

$$\begin{Bmatrix} u_1 \\ u_2 \\ u_3 \end{Bmatrix}_i = \begin{bmatrix} 1 & x_1 & x_1^2/2 \\ 1 & x_2 & x_2^2/2 \\ 1 & x_3 & x_3^2/2 \end{bmatrix} \begin{Bmatrix} u_{rb} \\ \varepsilon_x \\ \varepsilon_{x,x} \end{Bmatrix}_0$$

$$\begin{Bmatrix} u_1 \\ u_2 \\ u_3 \end{Bmatrix}_i = \begin{bmatrix} 1 & -L/2 & (-L/2)^2/2 \\ 1 & 0 & 0 \\ 1 & +L/2 & (+L/2)^2/2 \end{bmatrix} \begin{Bmatrix} u_{rb} \\ \varepsilon_x \\ \varepsilon_{x,x} \end{Bmatrix}_0$$

$$(\text{Eq. 3.8})$$

Equation 3.8 relates the nodal displacements and nodal locations of a three-node element to the rigid body displacement, the constant strain state, and the linear strain distribution that produce displacements in the element. The subscript 0 on the last term of Eq. 3.8 indicates that those quantities are being evaluated at the local origin, which in this case is located at node 2. The subscript i indicates a reference to the i-th element.

Equation 3.8 is an intermediate step in our development. The inverse of this equation contains the relationship being sought. This is the case because after a finite element problem is solved, we know the nodal displacements of each of the individual elements. As a result, we know the quantities on the left-hand side of Eq. 3.8 and we seek the quantities on the right-hand side of Eq. 3.8.

When Eq. 3.8 is inverted, we get the following:

$$
\begin{Bmatrix} u_{\mathrm{rb}} \\ \varepsilon_x \\ \varepsilon_{x,x} \end{Bmatrix}_0 = \begin{bmatrix} 0 & 1 & 0 \\ -1/L & 0 & 1/L \\ 4/L^2 & -8/L^2 & 4/L^2 \end{bmatrix} \begin{Bmatrix} u_1 \\ u_2 \\ u_3 \end{Bmatrix}_i
\qquad \text{(Eq. 3.9)}
$$

Equation 3.9 allows us to identify the contributions of the independent strain states that exist in a three-node bar element for this set of nodal displacements. The quantities on the left-hand side of Eq. 3.9 are the participation factors for the individual strain gradient components. The use of this capability is demonstrated in Section 3.8.

3.6   A COMMON BASIS FOR THE FINITE ELEMENT AND FINITE DIFFERENCE METHODS

We will use Eq. 3.9 to show that the finite element and finite difference methods have a common basis. The existence of a common basis for the two methods means that the finite difference method can be rationally extended to represent practically any domain or boundary condition that can be modeled by the finite element method.

On reflection, it is no surprise that these two powerful approximation techniques can be used to approximate the same set of problems. Both methods are formed from truncated Taylor series expansions. As discussed at length in Reference [1], the two methods solve the same problems with different approaches. The finite element method models the physical system and the finite difference method models the governing differential equation.

In later chapters, we will use the common basis of the two methods to provide the theoretical and practical foundations for point-wise error measures and for refinement guides based on first principles. The implications of the fact that these two powerful numerical methods have a common basis are discussed at length in Reference [1].

We will now show that Eq. 3.9 contains the standard form of the finite difference approximations for the first and second derivatives. This will be accomplished with a slight change of notation.

Figure 3.5 presents a one-dimensional, three-node finite difference template with even spacing. This figure is similar to Fig. 3.4 except that the notation is changed so that it matches the standard notation for a central difference template. In this notation, the nodal spacing is designated as h instead of as $L/2$. When Eq. 3.9 is rewritten with this notation, it becomes:

$$\begin{Bmatrix} u_{rb} \\ \varepsilon_x \\ \varepsilon_{x,x} \end{Bmatrix}_0 = \begin{bmatrix} 0 & 1 & 0 \\ -1/2h & 0 & 1/2h \\ 1/h^2 & -2/h^2 & 1/h^2 \end{bmatrix} \begin{Bmatrix} u_{n-1} \\ u_n \\ u_{n+1} \end{Bmatrix}_i \qquad \text{(Eq. 3.10)}$$

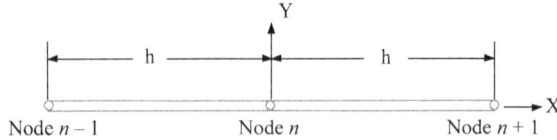

Figure 3.5. A three-node finite difference template.

The first row indicates that every node has a rigid body displacement that corresponds to the displacement of the local origin. When the equations contained in the second and third rows are extracted from Eq. 3.10, we have the following:

$$\varepsilon_x = \frac{du}{dx} = \frac{u_{n+1} - u_{n-1}}{2h}$$

$$\varepsilon_{x.x} = \frac{d^2u}{dx^2} = \frac{u_{n+1} - 2u_n + u_{n-1}}{h^2} \qquad \text{(Eq. 3.11)}$$

For those familiar with the finite difference method, the two components of Eq. 3.11 are identical to the standard form of the central difference template approximations of the first and second derivatives at the local origin.

The components of Eq. 3.11 are developed for evenly spaced nodes. The nodes can, in general, be unevenly spaced [1]. The use of unevenly spaced nodes reduces the order of the error in a finite difference template. However, it is not clear whether the idea of the order of error has much meaning in an adaptively refined problem where the introduction of smaller elements is the objective of the process.

It should be noted that Eq. 3.10 has been extended to multidimensions and uneven mesh spacing in Reference [1]. The ability to form finite difference templates for uneven meshes means that complex geometries can be treated and that adaptive refinement is easily available for the finite difference method. Furthermore, the idea of interelement jumps can be translated to nodal jumps in the finite difference method to provide termination criteria. These capabilities and the research possibilities they engender provide another important opportunity for computational mechanics by leading to a revival of the finite difference method in solving solid mechanics problems.

3.7 MODELING CAPABILITIES OF THE FOUR-NODE BAR ELEMENT

In Fig. 3.1, the ability of a four-node bar element to better represent a minimum point than a three-node element was shown. In this section, we will first identify the modeling capability of a four-node element that makes this improved result possible. Then, using this result, we will form a relationship that allows us to identify the *in situ* participation of the individual strain components of individual four-node elements. In the next section, we will use the capability developed in this section to identify the *in situ* performance of four-node bar elements.

When the displacement interpolation function for a four-node bar element is expressed in terms of the physically interpretable quantities, the result is the following:

$$u(x) = (u_{rb})_0 + (\varepsilon_x)_0 x + (\varepsilon_{x,x})_0 \frac{x^2}{2} + (\varepsilon_{x,xx})_0 \frac{x^3}{6} \qquad \text{(Eq. 3.12)}$$

where the subscript 0 refers to the local origin and the subscript following the comma indicates differentiation with respect to x.

This interpolation function is identical to the one for a three-node bar given by Eq. 3.6 with the addition of a cubic term. Equation 3.12 indicates that the displacements in a four-node element are produced by a rigid body displacement, a constant strain, a linearly varying strain, and a quadratically varying strain. That is to say, this element can represent a strain distribution that is one order higher than can be represented by a three-node element. The difference in the modeling capabilities produced by this additional term is shown graphically in Fig. 3.1.

We will now use Eq. 3.12 to form a relationship for a four-node element that is analogous to Eq. 3.9. This process is demonstrated for the four-node bar shown with even spacing in Fig. 3.6. The local origin is located at the center of the bar.

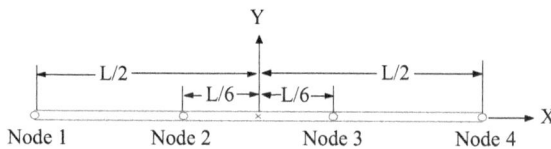

Figure 3.6. A four-node element with a local coordinate system.

When Eq. 3.12 is evaluated at the four nodal locations of a four-node bar, we have the following:

$$\begin{Bmatrix} u_1 \\ u_2 \\ u_3 \\ u_4 \end{Bmatrix}_i = \begin{bmatrix} 1 & x_1 & x_1^2/2 & x_1^3/6 \\ 1 & x_2 & x_2^2/2 & x_2^3/6 \\ 1 & x_3 & x_3^2/2 & x_3^3/6 \\ 1 & x_4 & x_4^2/2 & x_4^3/6 \end{bmatrix} \begin{Bmatrix} u_{rb} \\ \varepsilon_x \\ \varepsilon_{x,x} \\ \varepsilon_{x,xx} \end{Bmatrix} \qquad \text{(Eq. 3.13a)}$$

When Eq. 3.13a is evaluated at the four nodal locations shown in Fig. 3.6, we have the following:

$$\begin{Bmatrix} u_1 \\ u_2 \\ u_3 \\ u_4 \end{Bmatrix}_i = \begin{bmatrix} 1 & -L/2 & L^2/8 & -L^3/48 \\ 1 & -L/6 & L^2/72 & -L^3/1296 \\ 1 & L/6 & L^2/72 & L^3/1296 \\ 1 & L/2 & L^2/8 & L^3/48 \end{bmatrix} \begin{Bmatrix} u_{rb} \\ \varepsilon_x \\ \varepsilon_{x,x} \\ \varepsilon_{x,xx} \end{Bmatrix}_0 \qquad \text{(Eq. 3.13b)}$$

This equation relates the nodal displacements of a four-node element to the four physically interpretable strain components that exist at the local origin of the element being analyzed. The subscript i on the term on the left-hand side of Eq. 3.13b relates the nodal displacements to the ith element. The subscript 0 on the vector term on the right-hand side of Eq. 3.13b relates these quantities to the local origin of the ith element. The four nodal locations are identified in Fig. 3.6 as $x_1 = -L/2$, $x_2 = -L/6$, $x_3 = L/6$ and $x_4 = L/2$.

As was the case for Eq. 3.8, Eq. 3.13 is an intermediate step to the desired result. We are interested in finding the level of participation of the individual strain states that exist in an element for a given set of element displacements. The participation factors are found by inverting Eq. 3.13b.

When Eq. 3.13b is inverted, we get the following:

$$\begin{Bmatrix} u_{rb} \\ \varepsilon_x \\ \varepsilon_{x,x} \\ \varepsilon_{x,xx} \end{Bmatrix}_0 = \begin{bmatrix} -1/16 & 9/16 & 9/16 & -1/16 \\ -11/2\,L & 9/L & -9/2\,L & 11/2\,L \\ 9/2\,L^2 & -9/2L^2 & -9/2\,L^2 & 9/2\,L^2 \\ -27/L^3 & 81/L^3 & -81/L^3 & 27/L^3 \end{bmatrix} \begin{Bmatrix} u_1 \\ u_2 \\ u_3 \\ u_4 \end{Bmatrix}_i \qquad \text{(Eq. 3.14)}$$

Equation 3.14 allows us to identify the contribution of the independent strain quantities that a four-node bar element is capable of representing for a given set of nodal displacements.

3.8   IDENTIFICATION AND EVALUATION OF ELEMENT BEHAVIOR

In this section, we will demonstrate the use of Eq. 3.14 to identify the strain states that individual elements are actually representing in a finite element result. This result is obtained by using Eq. 3.14 to decompose the *in situ* displacements of individual elements into the contributions of the strain quantities that produce the nodal displacements. In later chapters, we use this ability to evaluate element performance and to formulate refinement guides.

We will identify the *in situ* strain modeling behavior of the individual elements for the results shown in Fig. 3.7. This figure presents the finite element strain representation and the exact strain distribution for a five-element model formed with four-node elements. This longitudinal bar problem is fixed at both ends and loaded with the symmetric load shown in Fig. 2.4a. This figure was originally presented as Fig. 2.7 and is reproduced here for the convenience of the reader.

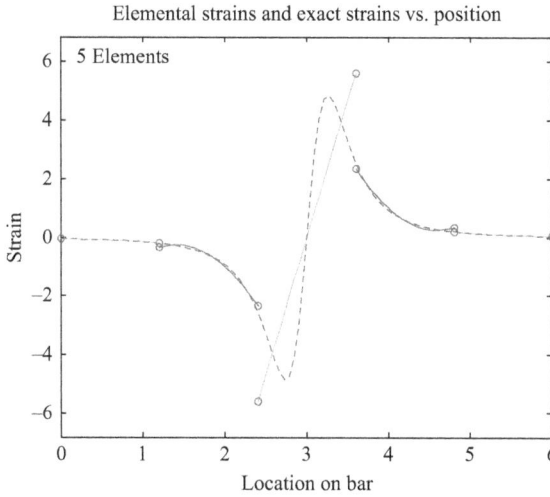

Figure 3.7. A five element finite element result and the exact result.

The participation factors for the linearly independent strain states that produce the displacements of the individual elements contained in Fig. 3.7 are presented in Table 3.1. These values are found by using Eq. 3.14 to decompose the nodal displacements of each of the five elements.

The quantities in the column labeled as $(u_{rb})_0$ identify the contribution of the rigid body displacement to the nodal displacements for the individual elements. The quantities in this column designate the total displacements of the local origin of the individual bars. This meaning can be seen in Eq. 3.12 because this quantity is the constant term in the Taylor series representation of the displacement interpolation polynomial. The other terms in the expansion do not contribute to the displacement of the local origin since they are multiplied by an x-term $(x, x^2, ...)$ which is zero at the origin.

Table 3.1. Physically interpretable components

Element no.	$(u_{rb})_0$	$(\varepsilon_x)_0$	$(\varepsilon_{x,x})_0$	$(\varepsilon_{x,xx})_0$
Element 1	−0.0390	−0.0897	−0.1204	−0.1903
Element 2	−0.3212	−0.5785	−1.6580	−4.2016
Element 3	−2.7992	0.0	9.3401	0.0
Element 4	−0.3212	0.5785	−1.6580	4.2016
Element 5	−0.0390	0.0897	−0.1204	0.1903

The column labeled as $(\varepsilon_x)_0$ contains the strain at the local origin of the individual elements. The meaning of this quantity can be correlated to Fig. 3.7 by comparing the strain at $x = 3.0$ in the figure to the value for $(\varepsilon_x)_0$ of Element 3 in Table 1. This comparison is possible because the point, $x = 3.0$, is also the local origin of Element 3. As can be seen, the strain at $x = 3.0$ is equal to zero in both the plot and in the table.

The quantities labeled as $(\varepsilon_{x,x})_0$ in Table 3.1 are the slopes of the strain at the center of the individual elements. These quantities measure the rate of change of ε_x in the x direction. The meaning of this quantity can be given substance by comparing the values in Table 3.1 to the slopes at the center of Elements 2 and 4 that are identified in Fig. 3.8. As can be seen, the slope of the strain in both elements is -1.6580.

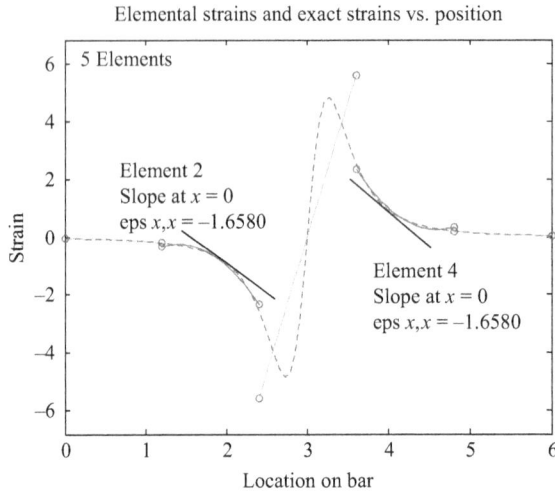

Elemental strains and exact strains vs. position

Figure 3.8. Slope at $x = 0$, $(\varepsilon_{x,x})_0$ for elements 2 and 4.

The slopes, $(\varepsilon_{x,x})_0$, of Elements 1 and 5 are negative and small relative to the other three elements in the model. This can be seen either by consulting Table 1 or by inspecting Fig. 3.8. The slopes of Elements 2 and 4 are both negative and significantly larger in magnitude than the slopes of Elements 1 and 5. Both Fig. 3.8 and Table 3.1 show that the slope of the center element, Element 3, is large and positive.

The quantities in the final column of Table 3.1 are the values of the second derivative of the strain at the local origin, $(\varepsilon_{x,xx})_0$. These quantities can be interpreted geometrically as the curvature of the strain representation at the center of the individual elements.

The meaning of curvature is illustrated in Fig. 3.9. In this figure, the radii of curvature for Elements 2 and 4 are shown. The radius of curvature indicates the size of the circle that has the same curvature as the function being analyzed at the point of tangency. The curvature is equal to the reciprocal of the radius of curvature.

An inspection of Fig. 3.9 shows that Elements 1 and 5 are nearly straight lines, so the magnitude of their curvature is small as is validated in Table 3.1. On the other hand, Fig. 3.9 and Table 3.1 show that that the magnitude of the curvature of Elements 2 and 4 is significantly larger

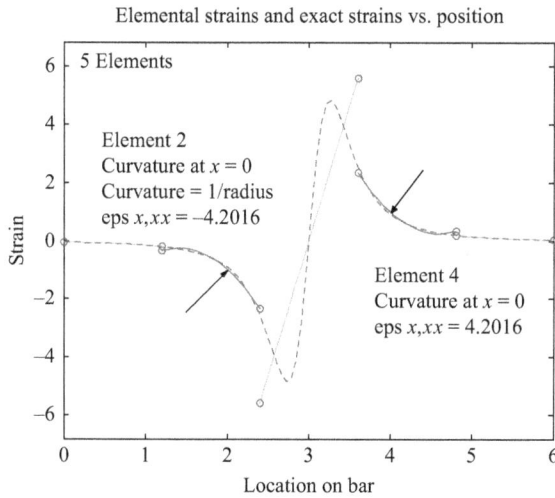

Figure 3.9. Curvature at $x = 0$, $(\varepsilon_{x,xx})_0$ for elements 2 and 4.

than the curvature in Elements 1 and 5. The signs of the curvature of these two elements are negative and positive, respectively. A negative sign indicates that the element is convex when viewed from above. That is to say, Element 2 will not "hold water." In contrast, the curvature of Element 4 will allow the element to "hold water." This is the same sign convention that is used for defining the signs of the bending moments that are applied to beams.

The curvature of Element 3 is given as 0.0 in Table 3.1. This means that the radius of curvature for this element approaches infinity. That is to say, the strain representation in Element 3 is a straight line, which can be seen in Fig. 3.9.

We have demonstrated the ability of Eq. 3.14 to identify the *in situ* participation of the individual strain states for individual elements. As mentioned in the introduction, we will use this capability to decompose the nodal displacements of an element into the independent strain states in order to form refinement guides in Chapter 11.

3.9 EVALUATION OF A TWO-DIMENSIONAL STRAIN MODEL

As noted earlier, this book focuses on one-dimensional problems as a way to introduce and extend the components that make up the adaptive refinement process. However, the focus on the one-dimensional problem does not exhibit a major contribution resulting from the use of strain gradient notation, namely, the identification of strain modeling errors in multidimensional finite elements.

The objective of this section is twofold: (1) to present the interpolation functions for a multidimensional case and (2) to demonstrate the advantages of being able to evaluate the modeling capabilities of an element by inspection that is provided by the physically interpretable notation.

This *a priori* evaluation of the modeling capability of an element is possible because the physically interpretable notation allows problem-specific knowledge to be brought to bear on the analysis.

The value of this capability will be shown by identifying strain modeling errors in four-node quadrilateral finite elements that are submerged when the standard notation is used.

This section and the one following demonstrate the power and usefulness of self-referential notation when it is applied to multidimensional solid mechanics problems. In this section, we will compare the strain representations that are produced by the standard notation to those produced by the physically interpretable notation for a four-node planar element, such as the one shown in Fig. 3.10. In the following section, the strain models that are formed in this section will be used to identify two types of strain modeling errors in the four-node planar element.

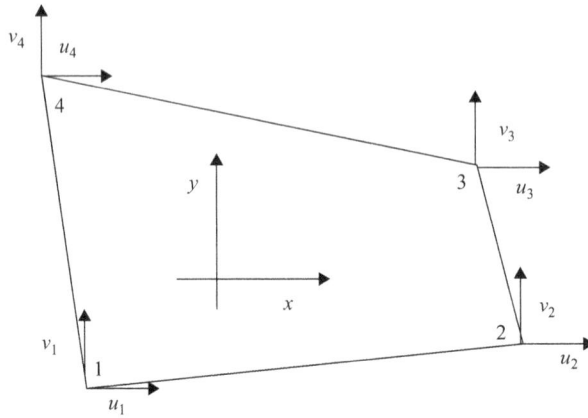

Figure 3.10. A four-node quadrilateral element.

The standard form of the displacement interpolation polynomials for a four-node element is the following:

$$u(x, y) = a_1 + a_2\, x + a_3\, y + a_4\, xy$$
$$v(x, y) = b_1 + b_2\, x + b_3\, y + b_4\, xy$$

(Eq. 3.15)

As can be seen, these functions are augmented linear polynomials because an xy term has been added to the complete linear polynomials in the two equations. The interpolation polynomials are not complete quadratic polynomials because they do not contain x^2 and y^2 terms. Note that the arbitrary coefficients—the a's and the b's—provide no direct information about the modeling capabilities of this finite element.

The displacement interpolation polynomials for the four-node quadrilateral expressed in physically interpretable notation are the following:

$$u(x, y) = (u_{rb})_0 + (\varepsilon_x)_0\, x + (\gamma_{xy}/2 - r_{rb})_0\, y + (\varepsilon_{x,y})_0\, xy$$
$$v(x, y) = (v_{rb})_0 + (\gamma_{xy}/2 + r_{rb})_0\, x + (\varepsilon_y)_0\, y + (\varepsilon_{y,x})_0\, xy$$

(Eq. 3.16)

These displacement polynomials contain the following eight linearly independent coefficients: two rigid body displacements, $(u_{rb})_0$ and $(v_{rb})_0$; one rigid body rotation, $(r_{rb})_0$; three constant strain terms, $(\varepsilon_x)_0$, $(\varepsilon_y)_0$ and $(\gamma_{xy})_0$; and two gradients of the normal strains, $(\varepsilon_{x,y})_0$ and $(\varepsilon_{y,x})_0$. That is to say, the four-node quadrilateral can represent the three rigid body motions, the three constant strain states, and two other deformation patterns. For a complete development of Eq. 3.16, see Reference [1].

The next step in identifying the advantages of the strain gradient notation is to form the strain models that are produced by the standard and the physically interpretable forms of the displacement interpolation polynomials given by Eqs. 3.15 and 3.16, respectively. Then these equations will be "interrogated" to determine what information concerning the strain modeling capabilities of a four-node quadrilateral is provided by direct inspection of these strain representations.

When the definitions of the three strain components are applied to the standard form of the displacement polynomial given by Eq. 3.15, we get the following strain models:

$$\varepsilon_x(x,y) = \frac{\partial u(x,y)}{\partial x} = a_2 + a_4\, y \qquad \text{(Eq. 3.17)}$$

$$\varepsilon_y(x,y) = \frac{\partial v(x,y)}{\partial y} = b_3 + b_4 x \qquad \text{(Eq. 3.18)}$$

$$\gamma_{xy}(x,y) = \frac{\partial u}{\partial y} + \frac{\partial v}{\partial x} = (a_3 + b_2) + b_4 x + a_4 y \qquad \text{(Eq. 3.19)}$$

This result shows that the two normal strains contain a constant term and one linearly varying term. Neither of the normal strain representations are complete linear polynomials, that is, they do not contain both an x and a y term. The shear strain component is a complete linear polynomial because it contains both an x and a y term. Nothing can be said about the detailed strain modeling capabilities of this element from visually inspecting Eqs. 3.17–3.19 because the coefficients of these equations have no intrinsic physical meaning.

The strain representations for the four-node quadrilateral element produced by the physically interpretable displacement polynomials given by Eq. 3.16 are the following:

$$\varepsilon_x(x,y) = \frac{\partial u(x,y)}{\partial x} = (\varepsilon_x)_0 + (\varepsilon_{x,y})_0\, y \qquad \text{(Eq. 3.20)}$$

$$\varepsilon_y(x,y) = \frac{\partial v(x,y)}{\partial y} = (\varepsilon_y)_0 + (\varepsilon_{y,x})_0\, x \qquad \text{(Eq. 3.21)}$$

$$\gamma_{xy}(x,y) = \frac{\partial u}{\partial y} + \frac{\partial v}{\partial x} = (\gamma_{xy})_0 + (\varepsilon_{x,y})_0\, x + (\varepsilon_{y,x})_0\, y \qquad \text{(Eq. 3.22)}$$

When Eqs. 3.20–3.22 are inspected, we see that the strain representations formed are expressed in terms of physically meaningful quantities. The three strain expressions given by Eqs. 3.20–3.22

contain five different quantities. The deformations produced by the three constant strain terms, $(\varepsilon_x)_0$, $(\varepsilon_y)_0$, and $(\gamma_{xy})_0$ are shown superimposed on the original shape of an element in Figs. 3.11a–3.11c, respectively.

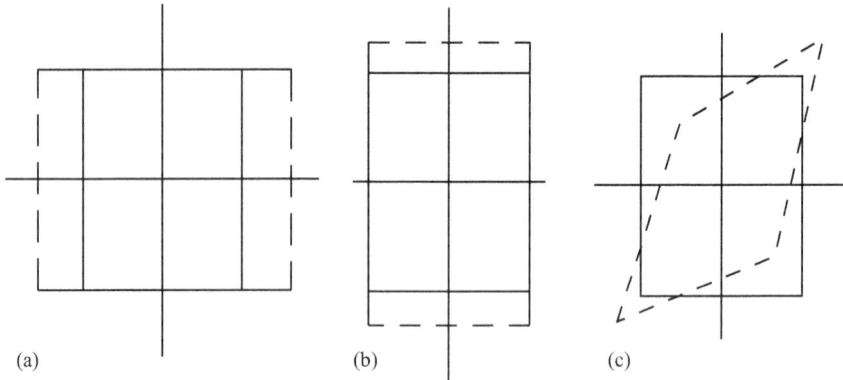

Figure 3.11. Constant strain states.

The displacements produced by the two strain gradient terms $(\varepsilon_{x,y})_0$ and $(\varepsilon_{y,x})_0$ are shown superimposed over the original shape of the element in Fig. 3.12. The two strain states shown in Fig. 3.12 can be interpreted as flexure terms, since they have the form of the strain distribution contained in beam elements. As can be seen in Fig. 3.12, one side of the element is in tension and the other is in compression.

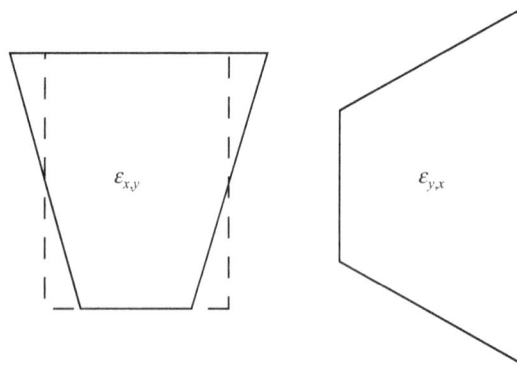

Figure 3.12. Flexural strain states.

As these two figures show, the coefficients of the strain expressions can be directly related to the physical problem being represented. In the next section, we will evaluate the strain models for the four-node planar element given by Eqs. 3.20–3.22 by comparing them to the expected representation from continuum mechanics.

3.10 ANALYSIS BY INSPECTION IN TWO DIMENSIONS

In this section, we will show that the physically interpretable notation provides a transparency to the finite element method that does not exist with the standard notation. We will demonstrate the value of this transparency by identifying several strain modeling errors that exist in the four-node quadrilateral by visually comparing the strain models given by Eqs. 3.20–3.22 to the Taylor series expansion of the strain representations from linear elasticity.

In order to simplify the comparison of the strain model in the four-node element to the expected result from continuum mechanics, we will put both strain representations in a similar form so they can be compared visually. When the strain expressions from continuum mechanics are expanded as Taylor series and put in vector form, we have the following:

$$
\begin{Bmatrix} \varepsilon_x \\ \varepsilon_y \\ \gamma_{xy} \end{Bmatrix} = \underbrace{\begin{Bmatrix} \varepsilon_x \\ \varepsilon_y \\ \gamma_{xy} \end{Bmatrix}_0}_{\text{Constant term}} + \underbrace{x \begin{Bmatrix} \varepsilon_{x,x} \\ \varepsilon_{y,x} \\ \gamma_{xy,x} \end{Bmatrix}_0 + y \begin{Bmatrix} \varepsilon_{x,y} \\ \varepsilon_{y,y} \\ \gamma_{xy,y} \end{Bmatrix}_0}_{\text{Linear terms}} + \cdots \qquad \text{(Eq. 3.23)}
$$

Only the constant and linear terms of this Taylor series expansion are explicitly presented. The higher order terms are not presented because the strain models for the four-node planar element do not contain any higher order terms. In Chapter 10, the role of the compatibility equation is explicated in this derivation.

In order to more easily evaluate the strain modeling capabilities of the four-node element, we will put Eqs. 3.20–3.22 in the following vector form with five separate components:

$$
\begin{Bmatrix} \varepsilon_x \\ \varepsilon_y \\ \gamma_{xy} \end{Bmatrix} = \underbrace{\begin{Bmatrix} \varepsilon_x \\ \varepsilon_y \\ \gamma_{xy} \end{Bmatrix}_0}_{\text{Constant term}} + \underbrace{\left[x \begin{Bmatrix} 0 \\ \varepsilon_{y,x} \\ 0 \end{Bmatrix}_0 + y \begin{Bmatrix} \varepsilon_{x,y} \\ 0 \\ 0 \end{Bmatrix}_0 \right]}_{\text{First linear term}} + \underbrace{\left[x \begin{Bmatrix} 0 \\ 0 \\ \varepsilon_{x,y} \end{Bmatrix}_0 + y \begin{Bmatrix} 0 \\ 0 \\ \varepsilon_{y,x} \end{Bmatrix}_0 \right]}_{\text{Second linear term}} \qquad \text{(Eq. 3.24)}
$$

The five components of this equation are arranged in three groupings. The first group of terms has one component. This component contains the constant term for the three strain components. These quantities match the constant term in the Taylor series expansion of the strain representations given by Eq. 3.23. This means that the four-node quadrilateral element correctly represents the constant strains in the bar.

The second grouping of terms is labeled as the "First Linear Term." This grouping identifies the element's representation of the normal strain components. The third grouping of terms is labeled as the "Second Linear Term." This grouping identifies the element's representation of the shear strain.

We will see that the quantity labeled as the First Linear Term contains errors of omission and the quantity labeled as the Second Linear Term contains errors of commission and omission. In other words, the normal strain expression is missing terms and the shear strain expression contains the wrong terms.

The errors of omission in the First Linear Term of Eq. 3.24 can be seen when it is compared to the Linear Term in Eq. 3.23. We see that the normal strain term in the First Linear Term of Eq. 3.24 contains two zeros. When this term is compared to the equivalent quantities in Eq. 3.23, we see that no similar zeros exist in the continuum representation. The two normal strain quantities missing from Eq. 3.24 are the following: $(\varepsilon_{x,x})_0 x$ and $(\varepsilon_{y,y})_0 y$. The absence of these terms can be viewed as imposing physical constraints on the individual four-node elements, which has the tendency of making the individual elements overly stiff.

The errors of commission are seen when the shear strain representation contained in the Second Linear Term of Eq. 3.24 is compared to the shear strain components in the Linear Term of Eq. 3.23. If we extract the incorrect shear strain representation contained in Eq. 3.24 for the convenience of the reader, we get the following equation that is identical to Eq. 3.22:

$$\gamma_{xy}(x,y) = (\gamma_{xy})_0 + (\varepsilon_{x,y})_0 x + (\varepsilon_{y,x})_0 y \qquad \text{(Eq. 3.25)}$$

When Eq. 3.25 is compared to the third row of Eq. 3.23, we see that they do not match each other. The constant terms are the same, but the linear terms are different. The coefficients of the two linear terms x and y in Eq. 3.25 are the normal strain quantities $(\varepsilon_{x,y})_0$ and $(\varepsilon_{y,x})_0$, instead of the expected shear strain quantities $(\gamma_{xy,y})_0$ and $(\gamma_{xy,x})_0$. The existence of the normal strain terms is an error of commission. These terms do not belong in the shear strain representation. The absence of the shear terms, $(\gamma_{xy,x})_0$ and $(\gamma_{xy,y})_0$, is an error of omission.

The fact that the two normal strain terms contained in Eq. 3.25 are in error can also be demonstrated with a physical argument. The two normal strain terms contained in the shear strain expression are incorrect because they violate the constitutive relationship for solid mechanics. There is no connection between the shear strains and the normal strains in linear elasticity. Thus, this element incorrectly engages a shear strain when the element is experiencing any linearly varying normal strain. This error is well known in finite element analysis and is identified as parasitic shear.

The perceptive reader will have deduced that planar eight- and nine-node elements also possess errors of the type identified for four-node elements. If the standard approach of using reduced-order Gauss quadrature integration is used in an attempt to eliminate parasitic shear from these elements, other errors are likely to be introduced. The errors in these elements can be minimized by identifying them and improving the strain model during the formulation process using this physically interpretable notation [1]. Because these higher order elements still contain errors even after they are improved, these elements are less effective than the six-node linear strain triangle. Thus, it is recommended that the six-node element be used exclusively in place of eight- and nine-node elements.

This discussion of the modeling characteristics of the four-node element is intended to demonstrate the direct connection between the approximate solution technique and the solid mechanics problem provided by the physically interpretable notation. This brief analysis has shown that the four-node quadrilateral element contains several strain modeling errors. Since the four-node quadrilateral element contains modeling errors in all three strain components, the four-node element is less capable of representing the continuum than is a three-node element except possibly in certain special problems [1]. In a nonconverged result, the four-node element is overly stiff because of the strain modeling errors.[1]

For completeness, it should be noted that the strain modeling errors in the four-node element result from the fact that the displacement interpolation polynomials for the element are not complete polynomials. As an aside, it should be noted that a six-node triangle, the linear strain triangle, is derived from complete displacement interpolation polynomials. The strain models for the six-node element match the complete linear representations presented in Eq. 3.23; hence, it contains no inherent strain modeling errors. Since the six-node element contains no inherent strain modeling errors, it is preferred to the four-node element in practically any application.

3.11 SUMMARY AND CONCLUSION

This chapter has demonstrated the value of the physically interpretable notation that provides a direct connection between the formulation of finite elements and the equations of continuum mechanics that are being represented by the finite element method. This notation expresses the interpolation polynomials that are used to form the finite element method in terms of strain quantities, the quantities that produce the displacements in the continuum. That is to say, this physically based notation provides a transparency to the finite element method that is not available with the standard notation. This self-referential notation was related to the arbitrary coefficients of the standard interpolation polynomials in Section 3.3.

We saw that this notation allows us to improve every facet of the finite element method. In Sections 3.4, 3.8, 3.9, and 3.10, this self-referential notation was used to evaluate the strain modeling capabilities of individual finite elements during the formulation process. In Sections 3.5, 3.7, and 3.8, we developed and applied the ability to evaluate the *in situ* performance of individual elements.

In Section 3.6, we showed that there is a direct relationship between the finite element and the finite difference methods. As we will see in later chapters, the identification of this relationship provides the theoretical foundation and practical basis for point-wise error measures, termination criteria, and refinement guides.

This is a disclaimer concerning the self-referential notation presented in this chapter. It should be noted that an understanding of the self-referential notation is only mandatory for researchers and practitioners who seek to understand the finite element method in depth. Once the insights provided by the physically based notation are embedded in the code used in application programs, the casual user never need know what errors have been eliminated from the analysis procedure as a result of using the techniques introduced here and developed in depth in Reference [1]. However,

---

[1] At the element level, the new ability to analyze the element formulation process eliminates the need to use the isoparametric formulation process, eliminates the need to consider spurious zero energy modes, provides a way to eliminate shear locking, and explains aspect ratio stiffening. These ideas and others for improving element development are presented in detail in Reference [1].

the understanding and appreciation of this notation opens the door for research in the application of the finite difference method to solid mechanics problems, in the development of mesh refinement strategies and in the development of error estimators.

3.12 REFERENCE

1. Dow, J. O. *A Unified Approach to the Finite Element Method and Error Analysis Procedures*, New York: Academic Press, 1999.
    This book puts the finite element method, the finite difference method, and error analysis on a common Taylor series basis. It develops and applies the physically based notation introduced in this chapter.

CHAPTER 4

The Source and Quantification of Discretization Errors

4.1 INTRODUCTION

When a finite element model cannot capture the exact solution, interelement jumps or discontinuities exist in the approximate strain representation. In this chapter, we will show that these jumps directly relate to the failure of the finite element representation contained in the individual finite elements to satisfy the governing differential equation being solved. The differences between the exact solution and the finite element result are called **discretization errors**. These errors get their name from the fact that they occur when the continuum is replaced by a discrete number of finite elements, which, in general, cannot represent the exact solution.

The relationship between the interelement jumps and the capability of the individual elements to represent the exact solution is demonstrated in Fig. 4.1. This figure presents the strain results for an initial finite element model that consists of five four-node elements and two sequential uniform refinements of 10 and 20 elements, respectively. As can be seen, the interelement jumps are reduced and the approximate solution gets closer to the exact result each time the model is improved by subdividing the elements that make up the model.

The reduction in the size of the interelement jumps occurs because each element is then required to represent a less complex portion of the exact solution as the size of the elements in the model is reduced. This, in turn, means that the interpolation polynomials of the individual elements are better able to represent the portion of the exact solution that exists on their domain.

The **objective of this chapter** is to show that the interelement jumps in the strain representations produced by a finite element model are a direct result of the failure of the finite element solution to satisfy the governing differential equation on the domain of the individual elements.

The direct connection between the discretization errors and the interelement jumps makes the interelement jumps an ideal error estimator for the adaptive refinement process. This is the case because the interelement jumps in strain express the errors in terms of the primary quantities that

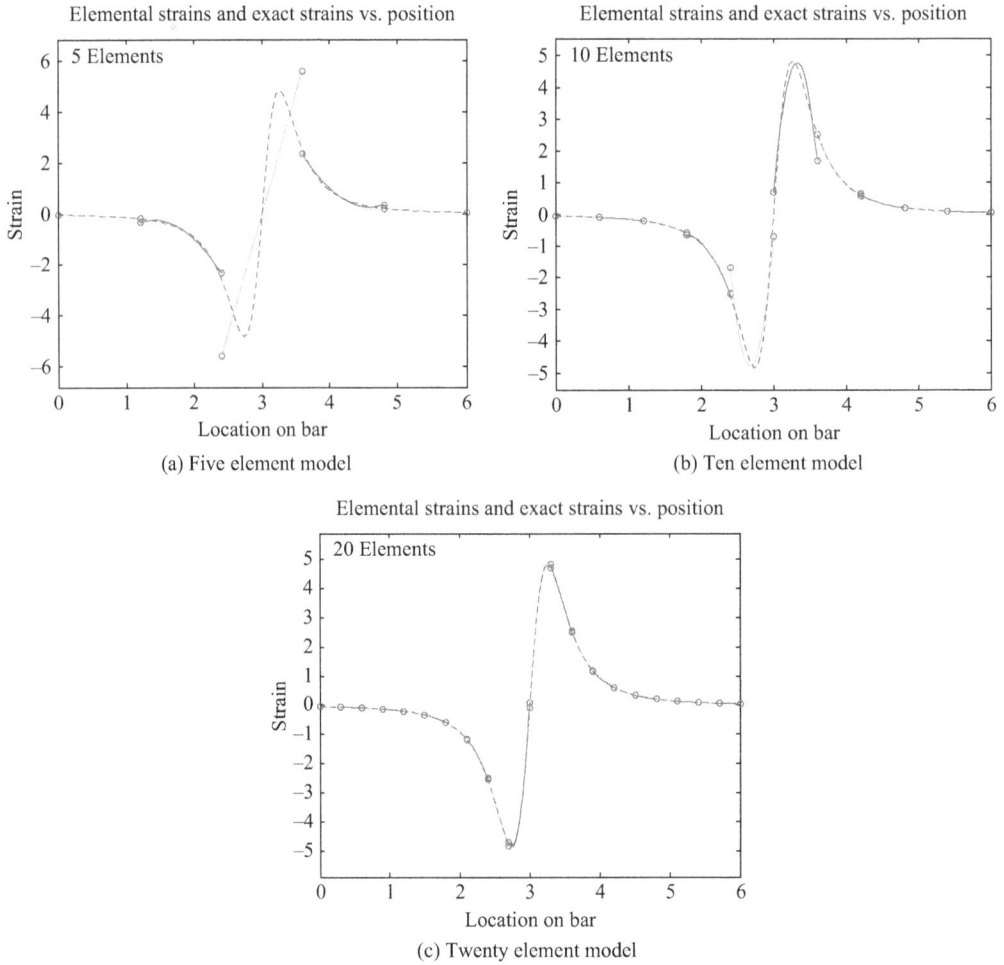

Figure 4.1. A uniformly refined model vs. the exact solution.

are sought in continuum mechanics, namely, strains and, indirectly, stresses. As a consequence of being expressed in terms of strain, a termination criterion for the adaptive refinement process based on this error estimator is intuitively satisfying.

For example, if the acceptable error in the strain result is defined as 4% of the critical strain for the material being analyzed, the use of a strain-based error estimator allows elements containing a higher level of error to be identified and subdivided. In other words, an error estimator based on strain allows the finite element result to be judged in terms of quantities (stress or strain) that are related to the major reason for performing finite element analyses.

In contrast, the first practical error estimator reported the error in an element as a percentage of the error in the strain energy content contained in an element. This metric is not of much use to an analyst because an estimate of the error in the strain energy content of an element does not tell an analyst much about whether the material will fail or if a region is overdesigned. Strain energy-based error estimators are developed in Chapter 7 for historical background and completeness.

The error estimator developed and presented here is classified as a **residual technique**. The failure of the finite element solution to satisfy the governing differential equation being solved is quantified by a residual that is produced when the displacements that exist in the individual elements for a solution are substituted into the governing differential equation. In physical terms, the existence of a nonzero residual means that the point-wise equilibrium is not satisfied by the finite element result.

As we will see in the sections that follow, the point-wise residual quantities can be interpreted as "fictitious" distributed loads on the individual elements. With this interpretation, the "fictitious" distributed loads can be used to form a set of "fictitious" equivalent nodal loads for the individual elements. When these fictitious nodal loads are treated in the same way as actual nodal loads and they are assembled in the usual manner, the resulting components of the "fictitious" applied load vector are found to be directly proportional to the interelement jumps in the strain.

4.2   BACKGROUND CONCEPTS—THE RESIDUAL APPROACH TO ERROR ANALYSIS

This section contains an overview of the residual approach to error analysis that was developed by Kelly [1]. This approach provides an intuitive interpretation of the discretization errors in a finite element model and shows that the interelement jumps in the stresses and/or strains quantify the discretization errors. The objective of this section is to outline this approach so that the individual steps presented in later sections can be seen in context.

It is easy to understand that discrete finite element models cannot represent the exact solutions for each of the infinite number of loading conditions that can be applied to a given continuous physical system. This is the case because a model with a finite number of degrees of freedom cannot represent the exact result for the infinite number of displacement patterns that are possible in the continuum. The innovation in Kelly's work consists of turning this point of view on its head.

Kelly postulates that an approximate finite element solution with its interelement jumps is the exact solution to an auxiliary problem that is closely related to the original problem. The auxiliary problem consists of the original problem that has been augmented with a set of additional loads that produce the interelement jumps in the strains. As we will see, the additional loads are directly related to the residuals that quantify the failure of the finite element solution to satisfy the governing differential equation on the domains of the individual elements.

We will demonstrate the formulation of the additional loads that are used to form the auxiliary problem for the case of the longitudinal bar problem with fixed ends, which is shown in Fig. 4.2a. The original problem, which was discussed in Chapter 2, consists of the five four-node elements with the symmetric load. The strain distribution produced by this finite element model is shown in Fig. 4.2b along with the exact solution. As can be seen, this finite element solution contains jumps in the strains between elements. The actual point-wise errors can be seen as the differences between the approximate finite element result and the exact solution.

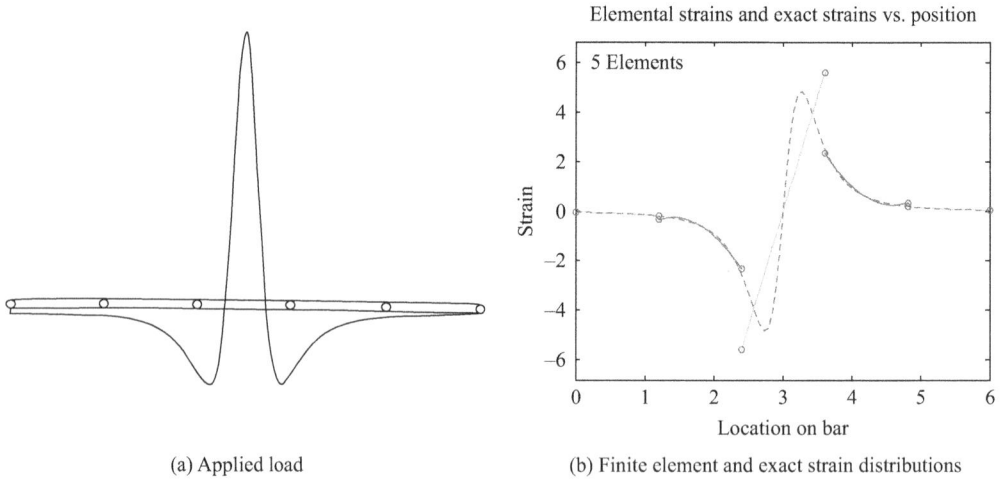

(a) Applied load   (b) Finite element and exact strain distributions

Figure 4.2. A finite element model and its strain result.

Note that each of the interelement jumps in this strain representation is similar in structure. In all four cases, the nodal strain of the right-hand element is below the nodal strain of the left-hand element. This means that the nodal loads that were added to form the auxiliary problem will point downward at each of the interelement nodes. In this case, the sign on each of the auxiliary loads is negative.

Figure 4.3 shows the auxiliary problem for which the finite element result presented in Fig. 4.2b is the exact solution. It consists of the original problem that was presented in Fig. 4.2a with the addition of a set of auxiliary nodal forces.

The additional forces are derived from the jumps in the strains that exist in the example problem. The auxiliary forces are equal to $F_{added} = AE\Delta_{strain}$. In this case, the forces are equal to the

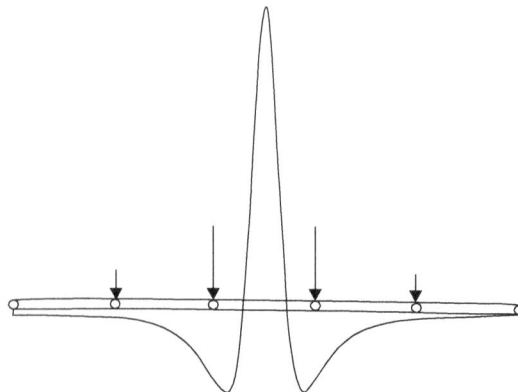

Figure 4.3. Finite element model with the augmented load.

following: −12.9414, −294.7040, −294.7040, and −12.9414, respectively. As is apparent in Fig. 4.2b, the jumps in the strains are proportional to the errors in the approximate solution.

In the next section, we will show that these interelement point loads are identical to an aggregation of the residuals that result from the failure of the individual elements to satisfy point-wise equilibrium on the individual elements. The point-wise residuals quantify the failure of the finite element solution to satisfy equilibrium. The interelement forces can be viewed as an aggregation of the point-wise failure to satisfy equilibrium. When these aggregated loads are applied to the original problem, they produce equilibrium at the nodes in the finite element model.

As a consequence of this relationship between the interelement jumps in strain and the satisfaction of local equilibrium, the interelement jumps decrease as the model is refined and the individual elements better represent the exact solution. Thus, the interelement jumps can serve as *a posteriori* (after the fact) error measures.

It should be noted that the interelement jumps in strains for two-dimensional problems consist of jumps in each of the three strain components on the boundaries between the elements, and not just at the nodes. However, the jumps on the boundaries between the elements are aggregated to point loads when the equivalent "fictitious" nodal loads are computed.

In addition to identifying the magnitudes and locations of the discretization errors, the interelement jumps in the strains can serve as termination criteria. This is the case because of the physical nature of the error measure. The jumps measure quantities that are directly related to the failure of the material in the continuum being modeled, for example, steel or aluminum. That is to say, if the jumps are relatively small with respect to the stresses on the interface or small with respect to the failure criterion, the model need not be refined further because the approximate results can be considered as accurate representations of the exact result.

4.3   QUANTIFYING THE FAILURE TO SATISFY POINT-WISE EQUILIBRIUM

This section will demonstrate that the interelement jumps in the finite element solutions are due to the failure of the approximate solution to satisfy point-wise equilibrium over the domains of the individual finite elements. We will accomplish this with the following three steps:

1. A residual function will be formed that quantifies the point-wise failure of the finite element solution to satisfy the governing differential equation over the domain of the individual elements.
2. Equivalent elemental nodal residuals will be formed by treating the elemental residual functions as fictitious or auxiliary distributed loads.
3. A global nodal residual vector will be formed by assembling the elemental nodal residuals into an auxiliary global load vector and then it will be compared to the fictitious loads that are produced by the jumps in the interelement strains.

We will see that the equivalent nodal loads formed from the residual quantities are identical to the loads formed from the jumps in the interelement strains. This allows us to conclude that the interelement jumps are produced by the failure of the finite element solution to satisfy point-wise equilibrium over the domain of the problem.

We will now demonstrate this process for the problem shown in Fig. 4.2 by computing the auxiliary loads shown in Fig. 4.3 using the step-by-step process just outlined:

Step 1—Residual Function Formulation: The failure of an approximate solution to satisfy point-wise equilibrium is quantified as a residual quantity. The residual computes the amount by which the finite element solution fails to satisfy the governing differential equation on a point-wise basis. The governing differential equation for a uniform bar is the following:

$$EA\frac{d^2 u(x)}{dx^2} = f(x) \qquad \text{(Eq. 4.1)}$$

where u is the displacement along the bar and $f(x)$ is the applied load.

If the approximate solution produced by the finite element model, $\tilde{u}(x)$, is substituted into Eq. 4.1, the equation may not be satisfied identically because the finite element result is not necessarily the exact solution. The amount by which the approximate solution fails to satisfy point-wise equilibrium is quantified with the following relationship:

$$r(x) = EA\frac{d^2 \tilde{u}(x)}{dx^2} - f(x) \qquad \text{(Eq. 4.2)}$$

where $r(x)$ is called the residual. The computation of the residual can be extended to higher dimensions.

Note that the residual has units that are identical to the applied force. Because of this similarity in units, the residual will be treated as a distributed load. That is to say, we will consider the residual as an auxiliary or a "fictitious" distributed force in the remainder of this development.

An example of a residual formed with Eq. 4.2 for the problem that was shown in Fig. 4.2a is presented in Fig. 4.4. The residual, which is shown as the solid line, is discontinuous on the full domain of the problem. However, the residual is continuous over the domains of the individual elements. This is to be expected when the two components on the right-hand side of Eq. 4.2 are considered. These two components are also shown in Fig. 4.4. Both components are continuous over the individual elements, but the contributions of the individual finite elements are discontinuous at the interelement nodes.

Step 2—Elemental Nodal Residual Formulation: In this step, the continuous residual functions on the domains of the individual elements are transformed to equivalent nodal quantities for each element. The equivalent nodal residuals are formed by using the standard finite element approach for including distributed loads into a finite element model. This procedure is applicable because, as was noted in the previous step, the residual function has the same units as a distributed load.

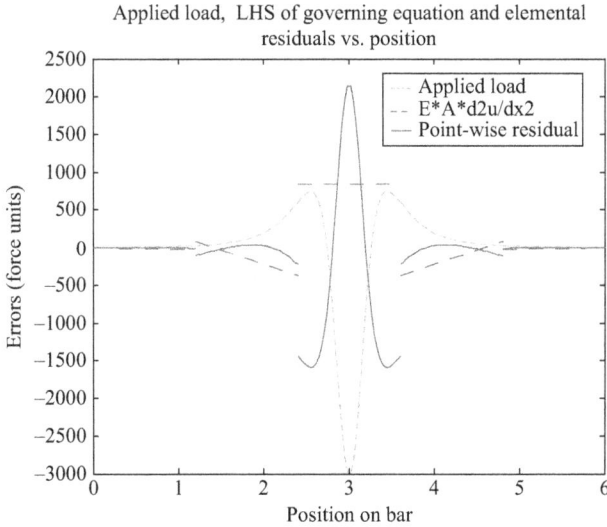

Figure 4.4. The elemental point-wise residuals and their components.

The formulation of these nodal quantities proceeds as follows. First, we form a function that is equivalent to the work function that comprises part of the potential energy expression for a bar element. That is to say, we form an integrand that consists of the residual function for a bar that is multiplied by the displacement interpolation function for the bar. We then integrate this result over the domain of the element. Finally, we take the derivatives required in the application of the principle of minimum potential energy to form the equivalent nodal residuals. The formulation of these elemental nodal residuals for this example is presented next.

The fictitious work done by the residual identified in Eq. 4.2 is produced when this fictitious force is moved by the displacement along the domain of the bar. The fictitious work function for the ith element is given as follows:

$$W_i = \int_{L_i} (r_i(x)\, u_i(x))\, dx \qquad (Eq.\ 4.3)$$

When Eq. 4.2 and the displacement interpolation polynomial for a four-node bar are introduced into Eq. 4.3, we have the following:

$$
\begin{aligned}
W_i &= \int_{L_i} (r_i(x)\, \tilde{u}_i(x))\, dx \\[2mm]
&= \int_{L_i} \left(EA\frac{d^2\,\tilde{u}(x)}{dx^2} - f(x) \right)_i \tilde{u}_i(x)\, dx \\[2mm]
&= \int_{L_i} \left(EA\frac{d^2\,\tilde{u}(x)}{dx^2} - f(x) \right)_i (N_1\ N_2\ N_3\ N_4) \begin{Bmatrix} u_1 \\ u_2 \\ u_3 \\ u_4 \end{Bmatrix}_i dx
\end{aligned}
\qquad (Eq.\ 4.4)
$$

When we take the derivatives necessary to form the equivalent nodal loads, we have the following:

$$\begin{pmatrix} F_1 & F_2 & F_3 & F_4 \end{pmatrix}_i = \begin{pmatrix} \dfrac{\partial W}{\partial u_1} & \dfrac{\partial W}{\partial u_2} & \dfrac{\partial W}{\partial u_3} & \dfrac{\partial W}{\partial u_4} \end{pmatrix}_i \tag{Eq. 4.5}$$

The four components of this row vector are the equivalent nodal residuals for the ith element.

When the equivalent nodal loads are found for the five elements contained in the problem shown in Fig. 4.2a using the residuals shown in Fig. 4.4, the results are presented in Tables 4.1–4.3.

Table 4.1 contains the contribution of the first term on the right-hand side of Eq. 4.2 or Eq. 4.4 to the equivalent nodal loads, namely, the portion of the residual due to the internal loads in the finite element solution for the five individual elements of the model.

Table 4.2 contains the contribution of the second term on the right-hand side of Eq. 4.2 or Eq. 4.4 to the equivalent nodal loads for the five individual elements of the model. As indicated in the caption of Table 4.2, the quantities in Table 4.2 are equal to the negative of the distributed loads that were applied to the original problem. This is the case because these quantities are formed in the same manner as the distributed load in the original problem. The only difference is the negative sign on the expression for the load.

Table 4.1. Internal load component of the nodal residuals, $EA\, d^2 \tilde{u} / dx^2$

El. no.	F_1	F_2	F_3	F_4
1	−0.4951	−2.1016	−7.6494	−2.7553
2	2.5747	−5.8890	−128.4081	−47.3405
3	126.0909	378.2728	378.2728	126.0909
4	−47.3405	−128.4081	−5.8890	2.5747
5	−2.7553	−7.6494	−2.1016	−0.4951

Table 4.2. Applied load component of the nodal residuals (−1*applied load)

El. no.	F_1	F_2	F_3	F_4
1	0.8959	2.1016	7.6494	3.3529
2	9.7690	5.8890	128.4081	67.2726
3	148.6809	−378.2728	−378.2728	148.6809
4	67.2726	128.4081	5.8890	9.7690
5	3.3529	7.6494	2.1016	0.8959

Table 4.3 contains the sum of Tables 4.1 and 4.2. These quantities are the equivalent nodal residuals for the **individual elements** produced by Eqs. 4.3–4.5. These residual quantities are treated as auxiliary forces in the analysis that follows.

When Table 4.3 is examined, we see that the residuals associated with the interior nodes of the individual elements are equal to zero. This result is consistent with the fact that that no discontinuities exist in the finite element strains on the interior of the four-node elements. This result reinforces the idea presented in Chapter 3 that the interelement jumps in the strain can be viewed as a failure to satisfy the "internal natural boundary conditions."

Table 4.3. Elemental equivalent nodal residuals

El. no.	F_1	F_2	F_3	F_4
1	−0.4008	0.0	0.0	−0.5976
2	−12.3438	0.0	0.0	−19.9321
3	−274.7718	0.0	0.0	−274.7718
4	−19.9321	0.0	0.0	2.5747
5	-0.5976	0.0	0.0	−0.4951

Step 3—Global Nodal Residual Formulation and Comparison: In this step, the elemental nodal residuals contained in Table 4.3 are assembled into the vector of global residual nodal loads that are applicable to the bar problem with two fixed ends. Then, these quantities are shown to be identical to the forces generated by the interelement jumps in the discontinuous finite element strain representation.

This applied load vector is formulated by assembling the elemental equivalent nodal residuals using the standard finite element procedure. We then eliminate the two end loads because of the fixed boundary conditions. These results are shown in Table 4.4. The zero loads on the interior of the bar shown in Table 4.3 are eliminated in order to highlight the nonzero elements. The columns labeled F_{Left} and F_{Right} are the two elemental components that are summed in the assembly process to give the global result contained in the final column of Table 4.4.

Table 4.4. Global nodal residuals

	F_{Left}	F_{Right}	F_{Total}
Load 1	−0.5976	−12.3438	−12.9414
Load 2	−19.9321	−274.7718	−294.7040
Load 3	−274.7718	−19.9321	−294.7040
Load 4	−12.3438	−0.5976	−12.9414

When the quantities in the last column of Table 4.4 are compared to the equivalent loads computed in Section 3.2 and shown in Fig. 4.3, they are found to be identical. This means that the jumps in the interelement strains do, indeed, quantify the failure of the individual finite element models to satisfy point-wise equilibrium.

In Chapter 2, we saw that as the finite element model is refined, either by uniform or adaptive refinement, two changes occurred in the approximate solutions: (1) the finite element representation better approximates the exact solution and (2) the interelement jumps in the strain are reduced.

This combination of results is consistent with Kelly's contention that the interelement jumps quantify the discretization errors in finite element models. The demonstration just presented shows that the interelement jumps are equivalent to the failure of the finite element model to satisfy point-wise equilibrium. This example explains Kelly's contention and gives it a solid theoretical foundation.

Since the residual quantities are related to quantities of interest in continuum mechanics, for example, strains, the residuals can serve as both an error measure and as a termination criterion. For example, if the residuals everywhere in the model are small with respect to the critical strain level, the approximate solution can be considered sufficiently accurate, and the analysis can be terminated. This is the case because the addition of a sufficiently small change in the strain level will not be responsible for the failure of material.

4.4 EVERY FINITE ELEMENT SOLUTION IS AN EXACT SOLUTION TO SOME PROBLEM

Although we have demonstrated that the interelement jumps in the strain representations are equivalent to the failure of the finite element model to satisfy point-wise equilibrium, we have not explicitly demonstrated Kelly's idea that every finite element solution is the exact solution to some problem. We will now demonstrate the validity of this idea.

This idea is valuable because of its intuitive appeal. If a solution has no interelement jumps, it is an exact solution. Analogously, if the interelement jumps are small with respect to the applied load, then the solution is sufficiently accurate.

We will modify the loading applied to a finite element model that has a result that contains interelement jumps so that the solution of the modified finite element model contains no interelement jumps. The result can be considered an exact result at the interelement nodes of the model. The new model is formed by adding a set of concentrated loads that are the negative of the "fictitious" loads shown in Fig 4.3. This situation is illustrated in Fig. 4.5.

The rationale for this approach is the following. If the "fictitious" loads in Fig. 4.3 produce the interelement jumps shown in Fig. 4.2b, then the opposite set of loads should eliminate these interelement jumps in the strains. When this modified problem is solved, the finite element strain representation is shown in Fig. 4.6 along with the exact solution.

As can be seen, the additional loads have eliminated the interelement jumps in the strain representation for this modified problem.

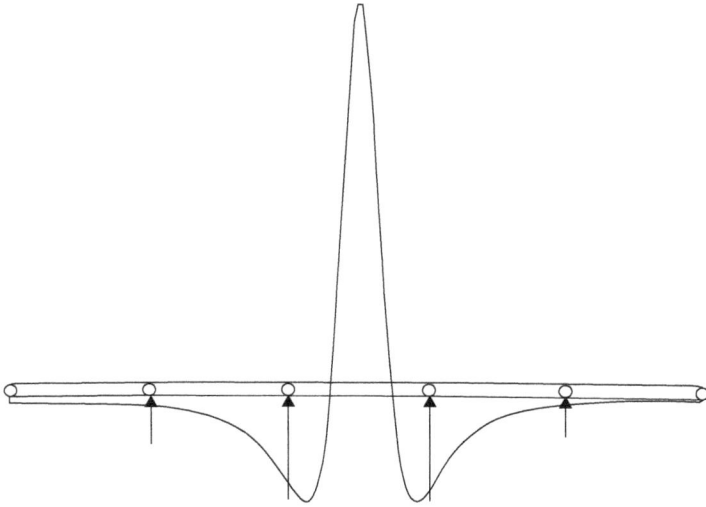

Figure 4.5. Finite element model with "strain smoothing" loads.

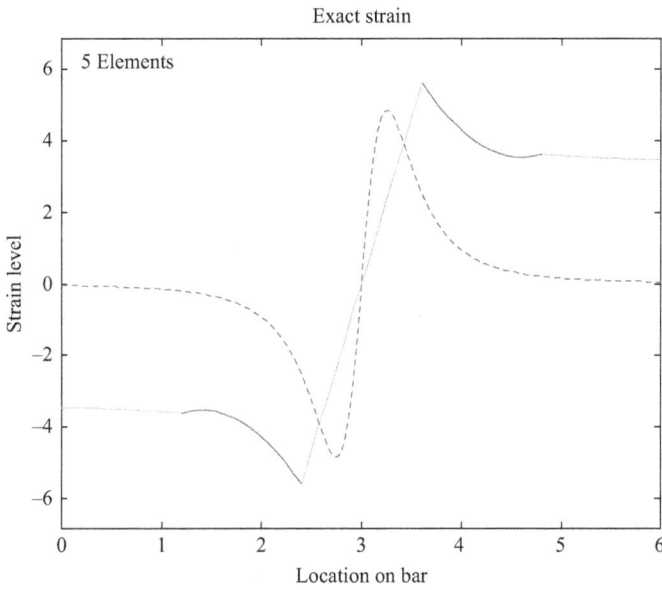

Figure 4.6. A smoothed strain result compared to the exact result.

The elimination of the interelement jumps by the introduction of these "fictitious" point loads shows that the interelement jumps are caused by the failure of the finite element solution to satisfy the governing differential equations on a point-wise basis. This is demonstrated by the results shown in Fig. 4.6. In this problem, the added point loads are exactly equal to this failure to satisfy the point-wise equilibrium and they are applied in a direction that compensates for this deficiency.

4.5 SUMMARY AND CONCLUSION

Two approaches for estimating errors in finite element results are identified in Chapter 1, namely, residual and recovery approaches. The developments and results presented in this chapter have introduced and validated the residual approach developed by Kelly for identifying the errors in finite element solutions. In this approach, the interelement jumps in the strains are found to be equivalent to the aggregated failure of the finite element solution to satisfy the governing differential equation being solved on the domain of the individual elements.

When this approach is applied to solid mechanics problems, the residuals can be interpreted as "fictitious" distributed loads that are applied to the finite element model. This interpretation gives the residual approach its intuitive feel. Since an exact finite element solution will produce no residuals, an exact solution will have no "fictitious" loads.

In contrast, a finite element solution that is not an exact solution will have a nonzero "fictitious" load. However, the closer the "fictitious" load is to zero, the better the solution. Furthermore, the magnitude of the "fictitious" load can be related to the size of the actual applied load in order to evaluate the accuracy of the finite element result. For example, if the fictitious load is small when compared to the applied load, the finite element solution can be considered to accurately represent the exact solution.

When the residuals are aggregated, they can be presented as equivalent to the interelement jumps in the strain components. In this interpretation, the interelement jumps can be directly related to failure criteria. If the normal strain that produces failure in the material being analyzed is known, the interelement jumps can be compared to this failure criterion. As a result, the acceptability of the finite element result can be judged against a meaningful standard.

The recovery approach that provided the first practical approach to error estimation is developed and applied in Chapter 7. The recovery approach has two disadvantages compared to the residual approach. Instead of using point-wise quantities that are readily available from the finite element results, a smoothed estimate of the actual strain is formed and the strain energy content of the difference between this estimate and the finite element result is computed. This process requires that integrations be performed over each element. However, the largest disadvantage is the metric used to estimate the strain. The error is reported as a percentage of the strain energy error in an element. The strain energy measure does not have the intuitive appeal of an error in the strain at given points in the finite element solution.

This chapter has introduced and validated Kelly's approach for identifying the errors in finite element solutions. We have seen that the interelement jumps quantify the discretization errors in the finite element result and that the discretization errors are due to the inability of the individual finite elements to satisfy point-wise equilibrium. In later chapters, we will use the interelement jumps as an error measure to drive an adaptive refinement process. We will also see that it works well as a termination criterion.

4.6 REFERENCE

1. Kelly, D. W. "The Self-Equilibration of Residuals and Complementary A-Posteriori Error Estimates in the Finite Element Method," *International Journal for Numerical Methods in Engineering* 20 (1984): 1491–506. DOI: http://dx.doi.org/10.1002/nme.1620200811

CHAPTER 5

MODELING INEFFICIENCY IN IRREGULAR ISOPARAMETRIC ELEMENTS

5.1 INTRODUCTION

The isoparametric procedure must currently be considered the standard approach for forming finite element stiffness matrices. However, the isoparametric approach is now obsolete. This is due to deficiencies in the formulation process that exist because of numerical shortcuts that were introduced in the 1960s to compensate for the limited computer capabilities that were available at that time. These computational shortcuts are no longer needed because computer capabilities have improved and the strain gradient approach for forming stiffness matrices has since been developed.

At the heart of the isoparametric method is the mapping that takes the actual geometry of an element onto a "regular" shape [1–3]. Figure 5.1 presents an example of an irregularly shaped or distorted two-dimensional element in the physical coordinate system, which is being mapped onto a regular domain in the "natural" coordinate system.[1]

This mapping of a distorted element onto a regular shape reduces the computational effort required to form an element stiffness matrix in two ways: **(1)** the interpolation functions are formed without a matrix inversion and **(2)** the necessary integrations are performed numerically using Gauss quadrature.

Although this mapping allows for computational convenience, **it introduces a significant flaw** into the modeling capabilities of distorted elements. When the initial physical configuration of the element is irregular, higher order strain distributions are not represented accurately. The deterioration

---

[1] The isoparametric formulation procedure is obsolete except possibly for certain applications in fracture mechanics that makes use of the alterations to the strain representations contained in irregularly shaped elements [4]. The strain gradient approach, the alternative to the isoparametric method, does not use a mapping and it significantly reduces the number of integrals that must be evaluated [5]. A 10-node strain gradient triangle is developed in detail in Chapter 10 in order to demonstrate the pedagogical and computational advantages of this element formulation procedure.

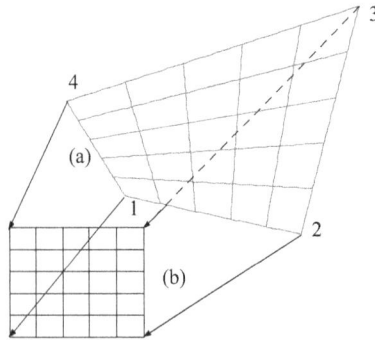

Figure 5.1. An isoparametric mapping in two dimensions.

of the strain modeling characteristics of distorted isoparametric elements is of interest here because of the inefficiencies these modeling errors introduce into finite element models.

The objective of this chapter is threefold: **(1)** to definitively identify the cause of the errors in the higher order strain distributions in distorted isoparametric elements, **(2)** to demonstrate that the errors introduced into the higher order strain distributions compromise the efficiency of finite element models, and **(3)** to identify a computationally competitive procedure for forming distorted elements that contains none of the deficiencies of the isoparametric approach.

Figure 5.2 demonstrates the corruption to the linear strain representation that occurs for increasing levels of distortion in a single three-node bar element. In the case of a three-node bar element, a distortion is produced by locating the interior nodal point away from the center of the element. The three-node isoparametric bar elements used to form these models have their interior nodes located at $x_2 = 0.50\,L$, $x_2 = 0.55\,L$, and $x_2 = 0.60\,L$, respectively.

As can be seen, the undistorted isoparametric bar element ($x_2 = 0.50\,L$) produces an exact linear representation. The other two strain distributions presented in Fig. 5.2 differ from the exact result. The amount of error in the strain representations increases as the distortion in the element increases.

As we will see in the developments that follow, the representation of higher order strain distributions in irregularly shaped isoparametric elements is corrupted when the ratio of the differential areas of the physical and the "natural" coordinate system is not constant. The differences in the ratios of the subareas produced by the mapping for a two-dimensional case are shown in Fig. 5.1.

In this figure, the distorted subdivisions of different sizes in the physical system are mapped onto equal-sized squares in the natural coordinate system. Since the unequal subdivisions in the physical system are mapped onto squares of equal size, the ratios of the physical areas to the mapped areas are different for each subdivision. As we will see in later developments, this difference in the ratio of the areas over the domain of the element produces the modeling errors in the higher order strain representations of isoparametric elements.

In the next section, we give a qualitative overview of the deterioration of isoparametric element performance as the element is progressively distorted. Next, we relate the errors in the strain

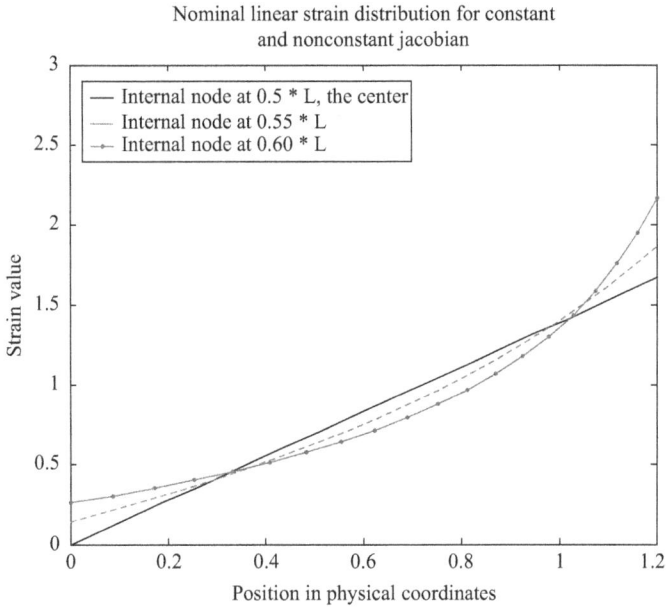

Figure 5.2. Strain distributions for distorted elements, $L = 1.2$.

representation to the distortion in an isoparametric element. Then, we present controlled examples that allow us to better understand the relationship between the strain modeling errors and the distortion of isoparametric elements. Finally, we show the existence of a computationally competitive alternative to the isoparametric element when the modeling inefficiency of distorted isoparametric elements is demonstrated [2].

5.2   AN OVERVIEW OF ISOPARAMETRIC ELEMENT STRAIN MODELING CHARACTERISTICS

Isoparametric elements are a holdover from the time when computer capacity was extremely limited by today's standards. As a result, there was a desire to reduce the computational effort required to evaluate the large number of integrals that exist in the isoparametric procedure [1–3]. In order to eliminate the need to integrate over the domain of elements with arbitrary sizes and shapes in two dimensions, the isoparametric formulation procedure maps every four-sided element onto a 2×2 square and every three-sided element onto an isosceles triangle.

When a finite element has an irregular or "distorted" shape, the coordinate transformation that performs this mapping introduces significant strain modeling errors in the higher order strain distributions. It should be noted that the mapping does not affect the rigid body motions or the constant strain states in an element. As a result, the distorted elements satisfy the convergence criteria for an element. Even though the distorted elements will converge toward a satisfactory result, the finite element model will contain more elements than necessary because of the errors in the higher order strain representations.

Figure 5.3 illustrates an example of the corruption of the strain modeling capabilities that the coordinate transformation induces in an irregular isoparametric element. This figure presents a sequence of results for the same problem when it is modeled using three-node isoparametric elements with increasing levels of distortion. The elements are progressively distorted by locating the interior node farther away from the center of the bar.

The negative effect of moving the interior node away from the center of the element on the higher order strain representations can be clearly seen by comparing the strain distribution produced by the three elements in the center of the model. As these elements are distorted, the linear strain

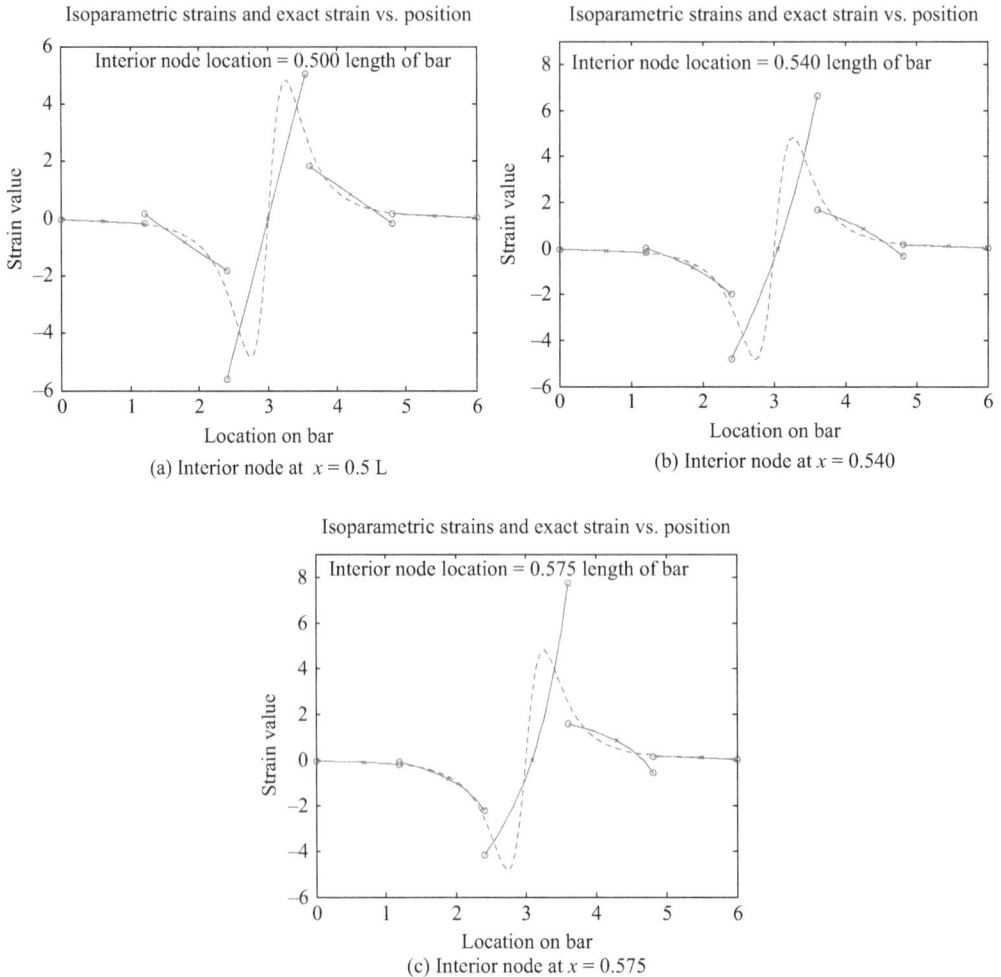

(a) Interior node at $x = 0.5$ L

(b) Interior node at $x = 0.540$

(c) Interior node at $x = 0.575$

Figure 5.3. Isoparametric strain representations.

representation is corrupted. Note that the strain representations in the two elements at the ends of the bar show little change as the elements are distorted. This is the case because the strains represented in these elements are nearly constant. As a consequence, they do not contain a significant linear strain component, so little or no alteration in the strain representation is produced because of a modeling error in this higher order strain state.

In Fig 5.3a, the strain distributions produced using undistorted three-node isoparametric elements are shown, that is, the interior node is located at the center of each element at $x_2 = 0.50\ L$. As can be seen, the strain distributions for the three elements in the center are clearly straight lines. That is to say, no error occurs in the linear strain representation.

In Fig. 5.3b, the same problem is solved with isoparametric elements whose interior nodes are located away from the center at $x_2 = 0.540\ L$. As can be seen, the three elements in the center of the model no longer represent linear strain distributions. The ability of this isoparametric element to represent a linear strain distribution has been corrupted by relocating the interior node away from the center of the bar.

In Fig. 5.3c, the three-node element is distorted even more by locating the interior node even farther from the center of the element at $x_2 = 0.575\ L$. As can be seen, the strain distributions produced by the three interior elements possess a higher level of curvature than those contained in Fig. 5.3b. This indicates that the ability of the isoparametric element to represent linear strain has been further corrupted.

Figure 5.3 has shown that the amount of error in the higher order strain representations is directly related to the amount of distortion in the isoparametric three-node bar. Similar results for a variety of two-dimensional isoparametric elements are presented in Reference [5].

5.3   ESSENTIAL ELEMENTS OF THE ISOPARAMETRIC METHOD

This section identifies the reasons why the isoparametric formulation procedure produces modeling errors in the higher order strain states when the element is distorted. As mentioned earlier, a coordinate transformation that maps any configuration of a physical element onto a fixed-size domain in the natural coordinate system is at the heart of the isoparametric method. The mapping is used to simplify the integrations involved in forming the element stiffness matrix.

A distorted three-node isoparametric bar element is illustrated in Fig. 5.4. A three-node bar element is distorted by locating the interior node away from the center of the element as shown in Fig. 5.4a. The three nodes in the physical system are mapped onto the three specified points in the natural coordinate system ξ at -1.0, 0.0, and $+1.0$ as shown in Fig. 5.4b.

As a result of locating the interior node away from the center, the ratios of the lengths in the two coordinate systems $\Delta x/\Delta \xi$ differ in the two portions of the bar, that is, $(\Delta x_1/\Delta \xi_1)$ differs from $(\Delta x_2/\Delta \xi_2)$. As we will see in the development that follows, if the differential form of this ratio is not a constant, the higher order strain representation in this isoparametric element is corrupted. For the case of a three-node bar, only the linear strain representation is affected because the rigid body and constant strain states are not affected by this source of modeling error.

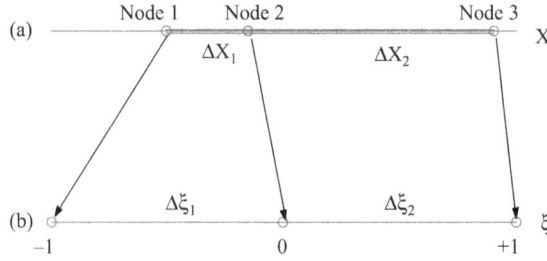

Figure 5.4. An isoparametric mapping.

In the isoparametric formulation procedure, the coefficients of the displacement interpolation function and the coefficients of the mapping function from the natural coordinates to the physical coordinates are the same. It is this characteristic that gives the isoparametric formulation procedure its name. The prefix *iso*, in Greek, means the same. As a consequence, the interpolation and the mapping functions for an isoparametric three-node bar element are given by the following:

$$u = N_1 u_1 + N_2 u_2 + N_3 u_3 \qquad \text{(Eq. 5.1)}$$

$$x = N_1 x_1 + N_2 x_2 + N_3 x_3 \qquad \text{(Eq. 5.2)}$$

$$\text{where } N_1 = \frac{\xi}{2}(\xi - 1); \quad N_2 = (1 + \xi)(1 - \xi); \quad N_3 = \frac{\xi}{2}(\xi + 1) \qquad \text{(Eq. 5.3)}$$

Equation 5.1 is the displacement interpolation function. It defines the displacements over the domain of the element in terms of the nodal displacements of the bar: u_1, u_2, and u_3.

Equation 5.2 is the transformation from the natural coordinates to the physical coordinates. Note that this is **not** a mapping from the physical system to the natural system. As we shall see, this transformation will have to be inverted in order to form the strain representations. It is this inversion that ultimately corrupts the strain representations.

The components of Eq. 5.3, the N's, are called shape functions or Lagrange interpolation polynomials. In Eq. 5.1, the shape functions force the displacements at $\xi = -1$ to equal u_1, the displacement at $\xi = 0$ to equal u_2, etc. In Eq. 5.2, the shape functions ensure that Node 1 in the physical system is mapped onto the location $\xi = -1$, Node 2 is mapped onto $\xi = 0$, etc. as shown in Fig. 5.3.

5.4   THE SOURCE OF STRAIN MODELING ERRORS IN ISOPARAMETRIC ELEMENTS

In order to demonstrate that the coordinate transformation given by Eq. 5.2 is responsible for the strain modeling errors in isoparametric elements, we will form the **strain-nodal displacement relation** for a three-node bar. When Eq. 5.1 is substituted into the one-dimensional strain-displacement relation, $\varepsilon_x = du/dx$, and the chain rule is applied, we get the following:

$$\varepsilon_x = \frac{dN_1}{d\xi}\frac{d\xi}{dx}u_1 + \frac{dN_2}{d\xi}\frac{d\xi}{dx}u_2 + \frac{dN_3}{d\xi}\frac{d\xi}{dx}u_3$$

$$= \left((\xi - 1/2)u_1 - 2\xi u_2 + (\xi + 1/2)u_3\right)\frac{d\xi}{dx} \qquad \text{(Eq. 5.4)}$$

When Eqs. 5.2 and 5.3 are consulted, we see that we cannot directly form the derivative $d\xi/dx$ that is introduced by the chain rule into Eq. 5.4. This is because ξ is not expressed in terms of x. What we do have is the opposite relationship: x is given as a function of ξ in Eq. 5.2.

We can form the expression for $d\xi/dx$ needed in Eq. 5.4 in a two-step process. First, we form an expression for $dx/d\xi$. Then, we invert the resulting derivative expression to obtain the desired result.

When we take the derivative of Eqs. 5.2 with respect to ξ, we get the following:

$$\frac{dx}{d\xi} = \frac{dN_1}{d\xi}x_1 + \frac{dN_2}{d\xi}x_2 + \frac{dN_3}{d\xi}x_3$$

$$= (\xi - 1/2)x_1 - 2\xi x_2 + (\xi + 1/2)x_3 \qquad \text{(Eq. 5.5)}$$

This expression is called the Jacobian of the coordinate transformation. As can be seen embedded in the notation, this equation identifies the point-wise ratio of the differential length dx in the physical coordinate system to the differential length $d\xi$ in the natural coordinate system. The discrete analog of Eq. 5.5 is addressed in the discussion concerning Fig. 5.4.

As the development proceeds, we shall see that the Jacobian is responsible for the corruption of the strain distributions in distorted isoparametric elements. In the cases of two- and three-dimensional elements, the Jacobian becomes a matrix and it identifies the point-wise ratios of the areas and the volumes in the two coordinate systems.

The idea of a nonconstant Jacobian is illustrated for a two-dimensional element in Fig. 5.1, which is repeated here as Fig. 5.5. The initial shape of the physical element differs significantly from the square domain in the natural coordinate system. In two dimensions, the Jacobian consists of the ratio of the differential areas in the physical system to those in the natural coordinate system. As can be seen in Fig. 5.5, the subdivisions in the distorted physical system all have different areas. In contrast, the subdivisions in the natural coordinate system all have the same area. As a result, the Jacobian for this distorted two-dimensional isoparametric element is not a constant.

In fact, an "irregular" or distorted isoparametric element can be identified by the form of the Jacobian. If the Jacobian associated with an isoparametric element is a constant, the element is "regular," and the higher order strain representations will not be corrupted. If the Jacobian is a polynomial, the element is irregular, and the higher order strain representations will be corrupted.

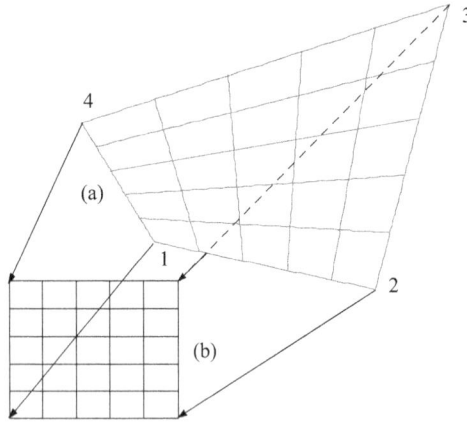

Figure 5.5. An isoparametric mapping.

The second and final step in forming the derivative needed to create an explicit expression for Eq. 5.4 is to invert Eq. 5.5. When this is done, we get the following:

$$\frac{d\xi}{dx} = \frac{1}{(\xi - 1/2)x_1 - 2\xi x_2 + (\xi + 1/2)x_3} \qquad \text{(Eq. 5.6)}$$

We will now demonstrate why the strain distribution in a three-node bar is affected by the location of the interior node. In order to make the development as transparent as possible, we will fix the locations of Nodes 1 and 3 and leave Node 2 as a parameter. This is done in order to reduce Eq. 5.6 to the following simpler equation.

When Node 1 and Node 3 are located at $x_1 = 0$ and $x_3 = 1$ in the physical system, Eq. 5.6 reduces to the following:

$$\frac{d\xi}{dx} = \frac{1}{(1 - 2x_2)\xi + 1/2} \qquad \text{(Eq. 5.7)}$$

The significant characteristic of Eq. 5.7 is that the derivative depends on the location of Node 2. When the interior node, Node 2, is located in the center of the bar at $x_2 = 0.5$, the Jacobian has a constant value equal to one-half. If the interior node is located anywhere else, the Jacobian is not a constant. It has a polynomial in the denominator.

We are now in a position to form an explicit expression for the strain in an isoparametric element. When Eq. 5.7 is substituted into Eq. 5.4, we have the following:

$$\varepsilon_x = \frac{(\xi - 1/2)u_1 - 2\xi u_2 + (\xi + 1/2)u_3}{(1 - 2x_2)\xi + 1/2} \qquad \text{(Eq. 5.8)}$$

When we inspect Eq. 5.8, we see that the strain representation has taken on the form of a rational polynomial. That is to say, one polynomial is divided by another polynomial.

As we saw in the discussion following Eq. 5.7, if the interior node, Node 2, is located at $x_2 = 0.5$, the Jacobian is equal to the constant value of one-half. When the denominator of Eq. 5.8 is a constant, the strain representation given by Eq. 5.8 is the correct linear representation for a three-node bar element. This explains why a three-node isoparametric element is able to represent a linear strain distribution when the interior node is located in the center of the element.

We will now identify the characteristic of a rational polynomial that corrupts the strain representations in distorted isoparametric elements. This is accomplished by analyzing a rational polynomial that is simpler than Eq. 5.8. This surrogate function has the same algebraic structure as Eq. 5.8, but the analysis is more compact and, hence, more transparent. We will analyze the following rational polynomial where c is a constant:

$$\frac{1+\xi}{c+\xi} \tag{Eq. 5.9}$$

The characteristic of a rational polynomial that negatively affects the strain representation in distorted isoparametric elements surfaces when Eq. 5.9 is divided synthetically. This division produces the following expression:

$$\frac{1+\xi}{c+\xi} = \frac{1}{c} + \frac{(1-1/c)}{c}\xi - \frac{(1-1/c)}{c^2}\xi^2 + \frac{(1-1/c)}{c^3}\xi^3 + \cdots \tag{Eq. 5.10}$$

Equation 5.10 is an infinite series. This result shows that when the Jacobian for an isoparametric element is not a constant, the strain representation will contain polynomial terms that do not belong in the strain representation. That is to say, the strain representation is corrupted.

The reformulation of the rational polynomial just presented shows that the existence of higher order terms in the strain representations of distorted isoparametric elements is responsible for the corruption of the strains that is produced by these elements. As was seen in Fig. 5.3, the errors in the strain representation are proportional to the level of distortion in the element.

5.5 STRAIN MODELING CHARACTERISTICS OF ISOPARAMETRIC ELEMENTS

This section demonstrates the effect of distortion on the ability of a **single** three-node isoparametric bar element to represent constant and linear strain distributions. Three finite elements with a length of 1.2 units that differ only by the location of the interior node in the individual isoparametric elements are subjected to loading conditions that produce constant and linear strain distributions in a continuous bar.

Figure 5.6 illustrates the strain distributions that are produced when a distributed load that has a constant magnitude is applied to the three elements. This loading condition produces a linear strain distribution in an exact solution. The three-node isoparametric bar elements used in this demonstration have their interior nodes located at $x_2 = 0.50\,L$, $0.55\,L$, and $0.60\,L$, respectively.

As can be seen, the undistorted three-node isoparametric bar element ($x_2 = 0.50\,L$) produces an exact result. The other two strain distributions presented in Fig. 5.6 differ from the exact result. The amount of error in the strain representations increases as the distortion increases.

We will now show that a distorted isoparametric element can accurately represent a state of constant strain. Figure 5.7 contains the strain distributions that are produced when the three different finite elements that are used to produce the results that were presented in Fig. 5.6 are subjected to a loading condition that produces a constant strain distribution in the continuum. In Fig. 5.7, the isoparametric elements with different levels of distortion produce the same result. This straight line shows that distorted isoparametric elements accurately represent constant strains.

In summary, this section has identified two significant characteristics of distorted isoparametric elements: **(1)** these elements are capable of accurately representing constant strains and **(2)** the level of the modeling errors in higher order strain representations is directly related to the amount of distortion in the initial configuration of the element. It is proven in Reference [5] that a three-node isoparametric element accurately represents rigid body motions and constant strains.

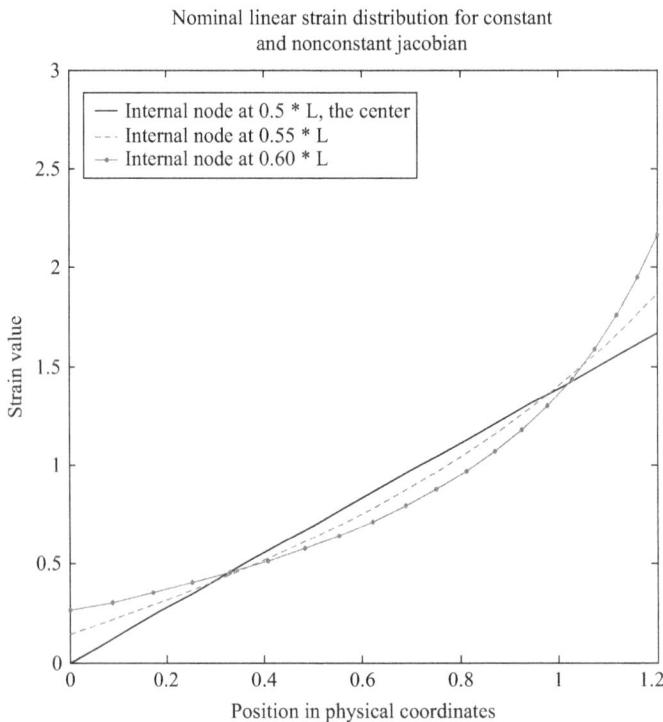

Figure 5.6. Strain errors due to element distortion.

Constant strain distribution for constant
and nonconstant jacobian

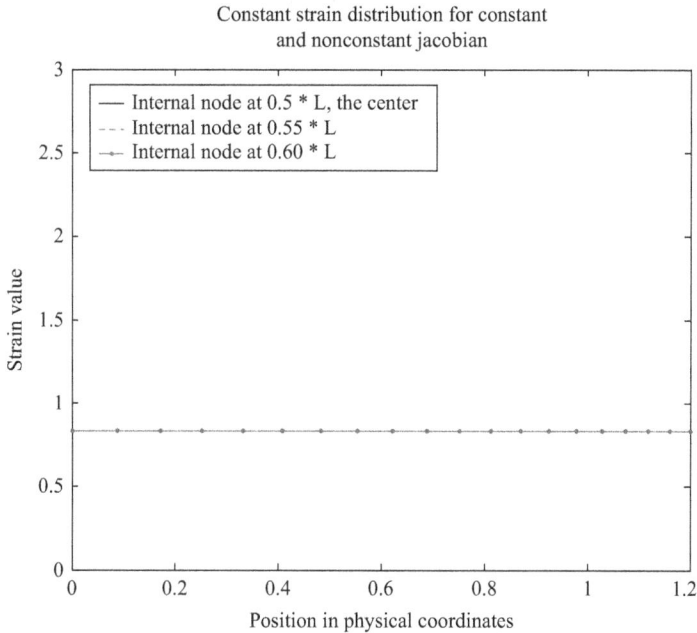

Figure 5.7. Constant strain representation with distorted isoparametric elements.

5.6 MODELING ERRORS IN IRREGULAR ISOPARAMETRIC ELEMENTS

The modeling inefficiency that exists when irregularly shaped isoparametric elements are used is demonstrated in this section. This demonstration compares the strain representations produced using distorted isoparametric elements to the strain representations produced by control models. The control models use distorted elements that are formed using the strain gradient approach developed in Reference [5] and demonstrated in Chapter 10. As we shall see, the elements formed using the alternative process do not contain the modeling deficiencies of isoparametric elements. In both models, the elements are distorted by locating the interior node away from the center of the element at 0.575 L. Each element has a length of 1.2 units.

The strain representations produced by both of the models formed with isoparametric elements and strain gradient elements are shown separately in Fig. 5.8. In Fig. 5.8a, the strain representation for the five-element control model is compared to the exact solution. In Fig. 5.8b, the strain representation produced by the five-element model formed using the distorted isoparametric elements is compared to the exact solution to the problem.

As can be seen in Fig. 5.8a, all five of the three-node elements in the control model, including the three elements in the center of the model, represent linear strain distributions. As noted earlier, a linear strain representation is the most complex strain distribution that a three-node bar can correctly represent. In other words, the strain approximation produced by the finite element model in Fig. 5.8a is as good as is possible to attain with this coarse mesh.

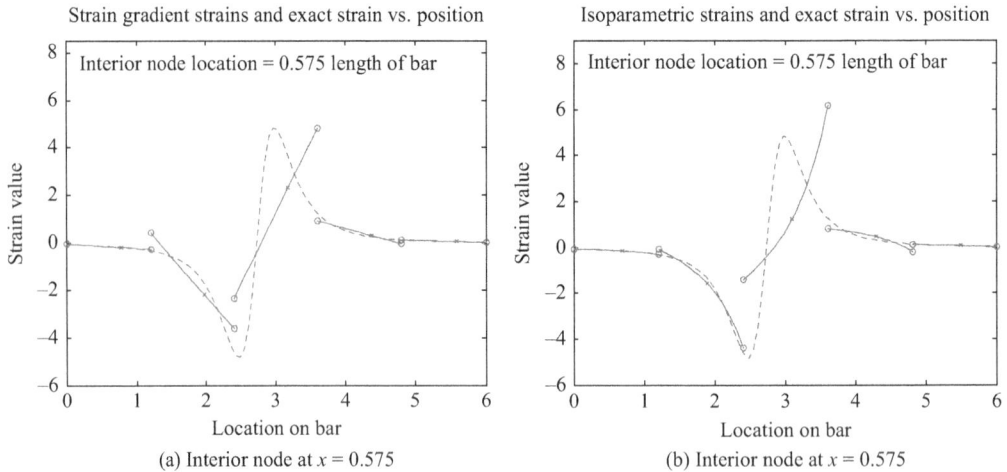

Figure 5.8. Results for two element types.

In Fig. 5.8b, the three distorted isoparametric elements in the center of the model have discernible curvatures. They do not represent linear strain distributions. The existence of this curvature in the strain representation is a strain modeling error that results from the distortion in this isoparametric element. The ability of these elements to represent linear strain distributions has been corrupted by relocating the interior node away from the center of the bar.

The discretization errors that are inherent in finite element models when a discrete representation cannot capture the exact solution are shown in Fig. 5.8. These errors are seen as the difference between the finite element representation and the exact solution. The errors, as quantified by the interelement jumps, are larger in the distorted isoparametric model except at the node located at $x = 1.2$. The curvature that exists in the strain representation of the distorted isoparametric element at this node reduces the level of error at this node.

The modeling inefficiency produced by the distorted isoparametric element is shown **quantitatively** in Fig. 5.9. This figure presents the magnitude of the interelement jumps for both models as a percentage of the converged maximum strain. The nodal errors shown are equal to the interelement jumps divided by the normalizing factor of 4.8648, which is the magnitude of the maximum strain existing in the exact result. This normalization factor is used so that the error estimator for the problem solved with the two different elements will have the same basis. The nodal errors are indicated by the x's in Fig. 5.9. The lines connecting the nodal errors are included for visual clarity. They do not possess any intrinsic meaning.

As can be seen, the errors contained in Fig. 5.9 correlate with the jumps in the interelement strains seen in Figs. 5.8a and 5.8b. The error in the isoparametric model at $x = 1.2$ is smaller than the error at this location in the control problem. As was noted earlier, the errors at every other node in the initial finite element model formed from the distorted isoparametric elements are higher than the errors in the control model. The difference in error is particularly large at the stress concentration on the right-hand side of the problem.

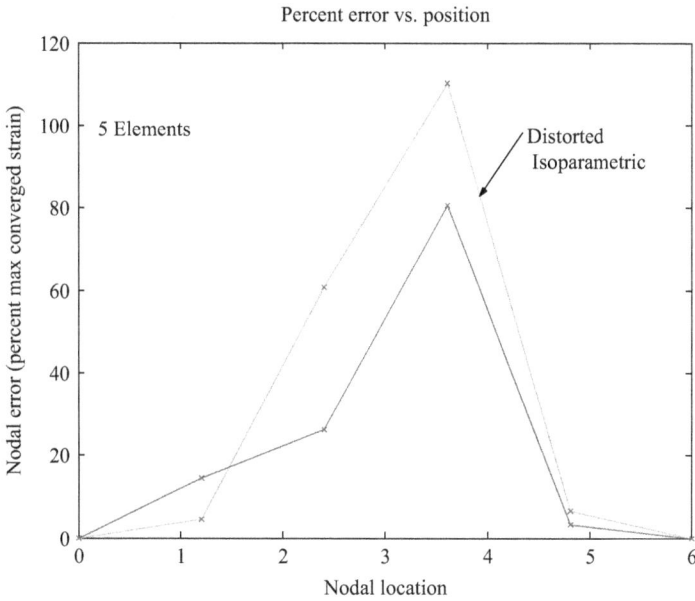

Figure 5.9. Distorted isoparametric error comparison.

This example has shown both qualitatively and quantitatively that the model formed using the distorted isoparametric elements contains significantly higher errors than does the control problem.

5.7 RESULTS FOR A SERIES OF UNIFORM REFINEMENTS

In this section, the inefficiency of distorted isoparametric elements is further demonstrated with a series of four uniform refinements of the problem that was solved in the previous section. The errors for the first two uniform refinements of the two models, which contain 10 and 20 elements, are shown in Fig. 5.10. As can be seen, the errors have been reduced in both the model formed with the distorted isoparametric elements and that with distorted control elements. However, the errors in the isoparametric models are still the larger of the two.

When the errors presented in Fig. 5.10 are compared to Fig. 5.9, the largest differences in the errors for the two models exist in the regions of rapidly changing strains. That is to say, the largest errors exist at the extreme points: the maxima, the minima, and the saddle points.

When the models are uniformly refined two more times so that they contain 40 and 80 elements, the resulting errors are shown in Fig. 5.11. These figures show the same pattern that was seen in Fig. 5.10. The errors for both refinements have been reduced for both models. However, the errors for the isoparametric model are higher than the errors for the model formed using elements that correctly represent the linear strain distribution.

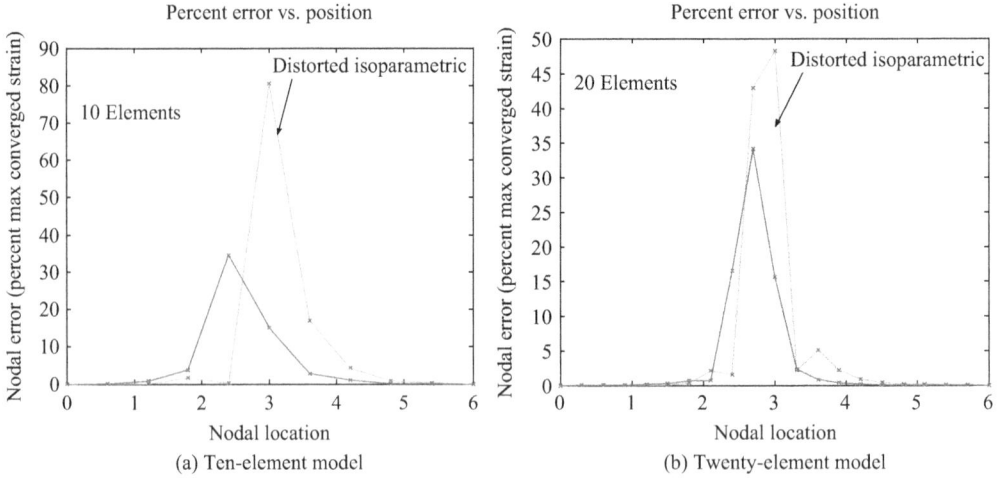

Figure 5.10. Contrasted errors in two uniformly refined models.

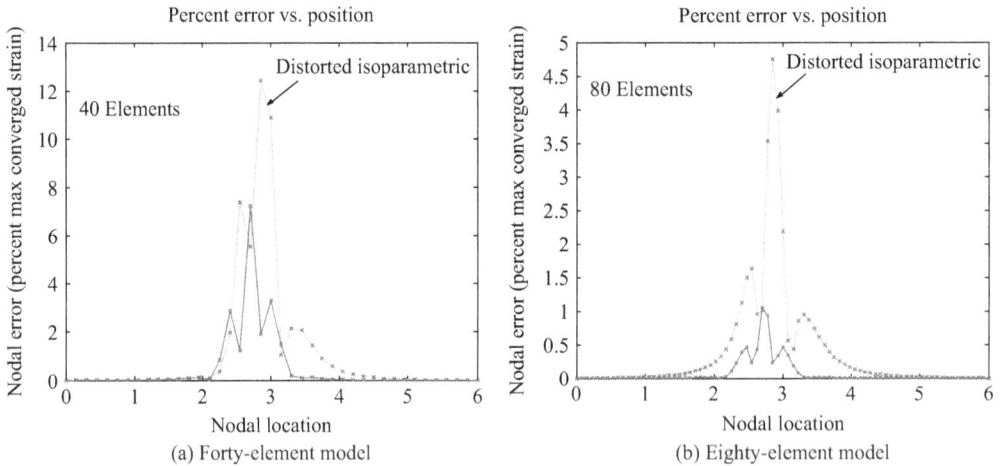

Figure 5.11. Contrasted errors in two uniformly refined models.

In these regions of rapidly changing strains, the linear strain modeling capabilities of the individual elements are called on in an attempt to represent the actual strain distribution. Since the ability of the irregularly shaped isoparametric elements to represent this higher order strain distribution has been corrupted, the errors are higher than for the element that correctly represents the linear strain distribution.

This series of uniform refinements shows that the use of distorted isoparametric finite elements produces inefficiencies in finite element models. This is the case because it would take more distorted isoparametric elements to reach the same level of accuracy that was reached with the control problem. It should be noted that the control problem is also assembled using distorted finite elements that were formed using strain gradient formulation procedure. Thus, an accurate alternative to the isoparametric method for forming distorted elements exists.

5.8   SUMMARY AND CONCLUSION

The results presented here have shown that the use of distorted isoparametric elements introduces inefficient finite element models. When distorted isoparametric elements are used, it takes more elements to achieve a given level of accuracy in a finite element model. This inefficiency is caused by the corruption of the representation of the higher order strain distributions in the individual finite elements that exist in distorted isoparametric finite elements.

Furthermore, we saw that the modeling errors in the higher order strain representations are caused by the mapping that is at the heart of the isoparametric formulation procedure. When an isoparametric element is distorted, the mapping introduces a polynomial into the denominator of the strain representation. As a result, the strain representation becomes a rational polynomial, which is equivalent to making the strain representation an infinite polynomial. The terms that are added implicitly to the strain representation corrupt the strain model.

Finally, it was shown that the formulation technique based on strain gradient notation accurately represents the strain distributions even when the finite elements are distorted. The strain gradient formulation technique, which is computationally competitive with the isoparametric procedure[2], is used to form a ten-node element in Chapter 10.

The isoparametric formulation procedure has a significant deficiency in a specialized application that has not been previously mentioned. Isoparametric elements are incapable of capturing the behavior of certain types of laminated plates. The isoparametric representation contains a **qualitative modeling error**. When the element is given a flexural displacement, the element rotates in the wrong direction. This modeling flaw does not diminish with refinement. Laminated plate elements formed using the alternative formulation procedure accurately capture this behavior [5].

In addition, the alternative formulation procedure has a pedagogical advantage over the isoparametric element. This alternative formulation procedure does not require the mapping that causes the strain modeling errors in isoparametric elements. Consequently, Gauss quadrature integration techniques are not used. The removal of these two auxiliary steps simplifies the finite element method. As a result, the physically based approach can be learned in a significantly shorter amount of time than the isoparametric approach. This means that the finite element method can be introduced sooner in the sequence of courses.

---

[2] In the alternative formulation procedure, the computational efficiency results from the fact that significantly fewer integrals must be evaluated in the formulation of the element stiffness matrix. In the case of a six-node linear strain triangle, only six simple integrals must be evaluated. These integrations are the area, the two first moments of area, and the three second moments of area [5]. In a six-node isoparametric element, 78 relatively complex integrals must be evaluated.

5.9   REFERENCES

1. Hughes, Thomas, R. J. *The Finite Element Method*, Mineola, New York: Dover Publications, 2000, p. 112. References to the origin of the isoparametric method.
2. Argyris, J. H. and Kelsey, S. *Energy Theorems and Structural Analysis*, London: Butterworths, 1960. (Originally published as a series of articles in *Aircraft Engineering*, 1954–55.)
3. Tiag, I. C. *Structural Analysis by the Matrix Displacement Method*, English Electric Aviation Report No. S017, 1961.
4. Mohammadi, S. *Extended Finite Element Method*, New York: Blackwell Publishing, 2008.
5. Dow, J. O. *A Unified Approach to the Finite Element Method and Error Analysis Procedures*, New York: Academic Press, 1999.

   This book puts the finite element method, the finite difference method, and error analysis on a common Taylor series basis. It develops the alternate element formulation procedure mentioned and applied here.

CHAPTER 6

INTRODUCTION TO ADAPTIVE REFINEMENT

6.1   INTRODUCTION

The initial finite element model for a problem rarely provides a solution that is accurate enough for use in the design process. The obvious strategy for improving the model is to repeatedly subdivide every element in the model until the change in two successive results is acceptably small. However, this brute force approach for reaching a converged solution leads to finite element models that are unmanageably large because elements are needlessly introduced into regions where little or no error exists.

The excessive growth produced by uniformly refining a finite element model can be eliminated by selectively improving the model only in regions containing unacceptable levels of error. A procedure for identifying such regions and improving the model in these regions is shown schematically in Fig. 6.1.

This procedure, known as adaptive refinement, begins by forming and solving an initial finite element model. The errors in the solution of the initial finite model are then estimated and evaluated with a termination criterion. If the specified level of accuracy is not achieved, the model is improved

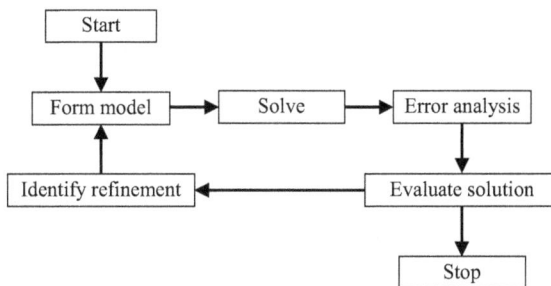

Figure 6.1. Adaptive refinement schematic.

by refining the mesh in regions of unacceptable error. The process is repeated, starting with the improved model, until an acceptable solution is obtained.

The **primary objective** of this chapter is to demonstrate the effectiveness of the adaptive refinement process. The **secondary objective** is to show that error estimators can be designed to produce finite element results with specified characteristics.

These objectives are achieved by developing error estimators based on physically interpretable quantities, namely, the interelement jumps in the strain. Since strains are quantities of direct interest to the analyst, error estimators can be given a specialized, strain dependent focus. For example, the error estimators can be based on a specific failure criterion, such as introducing a bias on shear strains for a brittle material.

In the first demonstration, an error estimator that focuses the refinements of the finite element model on regions containing stress concentrations guides the adaptive refinement process. The second demonstration produces a final result with a uniform level of error over the domain of the finite element problem.

The model refinement applied here is simplistic. If an element does not meet the termination criterion it is simply subdivided into two elements. This is done to demonstrate the efficacy of the adaptive refinement process with as few complications as possible. In contrast, refinement guides based on first principles of continuum mechanics are developed and applied in Chapter 11. In these approaches, an estimate of the exact solution that is emerging from the finite element model is used to identify a level of refinement that will produce the rapid convergence of the finite element model.

6.2   PHYSICALLY INTERPRETABLE ERROR ESTIMATORS

This chapter demonstrates that error estimators based on the jumps in interelement strains can be designed to produce finite element results with given characteristics. Specializing the error estimators is possible because they are based on physically interpretable quantities that are sought in the analysis.

Two error estimators are applied to problems modeled with three-node bar elements to demonstrate that error estimators can be designed to produce results with specified characteristics. The first error estimator focuses on improving the model in regions of high strain and the second is designed to produce a uniform level of error everywhere in the problem.

The two error estimators demonstrated here have the following general form:

$$\eta_i = \left| \frac{(\Delta \varepsilon)_i}{(\varepsilon)_{Normalizing}} \right| \times 100 \qquad \text{(Eq. 6.1)}$$

where η_i = percent of estimated error at the ith node.

$(\Delta\varepsilon)_i = (\varepsilon_{\text{Right-Hand Element}} - \varepsilon_{\text{Left-Hand Element}})_i$

$(\varepsilon)_{Normalizing}$ = the normalizing factor that focuses the model refinement.

The error estimator defined by Eq. 6.1 estimates the error given by the interelement jumps in the strain at the ith node as a percentage of the normalizing factor.

The numerator of Eq. 6.1, $(\Delta\varepsilon)_i$, is equal to the jump in the interelement strain at the ith interelement node. This jump is directly related to the errors in the two elements adjacent to the node. The choice of the normalizing factor, $(\varepsilon)_{Normalizing}$, defines the focus of the error estimator and, ultimately, the characteristics of the final finite element model.

The first error estimator demonstrated here is designed to focus on regions containing stress concentrations. This focus is provided by choosing a normalizing factor, the denominator of Eq. 6.1, that has a constant value. As a result, regions of high error are highlighted.

This globally normalized error estimator focuses on stress concentrations for the following two reasons: **(1)** stress concentrations are characterized by rapidly changing strain distributions and **(2)** finite element models based on low-order polynomials contain large errors when they attempt to represent complex functions unless they are highly refined. This version of the error estimator is demonstrated in Section 6.5.

On the other hand, if we choose a normalizing factor that is closely related to the strain levels in the elements adjacent to ith node, the errors in the final model are equally distributed over the domain of the problem regardless of the variations in the strain levels. That is to say, this locally normalized error estimator produces results in which the errors in regions of low strains and in regions of high strain are nearly equal. This version of the error estimator is demonstrated in Section 6.6.

6.3   A MODEL REFINEMENT STRATEGY

Now that we have an error estimator available to us, we must implement a strategy for improving the finite element model based on these error estimates. In the demonstrations of adaptive refinement presented in Sections 6.5 and 6.6, the decision on whether to subdivide an element is based on the average of the error estimates at the two ends of the finite element. If this average value for an element exceeds some threshold value, the element is divided into two equal sized elements.

This simple strategy is chosen because the purpose of these demonstrations is to present the efficacy of the adaptive refinement process, not to optimize the process. We will revisit the issue of refinement strategies in Chapter 11. We will see that the possibilities of refinement strategies are limited only by our imaginations.

6.4   A DEMONSTRATION OF UNIFORM REFINEMENT

In preparation for demonstrating the adaptive refinement process, a sequence of uniform refinements is applied to a sample problem in this section. The models that result from these uniform refinements serve two purposes: **(1)** they show the explosive growth in the number of elements in a uniformly refined model that necessitates the use of adaptive refinement and **(2)** the approximate strain distributions produced by these models provide a baseline against which to compare the results of the adaptive refinement process.

The finite element strain approximations and the error estimates produced by the initial model formed with five three-node elements are shown in Fig. 6.2. Figure 6.2a compares the strains in the finite element model to the exact solution. Figure 6.2b contains the error estimates at the nodes of this model. The error estimate at each node is normalized with the magnitude of the largest strain in the finite element model. The error estimates at the nodes are connected by straight lines for visual convenience. The lines have no significance.

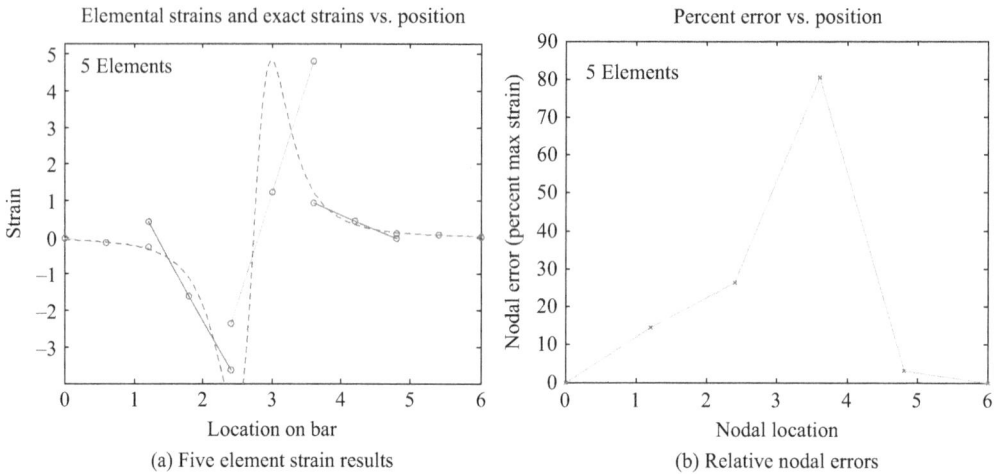

Figure 6.2. Initial mesh for both uniform refinement and adaptive refinement.

The error estimates at the two fixed ends of the finite element model are taken to be zero. These estimates are close to the actual error for this problem. The treatment of the boundary errors is postponed until Chapter 8 because the estimation of the errors on the boundaries is closely related to the point-wise error measures presented in Chapter 8.

When the error estimates shown in Fig. 6.2b are compared to the actual errors that can be seen in Fig. 6.2a, we see that the error estimates correlate well with the actual errors in the finite element strain distribution. The large errors exist where an attempt is being made to represent a complex strain distribution with a small number of elements that are, at most, capable of representing linear strain.

The finite element representation of the exact result will be improved by adding elements to the regions showing high error. This means that the error estimator given by Eq. 6.1 will successfully guide the adaptive refinement process. This contention is validated in the next section.

The results of the first uniform refinement of the initial model are shown in Fig. 6.3. As can be seen in Fig. 6.3a, the strain distribution for the uniformly refined finite element model more closely resembles the exact solution than does the result shown in Fig. 6.2a. The finite element approximation is smoother and the strain representations in the regions of the stress concentrations, while still discontinuous, are located near the point of maximum magnitudes in the exact solution.

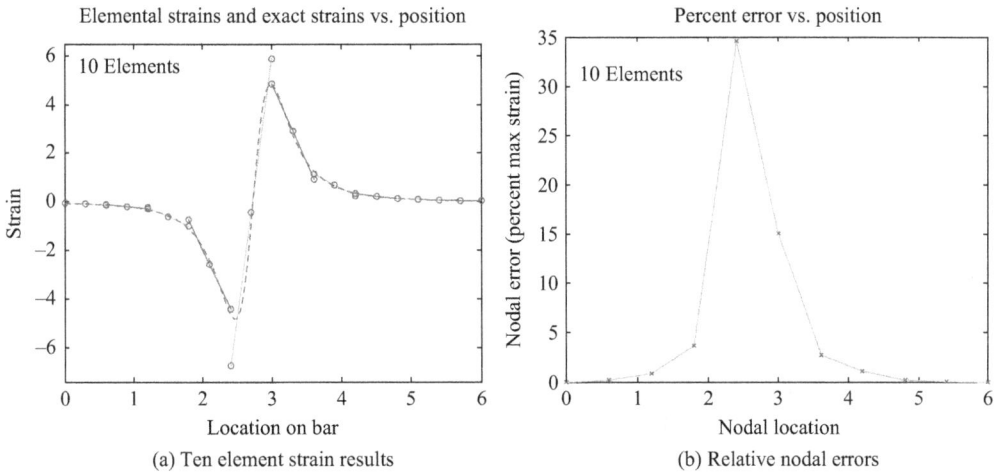

Elemental strains and exact strains vs. position

Percent error vs. position

(a) Ten element strain results

(b) Relative nodal errors

Figure 6.3. First uniformly refined finite element model.

These improvements in the finite element representation are captured by the error estimator as shown in Fig. 6.3b. When Fig. 6.3b is compared to Fig. 6.2b, we see that the error estimates over the whole model are reduced. The maximum error estimate has been reduced to approximately 35% from the previous estimate of approximately 80%. Again, the highest levels of error exist in regions of the exact strain distribution that are too complex to be captured by the linear strain representation of a single finite element.

The results of the second uniform refinement are shown in Fig. 6.4. The resulting error estimates contain an interesting result. The maximum error estimate has not changed much from the maximum estimate in the previous refinement. It is approximately 33% versus the approximate value of 35% in the previous refinement. However, the nature of the strain in the location of the maximum error estimate has changed markedly. Instead of being associated with a maximum point in the strain distribution, the maximum error estimate occurs at an inflection point in a region with a very small strain value in the exact solution. That is to say, the maximum error exists in a region of a stress concentration with a rapidly changing strain distribution. Again, the error estimator has identified a region in a strain distribution that cannot be easily represented by a linear strain representation.

Figure 6.5 presents the result of the third uniform refinement. In this refinement, the maximum error estimate has been significantly reduced from approximately 33% to approximately 7%. However, the error estimate in the region of the inflection point still exceeds the estimated error at the stress concentration.

The primary deficiency with uniform refinement is seen when Figs. 6.4 and 6.5 are compared. In Fig. 6.4b, we see that the error estimates are close to zero on the two outside portions of the model. In the uniform refinement shown in Fig. 6.5, the majority of the new elements are introduced into these regions of low error. That is to say, most of the new elements are introduced into regions where they are not needed.

Elemental strains and exact strains vs. position   Percent error vs. position

(a) Twenty element strain results   (b) Relative nodal errors

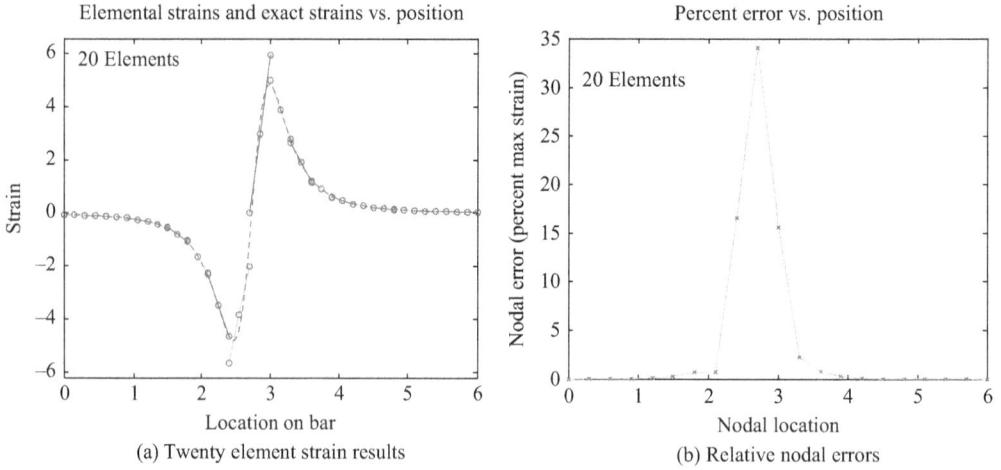

Figure 6.4. Second uniformly refined finite element model.

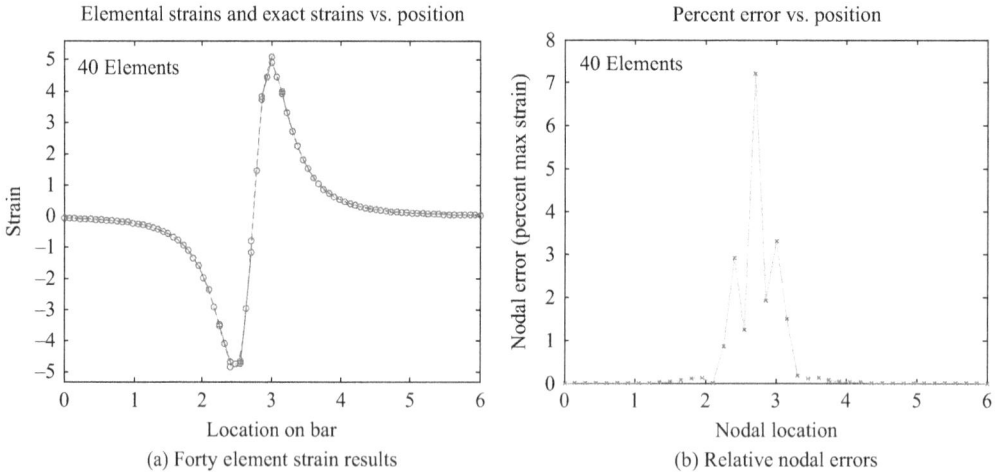

Elemental strains and exact strains vs. position   Percent error vs. position

(a) Forty element strain results   (b) Relative nodal errors

Figure 6.5. Third uniformly refined finite element model.

The introduction of these unneeded elements into the model characterizes the inefficiency found in the process of uniform refinement. It should be noted that the inefficiency seen in this one-dimensional case is minor when compared to the number of unneeded elements that can be introduced into two- and three-dimensional problems.

Figure 6.6 presents the results of the fourth uniform refinement of the model. In this refinement, the errors have the same general distribution as that seen in the previous refinement shown in Fig. 6.5b. However, the magnitudes of the errors have been significantly reduced. In fact, they are reduced to a level that would be considered accurate enough for practically any design process.

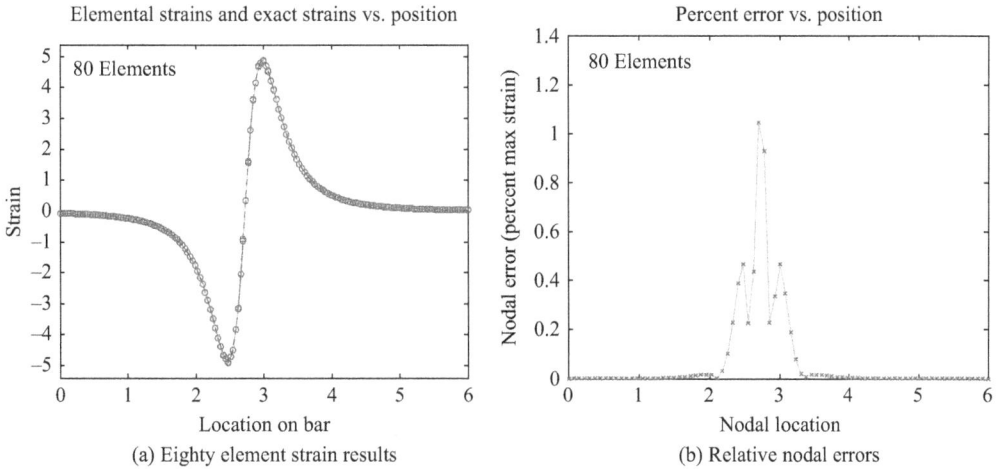

Figure 6.6. Fourth uniformly refined finite element model.

The inefficiency of the uniform refinement process is demonstrated again. Most of the new elements are introduced in regions of low error where no refinement is needed.

6.5 A DEMONSTRATION OF ADAPTIVE REFINEMENT

In this section, the effectiveness of the adaptive refinement process is demonstrated. The initial finite element model used in Fig. 6.2 is adaptively refined under the guidance of the error estimator defined in Eq. 6.1. This error estimator is normalized with respect to the maximum strain in the finite element model. This is the same error estimator that was used in the previous section to evaluate the results of the uniformly refined models. This choice for the normalizing factor means that the error estimator will focus on the stress concentrations.

The elemental error threshold for dividing an element in half is chosen as 4%. We will see that there are significantly fewer elements needed in the adaptively refined model to satisfy the termination criterion of 4% error or less at each node.

The adaptive refinement process achieved the desired level of accuracy after four iterations. The nodal error estimates for these five models are superimposed in Fig. 6.7. The most significant result shown in Fig. 6.7 is that the termination criterion was satisfied with only 19 elements. This compares to the 80 elements contained in the uniformly refined model that was required to achieve the same level of accuracy.

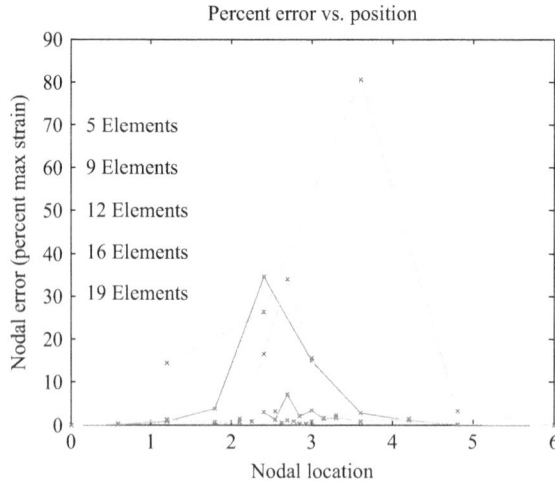

Figure 6.7. Nodal errors for four iterations of adaptive refinement.

To better see how the finite element model formed by the adaptive refinement process compares to the final model produced by uniform refinement, the result of the final iteration of adaptive refinement is shown in Fig. 6.8.

As can be seen, there are fewer elements in the regions of low error at either end of the model when Fig. 6.8a is compared to Fig. 6.6a. This is as would be expected. The elements added by the adaptive refinement process are located where they do the most good, namely, in the regions of high error. In other words, the adaptive refinement process adds elements to the region where the actual strain distribution is complex until there are enough elements to capture this complex strain distribution with linear strain representations.

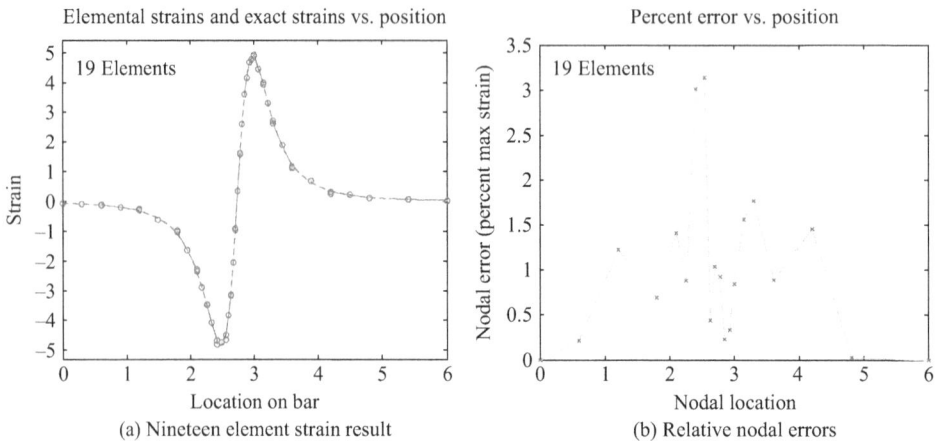

(a) Nineteen element strain result

(b) Relative nodal errors

Figure 6.8. Fourth adaptively refined finite element model.

Just to see what it takes in the way of adaptive refinement to reduce the estimated errors in the problem to a level below those in the final uniformly refined model, we will adaptively refine the problem with a lowered value for the criteria for subdividing an element and for the termination of the adaptive refinement process. The result for a threshold value of 1.5% is shown in Fig. 6.9.

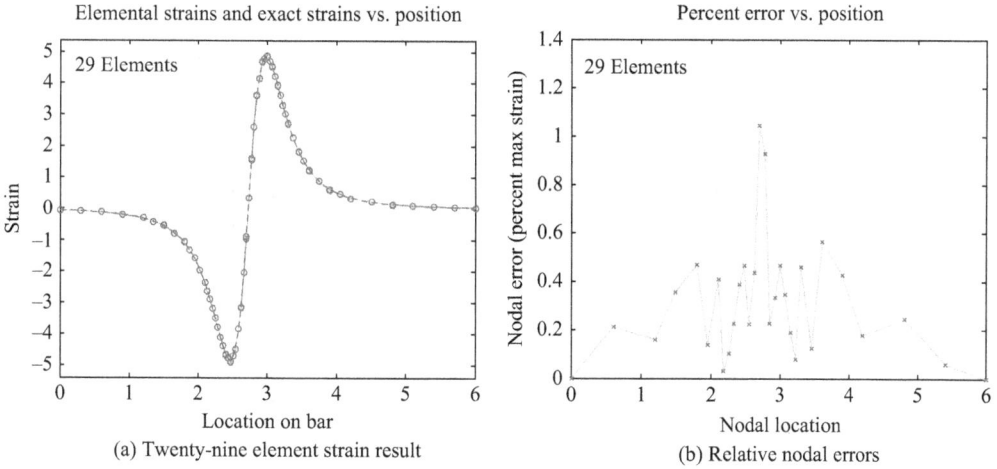

Figure 6.9. Adaptively refined finite element model—1.5% threshold.

When Fig. 6.9b is compared to Fig. 6.6b, we see that the error estimates in the central portion of the two figures are nearly identical. This result shows definitively that adaptive refinement produces finite element models that represent problems as well as a uniformly refined model with significantly fewer elements.

6.6   AN APPLICATION OF AN ABSOLUTE ERROR ESTIMATOR

In this section, the example problem is adaptively refined under the guidance of an error estimator that estimates the "absolute" error at an interelement node. That is to say, the error estimator defined by Eq. 6.1 is normalized with respect to a local value of the strain instead of with a large constant value of strain. The use of a local normalization of the error estimator is designed to distribute the absolute errors more uniformly over the domain of the problem. Specifically, the normalizing factor used in Eq. 6.1 is the following:

$$(\varepsilon)_{Normalizing} = \left| \varepsilon_{Local\ Average} \right|_i$$

$$= \frac{(\varepsilon_{Right\text{-}Hand\ Element} + \varepsilon_{Left\text{-}Hand\ Element})_i}{2}$$

(Eq. 6.2)

This normalizing factor is the local average of the strains at the ith node.

When the normalizing factor given by Eq. 6.2 is used, care must be taken that the average strain is not close to zero. If this is the case, division by zero can cause computational difficulties. In order to avoid this problem, a small value of strain is added to the denominator if the average of the two strains is close to zero. In the demonstration presented here, a small percentage of the maximum strain in the overall problem is added to the denominator if the existing average is close to zero.

The progression of the nodal error estimates is presented in Figs. 6.10–6.12. A sequence of figures is used to present the error estimates because the nodal errors do not decrease monotonically so it would be difficult to identify the progression of error estimates if they were all presented at once. In Fig. 6.10, the nodal error estimates for the initial finite element model and the first adaptive refinement are shown. As can be seen, the first application of adaptive refinement produces a model with 10 elements. This is equivalent to a uniform refinement. Each element in the initial five-element model had an estimated error that was larger than the threshold value of 4%. As a result, every element was subdivided.

In Fig. 6.11, the nodal error estimate for the second adaptively refined model is superimposed on the error estimates for the previous two models. The primary significance of this figure is that it clearly shows that the error estimates based on the locally normalized interelement jumps are not monotonic. In general, the error estimates for this third model are reduced, but the nodal error estimate for one point exceeds the maximum error estimate contained in the previous refinement.

The nodal error estimates for three more iterations of the adaptive refinement process are introduced in Fig. 6.12. As can be seen, 32 elements are required to satisfy the termination criterion of 4% used in this analysis. This contrasts with the 19 elements required with the error measure normalized with the maximum strain in the finite element model. This implies that the absolute error is equalized over the domain of the problems.

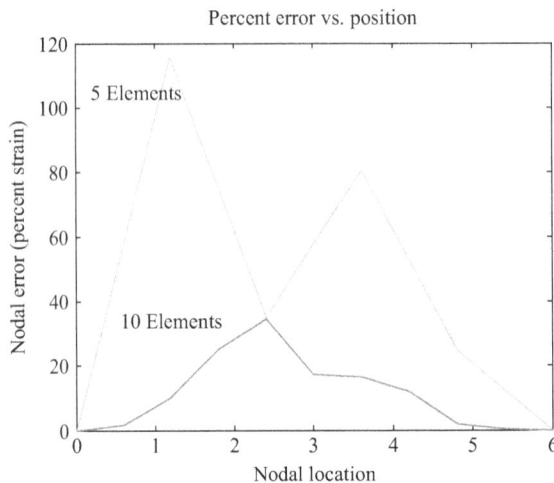

Figure 6.10. Nodal error estimate for the first two models.

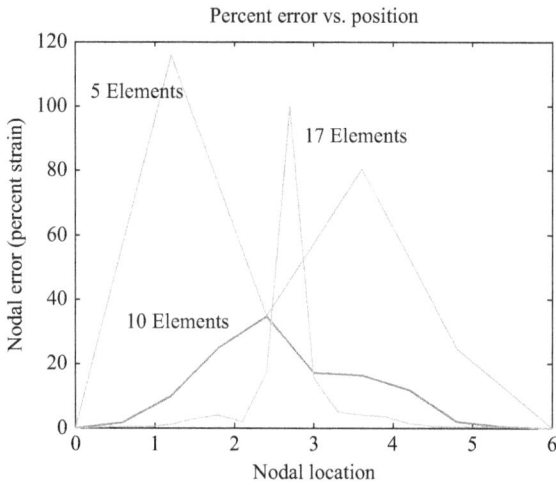

Figure 6.11. Nodal error estimate for the first three models.

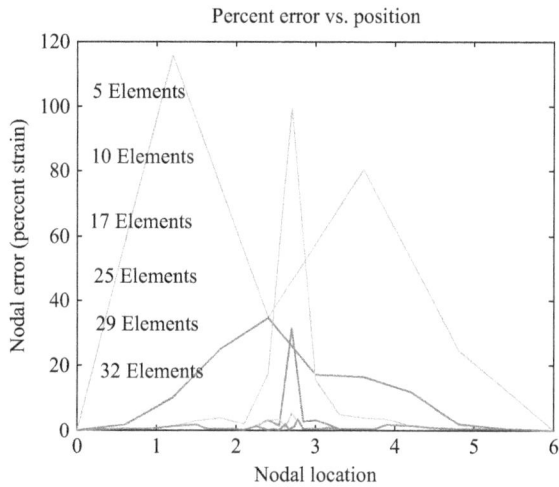

Figure 6.12. A sequence of nodal error estimates.

More detailed results for the final adaptively refined model are shown in Fig. 6.13. The finite element approximation is superimposed on the exact solution in Fig. 6.13a. As can be seen, the strain representation is a good match. This conclusion is confirmed in Fig. 6.13b where the nodal errors are significantly below the threshold of the termination criterion.

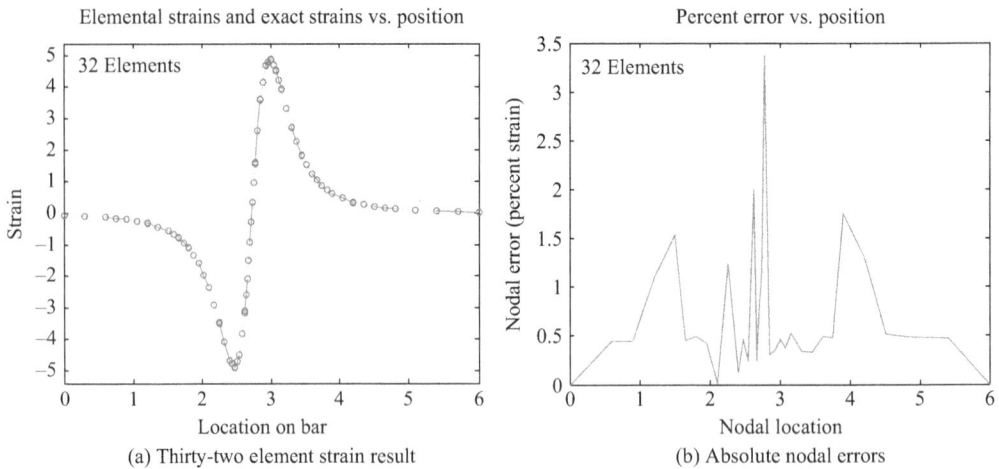

Figure 6.13. Finite element model after six iterations of adaptive refinement.

When Fig. 6.13a is compared to Fig. 6.8a, we see that the number of elements in the region of the stress concentration is equal for both cases. However, in Fig. 6.13a, we see that the finite element representation is smoother and there are more elements in the regions of low strain. In other words, the error estimator normalized with a local strain value did the job it is designed to do. It distributed the "absolute" errors more uniformly over the domain of the problem being solved.

6.7 SUMMARY

The idea of the adaptive refinement process has been introduced and its effectiveness has been demonstrated. We have seen that by refining a finite element model in regions of high error that the adaptive refinement process creates models that produce accurate representations of the exact result more efficiently than does uniform refinement.

In addition, it has been shown that the adaptive refinement process can be designed to produce a result with given characteristics. In one demonstration, the adaptive refinement process focused on regions of critical strain. In another demonstration, the model was refined so the errors were more uniformly distributed over the domain of the problem.

The use of the interelement jumps as the basis of the error estimates allows the errors to be interpreted in terms of quantities of interest to an analyst. Specifically, the errors are related to strain quantities so the termination criterion can be specified in terms of critical strain levels for the material being used.

Although the error estimators based on the interelement jumps are effective, they have the following deficiencies:

1. The interelement jumps disguise the level of errors in individual elements; when negative and positive errors exist in an element they cancel each other in the integration process.
2. The interelement jumps aggregate the errors in adjacent elements so the errors in critical elements are diluted, that is, the error resolution is reduced.
3. The interelement jumps submerge the severity of the errors at critical points because the jumps depend on the sum of integrated quantities.
4. The interelement jumps do not exist naturally on fixed boundaries so errors on these boundaries must be assessed differently.

In the following chapters, we will develop error estimators based on different theoretical bases that correct the deficiencies identified for the error estimators presented in this chapter.

6.8   REFERENCES

1. Dow, J. O. *A Unified Approach to the Finite Element Method and Error Analysis Procedures*, New York: Academic Press, 1999.
   This book puts the finite element method, the finite difference method, and error analysis on a common Taylor series basis. It develops the physically based elements described in the Introduction to Chapter 5.
2. Hughes, Thomas, R. J. *The Finite Element Method*, Mineola, New York: Dover Publications, 2000, p. 112. References to the origin of the isoparametric method.
3. Argyris, J. H. and Kelsey, S. *Energy Theorems and Structural Analysis*, Butterworths, 1960. (Originally published in a series of articles in Aircraft Engineering, 1954–55.)
4. Tiag, I. C. *Structural Analysis by the Matrix Displacement Method*, English Electric Aviation Report No. S017, 1961.

CHAPTER 7

STRAIN ENERGY-BASED ERROR ESTIMATORS—THE Z/Z ERROR ESTIMATOR

7.1 INTRODUCTION

In the previous chapter, the adaptive refinement procedure is applied to several problems. As was apparent in these examples and as can also be seen in the schematic diagram shown in Fig. 7.1, the critical component of the adaptive refinement process is **the error analysis operation**.

After the initial model is formed and solved, the errors in the finite element solution are quantified with an error estimator. Then, on the basis of the error analysis, the finite element solution is evaluated. If each element in the model satisfies the evaluation criterion, the analysis is stopped. If any elements do not pass the evaluation, the model is refined in the regions of high error and the process repeats.

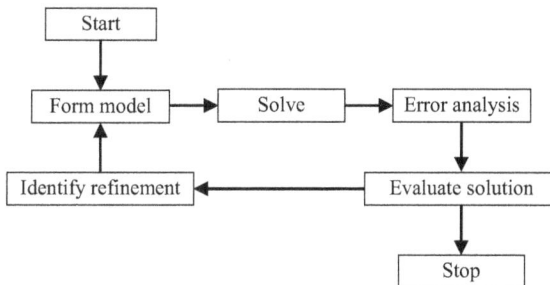

Figure 7.1. Adaptive refinement schematic.

Zienkiewicz and Zhu (Z/Z) developed the first practical error estimator in 1988 [1, 2]. This error estimator quantifies the error in each element by computing the strain energy contained in the difference between the discontinuous finite element strain distribution and an improved strain distribution that is closer to the exact solution than the finite element result.

The advantage of using the strain energy as a metric is that the positive and negative differences between the two strain representations cannot cancel each other. This is because the strain energy expression squares the differences so that the differences always contribute a positive quantity to the estimate of the error.

However, the use of strain energy as a metric has disadvantages when compared to the point-wise error estimators presented in Chapters 6 and 8 in this book. The Z/Z approach does not have the resolution or the computational efficiency of the point-wise approaches. Integration must be performed over the domain of each element in order to estimate the error in the strain energy. More importantly, the strain energy-based error estimator does not report the error in terms of a primary variable in continuum mechanics such as stress or strain.

Regardless of its deficiencies, the primary **objectives of this chapter** are to present the development of the Z/Z error estimator and to demonstrate its behavior in the adaptive refinement process. This error estimator is discussed in detail both because of its place in the history of error analysis and because concepts embedded in the Z/Z approach are used in the development of new approaches for forming refinement guides that are presented in Chapter 11.

This chapter proceeds as follows. For the sake of completeness, the evolution of the procedure for forming the smoothed solution in the Z/Z approach is presented in the next section. Next, the procedure for estimating the error in the elemental strain energy is outlined. In the following two sections, versions of strain energy metrics used as error estimators are developed. After that discussion, the error estimates produced by the Z/Z approach for uniformly and adaptively refined models are presented.

7.2 THE BASIS OF THE Z/Z ERROR ESTIMATOR—A SMOOTHED STRAIN REPRESENTATION

In the original version of the Z/Z approach, a smoothed solution is constructed over the domain of the entire problem. The smoothed solution is formed using a least squares approach with the following two constraints: **(1)** the smoothed solution is forced to have the same amount of strain energy as the finite element result and **(2)** the errors in the finite element result are distributed so the total error is minimized.

The computational effort required for this process is of the same magnitude as the effort to compute the solution of the finite element model. In later developments, the computational effort required to form the smoothed solution is reduced by forming the smoothed result on an **element-by-element basis.** The resulting error estimates are not compromised by the simpler approach to smoothing.

An overview of the Z/Z error estimator computed with **element-by-element smoothing** is shown in Fig. 7.2. In Fig. 7.2a, the strain distribution over a representative patch of three-node bar elements is illustrated along with the average of the interelement nodal strains, which are denoted by the x's.

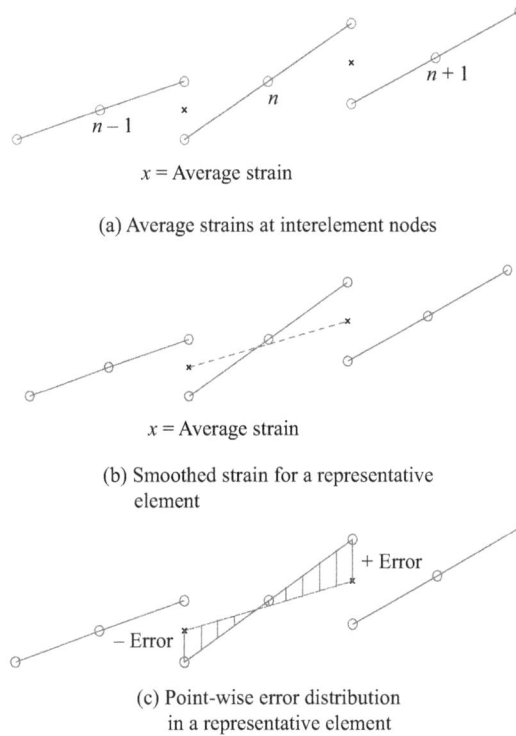

x = Average strain

(a) Average strains at interelement nodes

x = Average strain

(b) Smoothed strain for a representative
element

+ Error

− Error

(c) Point-wise error distribution
in a representative element

Figure 7.2. The basis of the Z/Z error measure.

Figure 7.2b shows the smoothed strain distribution that is formed by applying the displacement interpolation function to the average of the interelement jumps and the strain at the interior node of Element (n). In Fig. 7.3c, the estimated point-wise error in the finite element strain distribution is shown as the difference between the finite element and the smoothed strain representations.

As can be seen in Fig. 7.2c, the estimated errors can be either positive or negative depending on whether the finite element strain is larger or smaller than the smoothed strain. If the differences are summed, the error can be underestimated because the positive and negative differences will cancel each other. In order to eliminate this possibility, the error measure is computed as the strain energy contained in the difference between these two strain representations. Since strain energy is a function of the square of the strain, the magnitude of the error estimator is unaffected by the sign of the differences in the strains.

The Z/Z approach to error estimation is based on the assumption that the smoothed strain distribution that is formed by averaging the interelement nodal strains is closer to the exact solution than is the discontinuous finite element solution. This assumption is justified from an intuitive point of view, which recognizes that the smoothed solution is continuous instead of discontinuous. However, this assumption is given a solid theoretical foundation by the results presented in Chapter 4—The Source and Quantification of Discretization Errors.

In Chapter 4, the interelement jumps in the strains in a finite element result are shown to be due to the failure of the finite element solution to satisfy the governing differential equation being solved. In the smoothing process, a portion of the interelement jump is added to the discontinuous finite element solution on the domain of the individual elements. As a consequence, the smoothed strain representation is closer to the exact solution than the discontinuous finite element solution.

Procedures for forming the locally smoothed strain representations were refined in later versions of the Z/Z approach. These changes produced incremental improvements to the error estimates without changing the overall concept.

One modification developed the locally smoothed solution from the strains at the Gauss points of the element being evaluated and the strains at the Gauss points contained in surrounding elements. This change was made because the finite element strains at the Gauss points may be closer to the actual strains than at other points on the domain of an element. However, when adaptive refinement is being used, it is not clear if this subtlety needs to be considered.

The behavior of the strains at the Gauss points is outlined and demonstrated at the end of this chapter in Appendix 7A—Gauss Points, Super Convergent Strains, and Chebyshev Polynomials. One reason for including this Appendix is to introduce Chebyshev Polynomials to readers unfamiliar with these important functions.

7.3   THE Z/Z ELEMENTAL STRAIN ENERGY ERROR ESTIMATOR

As shown in Fig. 7.2c, the point-wise error in the finite element strain result for an individual element is assumed to be the difference between the finite element and the "smoothed" strain representations. This idea is given in equation form as:

$$\Delta\varepsilon_n(x) = \left[\left(\varepsilon_n(x)\right)_{\text{Finite Element}} - \left(\varepsilon_n(x)\right)_{\text{Smoothed}}\right] \qquad \text{(Eq. 7.1)}$$

where

$\Delta\varepsilon_n(x)$ = the estimated point-wise strain error in Element (n).

$\left(\varepsilon_n(x)\right)_{\text{Finite Element}}$ = the finite element strain distribution.

$\left(\varepsilon_n(x)\right)_{\text{Smoothed}}$ = the improved or smoothed strain distribution over Element (n).

The estimated error in the strain energy of Element (n), ΔSE_n, with a constant cross section is given as:

$$\Delta SE_n = EA/2 \int_0^L \left(\Delta\varepsilon_n(x)\right)^2 dx \qquad \text{(Eq. 7.2)}$$

where ΔSE_n = the estimated strain energy error in Element (n).

In mathematical terms, the error measure given by Eq. 7.2 is called a weighted L_2 norm because it is squared and, hence, always positive.

7.4   THE Z/Z ERROR ESTIMATOR

The Z/Z error estimator used in the adaptive refinement process is formed by normalizing the error measure given by Eq. 7.2 as follows:

$$\eta_n = \left[\frac{(\Delta SE_n)^{1/2}}{(SE)^{1/2}_{Normalizing}} \right] \times 100 \qquad \text{(Eq. 7.3)}$$

where η_n = the Z/Z elemental error measure for Element (n) as a percentage.

ΔSE_n = the estimated strain energy error in Element (n).

$(SE)_{Normalizing}$ = the normalizing factor that focuses the error estimator.

The numerator of the elemental error estimator is the square root of the estimate of the error in the strain energy of an individual element that is given in Eq. 7.2. The denominator can be chosen to make the error estimator perform in different ways. The square root of the components of Eq. 7.3 is taken to make it appear as if the error estimates represent the error in terms of strains, not strain energy.

In the first form of the Z/Z approach, the normalizing factor was identified as being equal to the total strain energy in the problem. Since the total strain energy depended on the size of the particular problem and the level of loading, this error estimator provided little, if any, physical insight into the meaning of the error estimate. As a result, it is difficult to define a termination criterion that is understandable in terms of the problem being solved.

In order to provide some physical insight into the meaning of the error estimator, a normalizing factor based on the strain energy content of the element being analyzed can be used. The most obvious candidate for such a localized normalizing factor is an estimate of the total strain energy in the element. This estimate is taken as the sum of the actual strain energy in the element plus the estimated error in the strain energy content of the element. This quantity is given as:

$$(SE)_{Normalizing} = (SE_n + \Delta SE_n) \qquad \text{(Eq. 7.4)}$$

where SE_n = the strain energy content of Element (n).

ΔSE_n = the estimated strain energy error in Element (n).

When the error estimator is normalized with Eq. 7.4, the result can be interpreted as a percentage of the total error in the Element (n).

However, this normalizing factor **can** cause the adaptive refinement process to diverge. The divergent behavior can occur because the numerator and the denominator of Eq. 7.3 each have a different nature. The numerator of Eq. 7.3 contains contributions from the errors in the adjacent elements as well as from the element being analyzed. However, the denominator contains only contributions from the element being analyzed.

As a result of these different characteristics, an element actually containing a low level of error that is adjacent to a high error element will have an error estimate that is excessive. This, in turn, may cause unneeded refinement of this low error element. The refinement of this low error element exacerbates the problem, and the refinement can diverge. An example of this divergence when the Z/Z error estimator is normalized with Eq. 7.4 is presented in Appendix 7B—An Unsuccessful Example of Adaptive Refinement.

7.5   A MODIFIED LOCALLY NORMALIZED Z/Z ERROR ESTIMATOR

The divergent behavior identified in the previous section can be eliminated by modifying the denominator of Eq. 7.3 so that it has the same characteristics as the numerator. This change is accomplished by including strain energy contributions from the elements adjacent to the element being evaluated into the denominator of the error estimator. When this is done, the modified version of the locally normalized Z/Z error estimator becomes:

$$\eta_n = \left[\frac{\left(\Delta SE_n\right)^{1/2}}{\left(SE\right)^{1/2}_{Normalizing}} \right] \times 100 \qquad \text{(Eq. 7.5)}$$

where η_n = the Z/Z elemental error measure as a percentage.

ΔSE_n = the estimated strain energy error in Element (n).

$(SE)_{Normalizing} = SE_n + \Delta SE_n + C(SE_{n-1} + SE_{n+1})$

C = A participation factor for the elements adjacent to Element (n).

When the denominator of Eq. 7.5 is compared to the denominator given by Eq. 7.4, we see that it contains a contribution that is due to the strain energy content of the elements adjacent to the element being evaluated. The amount of strain energy added to the denominator is controlled by the participation factor, C. The participation factor for these strain energy quantities is arbitrarily chosen as 0.5 in order to demonstrate the process for the examples presented in Sections 7.6 and 7.7.

The modification just introduced into the locally normalized Z/Z error estimator addresses the situation that caused the example presented in Appendix 7B to diverge. The overly high error estimates for lightly strained elements are reduced because the denominator is no longer close to zero. We will see the effect of this improvement in Section 7.7 when Eq. 7.5 successfully guides the adaptive refinement process for the problem that diverged in Appendix 7B.

7.6   A DEMONSTRATION OF THE Z/Z ERROR ESTIMATOR

In this section, the Z/Z error estimator defined by Eq. 7.5 is applied to a sequence of uniform refinements beginning with the five-element control problem that is formed with three-node elements. The result of the analysis of the same problem with the residual approach that was developed in Chapter 4 is also presented. This is done to show that the two approaches possess the same characteristics. The similarity in results means that the point-wise error estimator is at least as effective as the Z/Z approach for evaluating finite element results.

The results for the initial five-element model are shown in Fig. 7.3. Figure 7.3a compares the strains in the finite element model to the exact solution. Figure 7.3b contains the Z/Z error estimates for the individual elements. Figure 7.3c contains the nodal error estimates that are produced by the residual approach.

In the plots of the Z/Z elemental error estimates, the estimates are arbitrarily presented as if they apply at the center points of the elements. The errors are presented in a continuous curve by connecting the error values with straight lines and by taking the errors at the ends of the bar to be zero. For this problem the errors are close to zero at the end points. The lines connecting the points where the errors are estimated have no real meaning. The lines are presented for visual convenience.

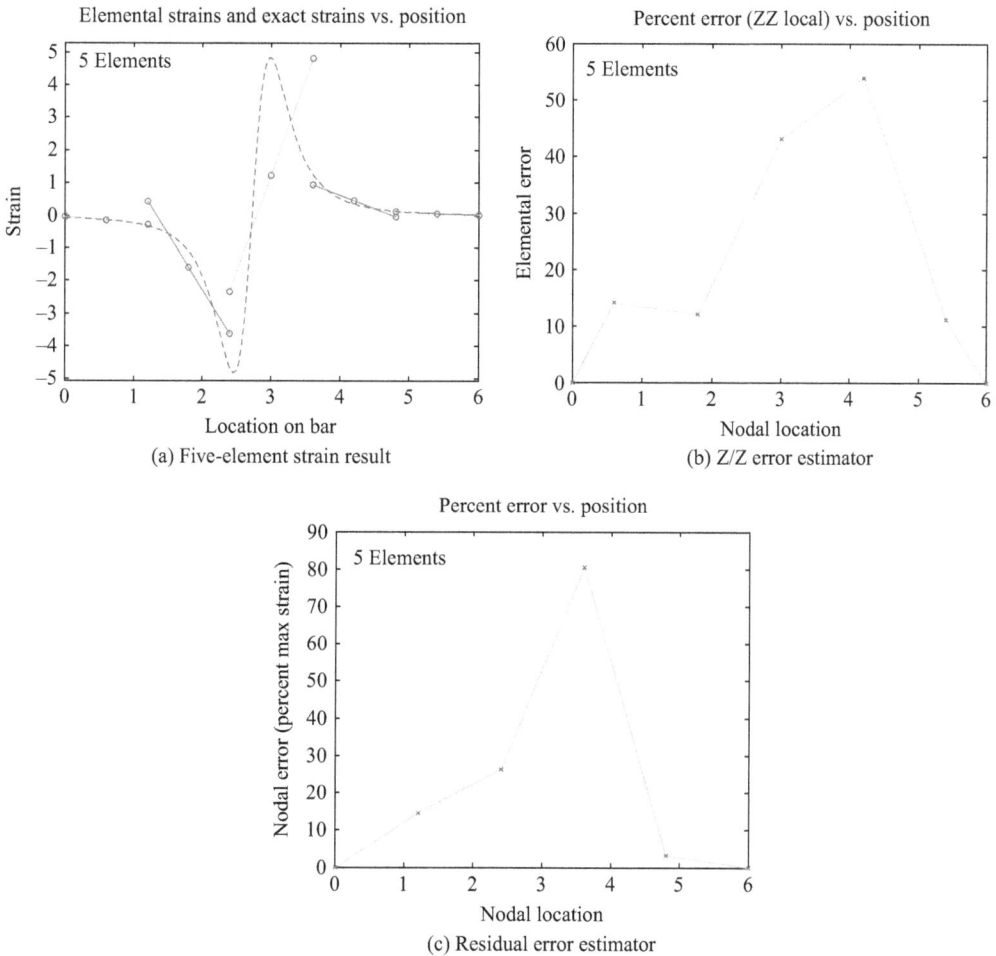

(a) Five-element strain result

(b) Z/Z error estimator

(c) Residual error estimator

Figure 7.3. A five-element model.

The error estimates produced by the residual approach in Fig. 7.3c are nodal quantities. Since the Z/Z error estimates are elemental quantities, the two error estimates do not correspond one to one. However, when Figs. 7.3b and 7.3c are compared, we see that the two error estimators identify the same regions as having large values of error. The residual result has a higher maximum value for the error estimator. This difference in magnitude is not significant for this discussion because the objective of this chapter is to demonstrate the characteristics of the Z/Z approach and to show that the approach successfully guides the adaptive refinement process.

When the five-element model is modified with two uniform refinements, the results for the 10- and 20-element models are shown in Figs. 7.4 and 7.5.

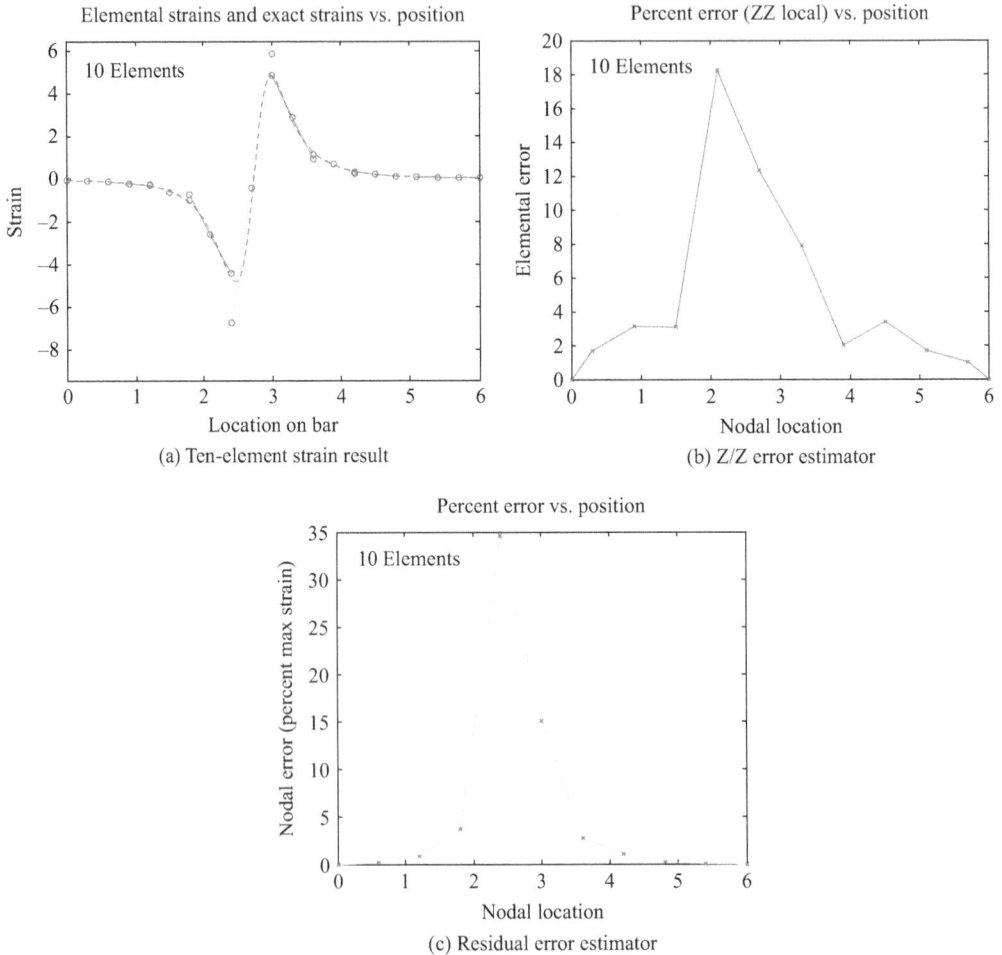

(a) Ten-element strain result

(b) Z/Z error estimator

(c) Residual error estimator

Figure 7.4. A 10-element model.

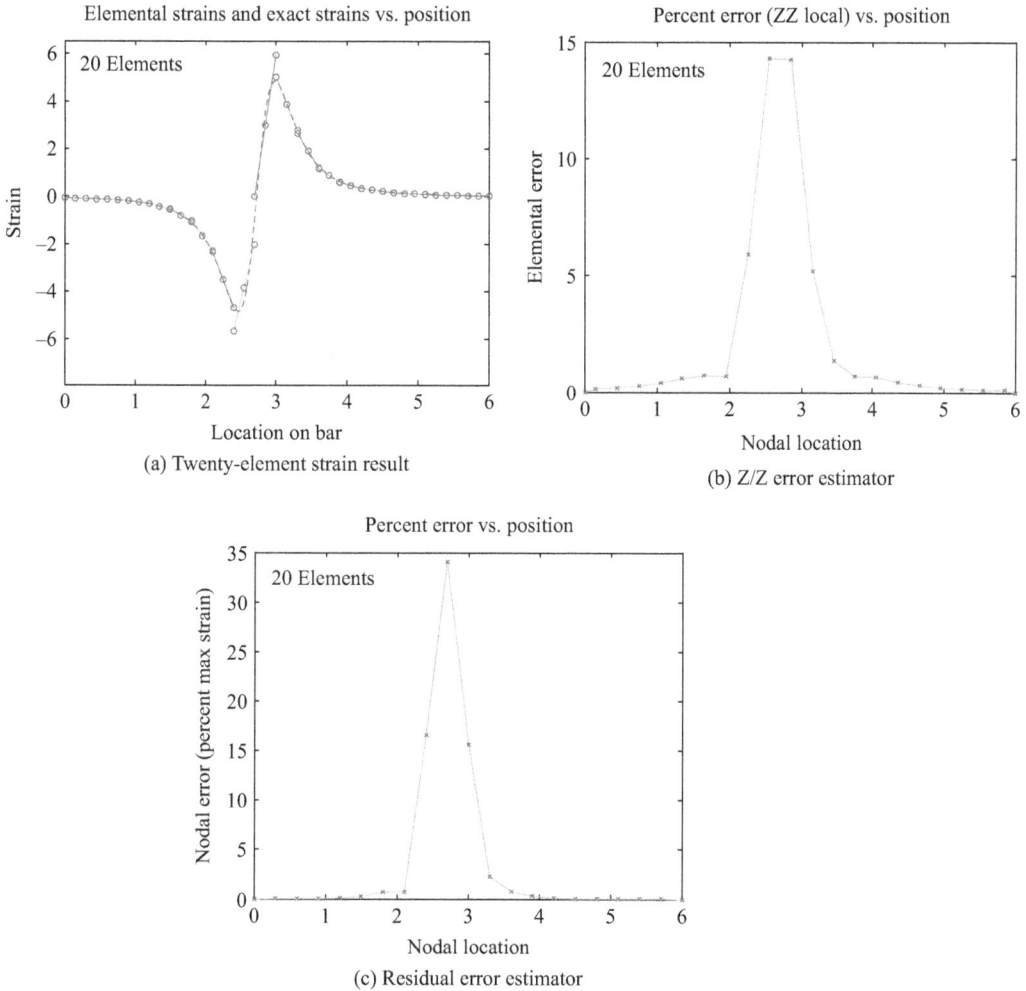

(a) Twenty-element strain result

(b) Z/Z error estimator

(c) Residual error estimator

Figure 7.5. A 20-element model.

As can be seen in Figs. 7.4 and 7.5, the qualitative characteristics of the two error estimators are similar. That is to say, both error estimators identified the same region as having the maximum error in the first refinement. In the second refinement shown in Fig. 7.5, both approaches identified the same new location as having the maximum error. Furthermore, the regions identified with the highest level of error correspond to the regions that actually contain the maximum error.

When the magnitudes of the error estimates are compared, we see that both were reduced in a similar manner by the uniform refinements. In addition, the differences between the maximum and the minimum estimated errors were similar in the two error estimators. That is to say, they had approximately the same resolution between high and low errors. These results indicate that both approaches will successfully identify regions in the finite element model where the model must be refined in order to improve the results.

The results for uniformly refining the model two more times are shown in Figs. 7.6 and 7.7, respectively. As can be seen, the uniform refinements significantly reduce the maximum error for both approaches to error analysis. Furthermore, the error estimates for both approaches correlate well with the actual errors along the length of the bar. The actual errors are seen as the difference between the finite element results and the exact solution, which are shown in Figs. 7.6a and 7.7a, respectively.

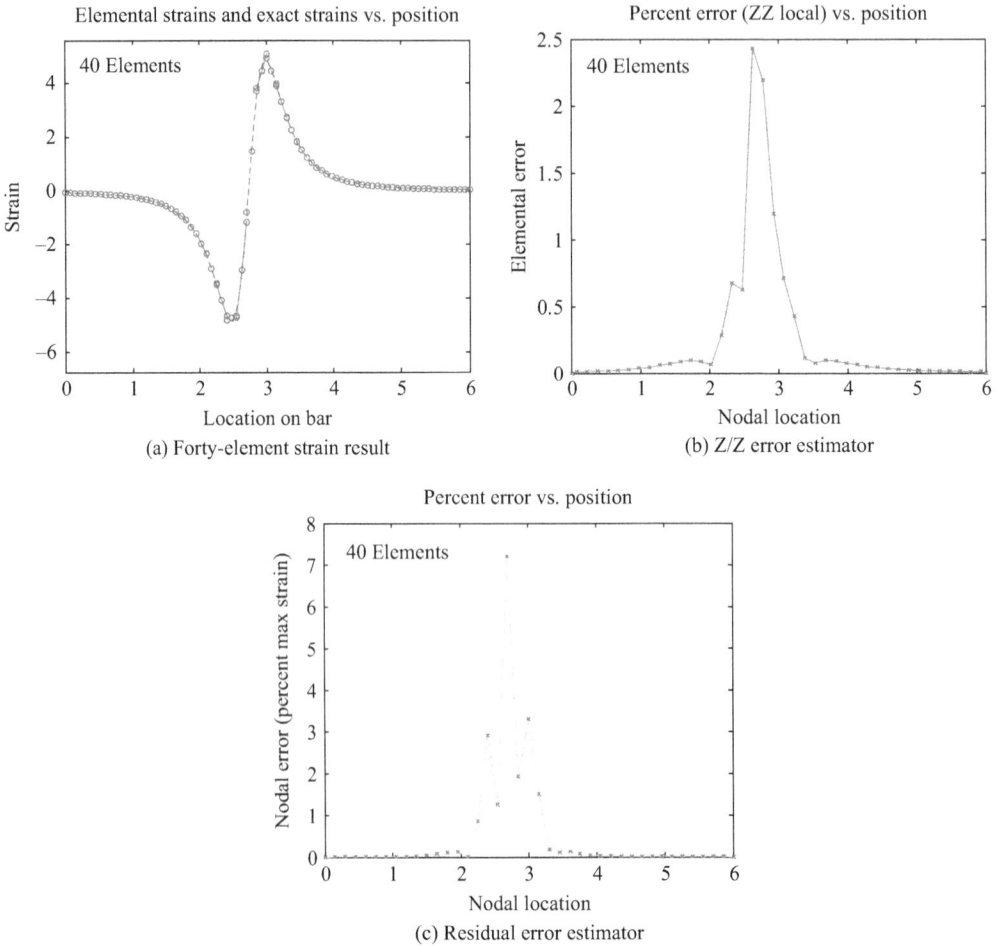

(a) Forty-element strain result

(b) Z/Z error estimator

(c) Residual error estimator

Figure 7.6. A 40-element model.

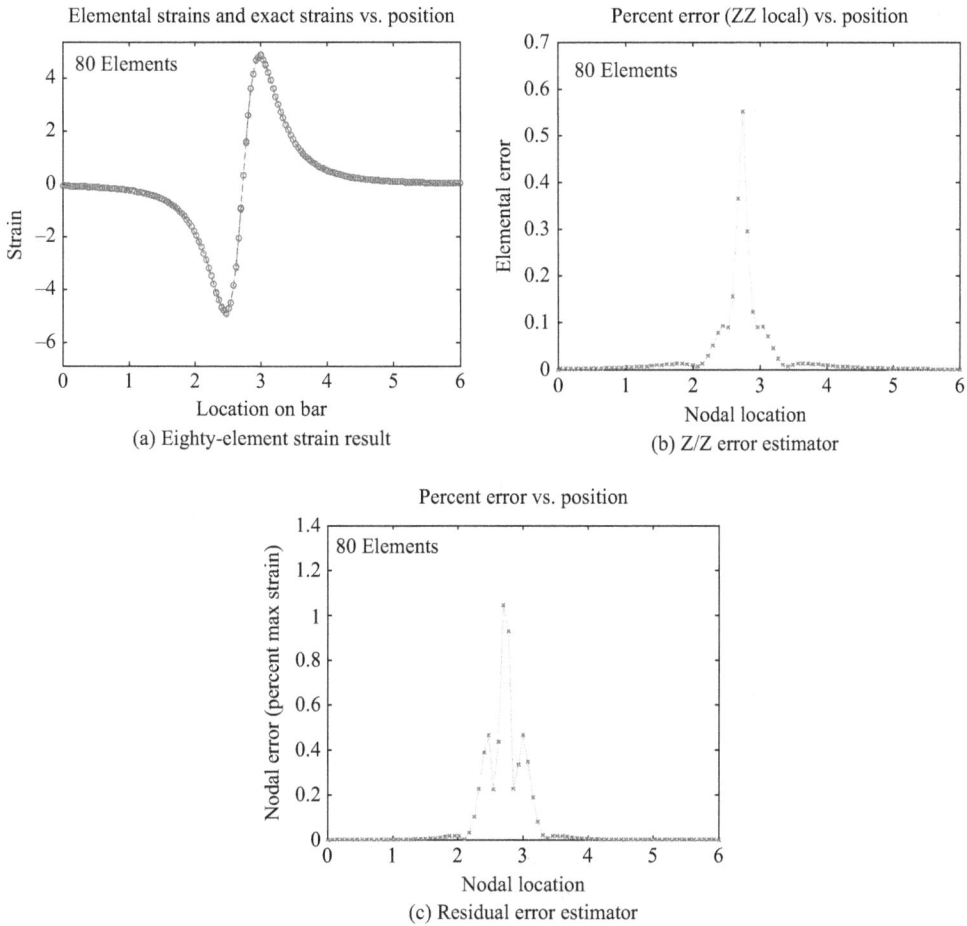

Figure 7.7. An 80-element model.

It should be noted that this result validates the desirability of using adaptive refinement to improve finite element models instead of uniform refinement. That is to say, in uniform refinement, a majority of the elements added to the model are introduced in regions of low error.

7.7   A DEMONSTRATION OF ADAPTIVE REFINEMENT

In this section, the Z/Z error estimator defined by Eq. 7.5 is used to guide the adaptive refinement of the five-element demonstration problem. Here, an element is subdivided into two equal length elements if the refinement guide given by Eq. 7.5 is greater than or equal to a value of 4%. This is the same problem that diverged when the denominator given by Eq. 7.4 is used in the error estimator given by Eq. 7.3 for the problem solved in Appendix 7B.

Figure 7.8 presents the results of solving the five-element model and its corresponding error analysis. As can be seen in Fig. 7.8a, the three lightly strained elements, Elements (1), (4), and (5), represent the exact solution reasonably well. In contrast, the two elements with the highest rates of change in strain (high strain gradients) and relatively high magnitudes of strains that are attempting to represent the regions of rapidly changing strains do not closely represent the exact solution.

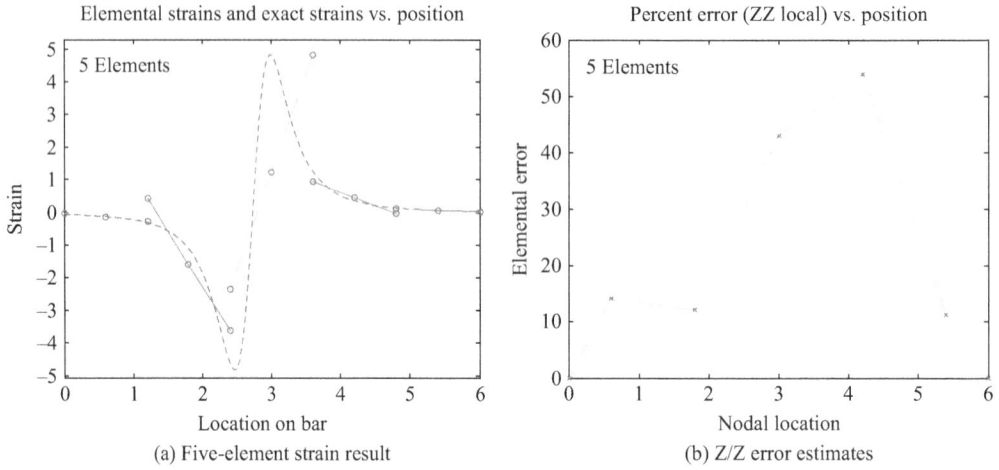

(a) Five-element strain result

(b) Z/Z error estimates

Figure 7.8. The initial five-element mesh.

The approximate error estimates presented in Fig. 7.8b do not completely agree with the observations just made about the accuracy of the strain representations of the individual elements. The element with the highest estimated error represents the exact solution reasonably well. The error estimate for this element does not coincide with the actual error for two reasons: **(1)** since it is located next to an element that is incapable of capturing the shape of the exact result with any degree of accuracy, the interelement jump between the two elements is large and **(2)** this element and the element located to its right are low strain elements.

As a result, the numerator of the error estimator is overly large and the denominator is relatively small. This leads to an error estimate that is higher than the actual error in the element. Consequently, this element with an acceptable level of error is needlessly subdivided. This will add computational effort, but the improvement of the model is not negatively affected.

Since the error estimate for every element in this model exceeds the refinement threshold of 4%, every element in the model will be subdivided to form the new model. As a result, the initial refinement is equivalent to a uniform refinement. It is worth noting that the error estimate for every element is also larger than 10%. This means that if the refinement threshold is relaxed, the refined model would not possess fewer elements.

The results of the first refinement are shown in Fig. 7.9. As can be seen, the maximum error has been reduced from approximately 55% to approximately 19% and the error estimates have been significantly reduced for every element. In addition, the error estimates for three of the 10 elements exceed the threshold level of 4%. All three of these elements either encompass the region of actual maximum error or are adjacent to this region. That is to say, this set of error estimates will refine the model where it needs refining.

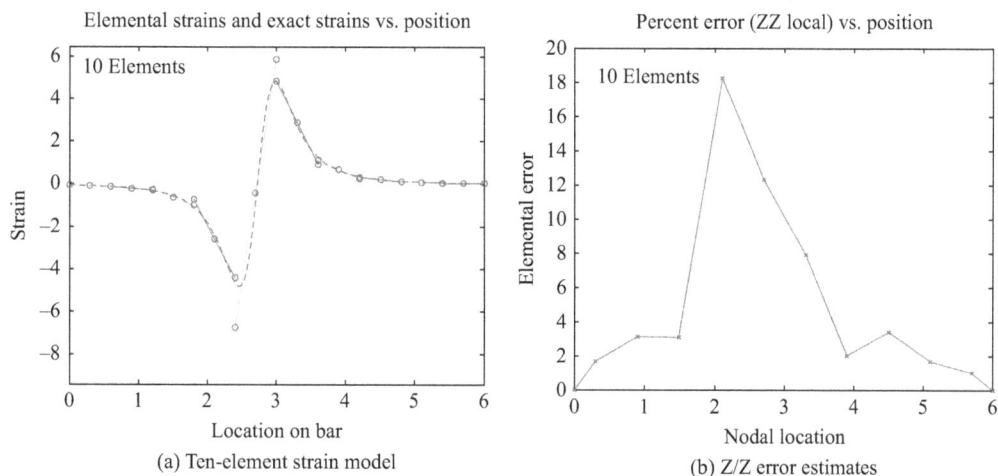

Figure 7.9. The first adaptively refined mesh of 10 elements.

Figures 7.10–7.12 contain the subsequent three adaptive refinements required to satisfy the termination criterion of 4%. The only unexpected feature of these refinements is the fact that the maximum error increased in the second refinement. Figure 7.10b shows that the maximum error increased from approximately 18% in the first refinement to approximately 21% in the second refinement.

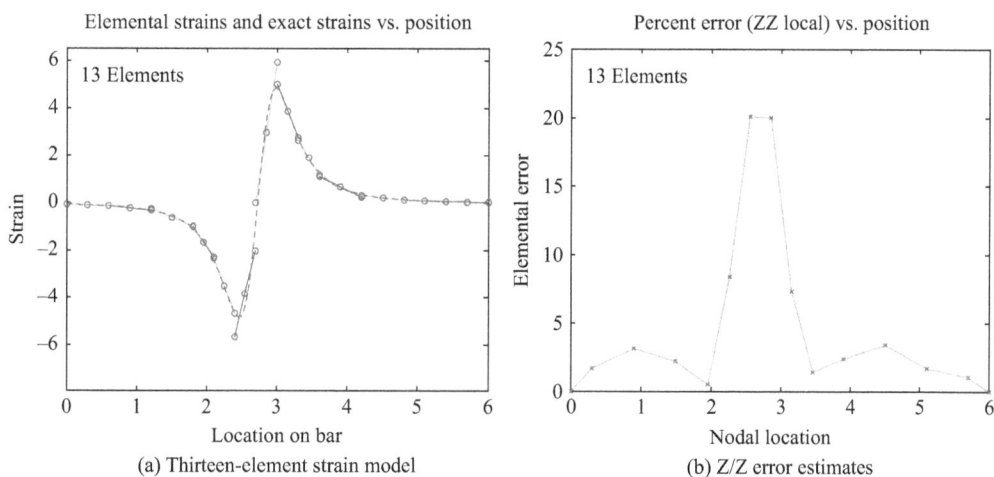

Figure 7.10. The second adaptively refined mesh of 13 elements.

The reason for this increase can be seen by comparing Figs. 7.9a and 7.10a. The single element that attempts to represent the two critical stresses and the inflection point between the positive and negative stress concentrations in Fig. 7.9a has been subdivided in the second refinement. As can be seen in Fig. 7.10a, a significant interelement jump has been introduced by subdividing the element in the region of the inflection point in the center of the bar. That is to say, a modeling deficiency that was submerged on the interior of a single element has been made explicit on the boundary between two elements.

The refinements shown in Figs. 7.11a and 7.12a introduce elements into the regions where the error estimates exceed the refinement threshold and the finite element model fails to accurately represent the exact solution. Thus, it can be concluded that the locally normalized Z/Z error estimator given by Eq. 7.5 produces a convergent result as it is designed to do.

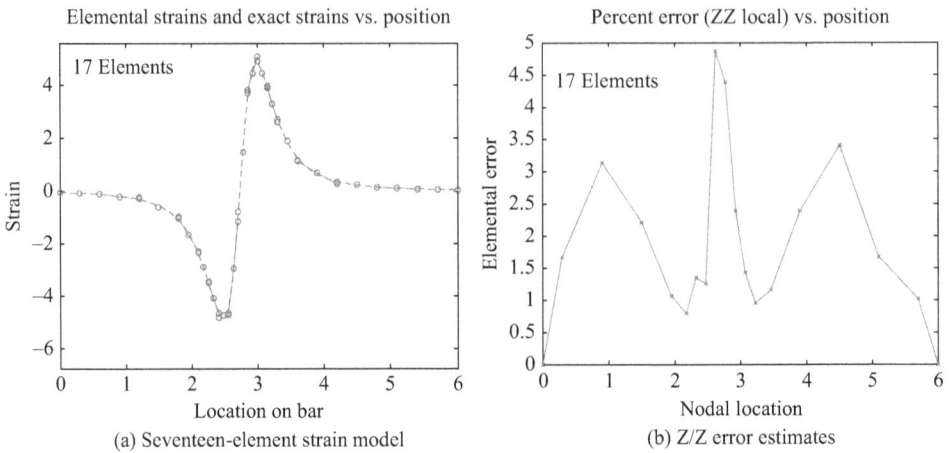

(a) Seventeen-element strain model

(b) Z/Z error estimates

Figure 7.11. The third adaptively refined mesh of 17 elements.

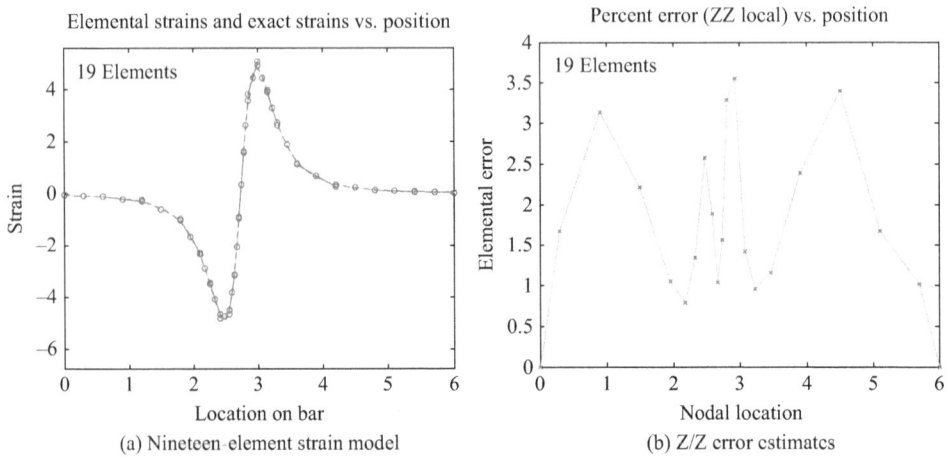

(a) Nineteen-element strain model

(b) Z/Z error estimates

Figure 7.12. The fourth adaptively refined mesh of 19 elements.

7.8 SUMMARY AND CONCLUSION

This chapter has presented and demonstrated the Z/Z approach to error estimation. This intuitively satisfying procedure estimates the error in the strain energy content of an individual element by comparing the discontinuous finite element strain results to a smoothed solution over the domain of the element. This approach is intuitively satisfying because the smoothed solution is taken to be closer to the exact solution than the discontinuous finite element result because it is continuous.

In addition to being intuitively satisfying, the Z/Z approach now has a solid theoretical foundation. This foundation was provided by the contents of Chapter 4, which showed that the interelement jumps in the finite element solution are a direct quantification of the discretization errors in the finite element solution. As a result, it can be shown that the smoothed solution used in the Z/Z approach is, indeed, an improved solution.

The Z/Z smoothed strain representation is formed by adding a strain component on the domain of an element due to the interelement jumps to the finite element solution. As mentioned above, the interelement jumps in the strain are due to the discretization errors in the finite element model. Thus, the smoothed solution used in the Z/Z error estimator is an improvement over the finite element strain representation because the discontinuous finite element strain distribution is augmented with a portion of the error that exists in the finite element result. As a result, the smoothed strain representation is closer to the exact result than is the discontinuous finite element result.

The recognition that this smoothed solution is closer to the exact solution than the finite element result suggests a new way to develop refinement guides for the adaptive refinement process. This innovative approach for estimating the level of refinement to be given to each of the high error elements is developed in Chapter 11 and proceeds as follows. First, the existing finite element result for a group of nodes close to and including the element being considered are used to form an improved solution. This improved solution is then decomposed into the coefficients of the physically interpretable strain gradient notation discussed in various locations in this book.

This decomposition contains coefficients of a higher order than are contained in the strain representation produced by the element being analyzed. This is the case because information from more nodes is used in this decomposition than exist in the individual elements. This higher order representation is taken to be an approximation of the exact solution that is emerging from the finite element model. Then, the number of subdivisions that must be given to an element so the refinement can represent the complexity of the emerging solution is identified. This is accomplished by finding the number of elements that are needed to represent the highest order strain gradient component contained in the emerging solution to within the accuracy defined by the termination criteria. As we will see in Chapter 11, this approach produces a rapidly converging adaptive refinement process.

This approach to model refinement is an improvement over previous approaches used in the adaptive refinement process. The previous approaches identify the level of refinement to be given to an element with a correlation based on the magnitude of the error estimated for the element. However, it is not always clear how this correlation is determined.

The availability of a refinement guide that uses a rational approach instead of a correlation reduces the demands on the error estimator. The error estimator needs only to function as a termination criterion. It need not be directly involved in the model refinement process. As a consequence, the ideal characteristics of an error estimator can be defined as follows.

The first priority for an ideal error estimator is that it can be expressed in terms of a metric that allows a meaningful termination criterion to be defined. The second desirable characteristic is that the computation of the error estimator requires a small amount of effort.

The point-wise error estimator developed and presented in the previous chapter possesses these two ideal characteristics. The error estimator based on the interelement jumps in the strain representations estimates the error in terms of a quantity of primary interest in solid mechanics. The error is estimated as a percentage of a critical strain value. This can easily be transformed to a metric expressed in terms of stress if so preferred. As a result, the termination criterion can, for example, be related to a failure model for the material being analyzed.

In contrast, the Z/Z approach estimates the error with a metric based on the strain energy content of the element. The strain energy is not a quantity that is of primary interest in solid mechanics. In other words, the use of strain energy as the metric in the error estimator does not allow a termination criterion to be defined that gives a clear indication that the quantities desired in the analysis are accurately represented.

In addition, the point-wise error estimator presented in the previous chapter does not require the evaluation of an integral to estimate the error in an element. It can be computed from quantities that are readily available in the finite element analysis, namely the differences in the strains at the interelement nodes. However, the most important feature of the point-wise error estimator is the qualitative advantage identified in the previous paragraph. Namely, the error is estimated in terms of a quantity or metric that be used to identify a useful criterion for defining when a finite element result is accurate enough for the problem being solved.

7.9 REFERENCES

1. Zienkiewicz, O. C. and Zhu, J. Z. "A Simple Error Estimator and Adaptive Procedure for Practical Engineering Analysis," *International Journal for Numerical Methods in Engineering* 24 (1988): 337–57. DOI: http://dx.doi.org/10.1002/nme.1620240206.
2. Zienkiewicz, O. C. and Zhu, J. Z. "The Superconvergent Patch Recovery and *A-Posteriori* Error Estimates, Parts I and II," *International Journal for Numerical Methods in Engineering* 33 (1992): 1331–82. DOI: http://dx.doi.org/10.1002/nme.1620330702.

GAUSS POINTS, SUPER CONVERGENT STRAINS, AND CHEBYSHEV POLYNOMIALS

7A.1   INTRODUCTION

As mentioned in the main text, one of the modifications to the Z/Z approach formed a smoothed solution on an element-by-element basis by interpolating the strains at the Gauss points of the elements that surround the element being evaluated. The modification was introduced because the strains at the Gauss points **can** be more accurate than the strains at the nodes of the element. Furthermore, the strains at the Gauss points are readily available in isoparametric elements. It should be noted that the use the average of the strains at the interelement nodes to form a locally smoothed strain distribution was shown in the main text to be closer to the exact result than the discontinuous finite element result.

The objective of this Appendix is threefold: (1) to demonstrate the behavior of the strains at the Gauss points, (2) to identify the fact that the Gauss points are the zeros of the Chebyshev polynomials (of the first kind), and (3) to introduce the concept of Chebyshev polynomials and their significance to numerical analysis [A1].

The first objective is accomplished by presenting the modeling characteristics of the following three finite element models constructed with three-node bar elements:

1. A finite element model consisting of five equal length elements loaded with a constant load.
2. A finite element model consisting of five equal length elements loaded with a linearly varying load.
3. A finite element model consisting of five **unequal** length elements loaded with a linearly varying load.

The second and third objectives are included simply for the edification of the reader. The idea of Chebyshev polynomials will be discussed briefly after the demonstration of the characteristics of the finite element strain representations at the Gauss points.

7A.2 MODELING BEHAVIOR OF THREE-NODE ELEMENTS

We will now demonstrate the ability of a three-node element to represent a linearly varying strain distribution with an example problem. A finite element model formed from five three-node elements is loaded with a constant load. The strain result for this case is a linearly varying strain distribution. The finite element result and the exact results are shown in Fig. 7A.1. As can be seen, the two results are identical. Thus, the element accurately represents the linear strain distribution. The three-node element behaves as designed.

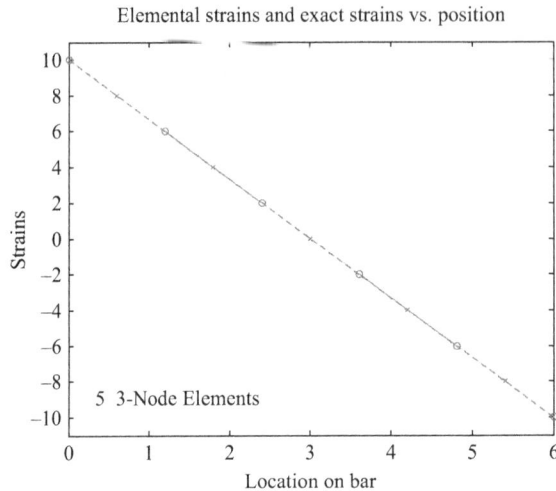

Figure 7A.1. A linear strain distribution.

We will now demonstrate the ability of a three-node bar model to represent a parabolically varying strain distribution with two finite element models. Loading the finite element model with a linearly varying load produces the parabolic strain distribution. The first model will be formed with five three-node elements of equal length. The second model will consist of five elements of unequal length. A three-node element cannot **explicitly** model a parabolic strain distribution because the most complex strain distribution that a three-node bar element can represent is a linear variation. However, we will see that the finite element result matches the exact strain distribution at the Gauss points for a second-order Gauss quadrature numerical integration scheme.

The linearly varying load that was used to load these two example problems is shown in Fig. 7A.2a. The resulting strain distribution for the finite element model formed with five equal length elements is shown in Fig. 7A.2b. The finite element strain distribution is superimposed on the exact result in Fig. 7A.2b.

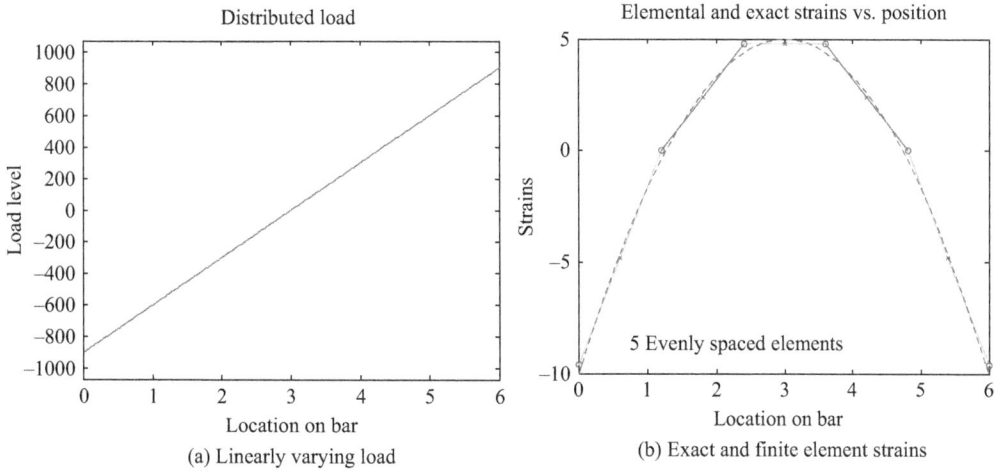

Figure 7A.2. An evenly spaced model representing a quadratic strain.

As can be seen in Fig. 7A.2b, this finite element model does not capture this strain distribution exactly. However, the finite element strain result matches the exact result at two points on the interior of each of the three-node bar elements.

The exact and the finite element strain distributions shown in Fig. 7A.2b are reproduced in Fig. 7A.3 with the Gauss points for a second-order Gauss quadrature shown as the vertical dashed lines for each element. As can be seen, the finite element strains and the strains in the exact result are identical at the Gauss points. This is what is meant when it is said that the strains at the Gauss points are super-convergent. The strain results at the Gauss point are closer to the exact results than at other points on the domain of the element.

The strain distribution for a finite element model formed with five elements of **unequal** length is shown in Fig. 7A.4. The finite element strain distribution is superimposed on the exact result in this figure. As can be seen, this finite element model does not capture this strain distribution exactly. However, the finite element strain result again matches the exact result at the two Gauss points on the interior of each of the three-node bar elements of unequal length.

However, when Figs. 7A.3 and 7A.4 are compared, we see a significant difference in the finite element strain distribution between the two cases. There are no interelement jumps in Fig. 7A.3 but there are interelement jumps in Fig. 7A.4. The result in Fig. 7A.3 means that the nodal equivalent errors in one element cancel the nodal equivalent errors in the adjacent element. Since the elements are of different lengths in Fig. 7A.4, the errors in one element are not the same size as they are in the other element. As a result, the errors in one element do not cancel those in the adjacent element. This result is noted because it could explain some anomalous error estimates in some problem.

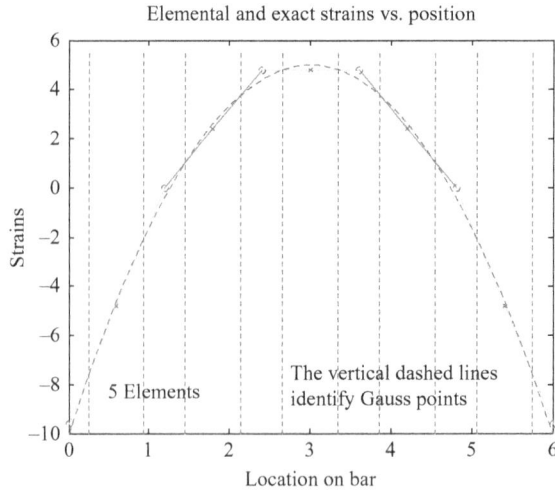

Figure 7A.3. Strains at Gauss points.

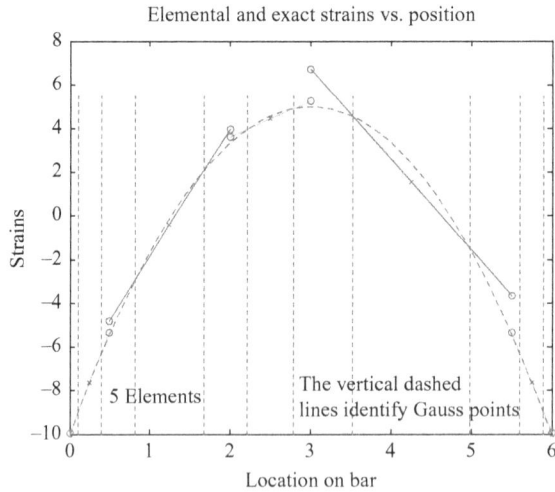

Figure 7A.4. Strains at Gauss points.

7A.3 GAUSS POINTS AND CHEBYSHEV POLYNOMIALS

The reader might want to explore the efficacy of using the Gauss points in a lower order element when the element is attempting to represent higher order strain distributions. This would be a way to establish if the determination of the strains at the Gauss points is primarily used as a convenience in the isoparametric approach or whether it actually improves the strain results for high-order strain representation.

The idea of Gauss points surfaces when the idea of approximate numerical integration using Gauss quadrature is introduced. When the Gauss quadrature procedure is introduced, it is shown that a Gauss quadrature will accurately integrate a polynomial expression with an order of ($2n$-1) where n is the number of Gauss points. The locations of the Gauss points can be found through an optimization process that forces the error to be zero for the highest-order polynomial that is accurately integrated.

However, it turns out that the Gauss points are also the zeros of the Chebyshev polynomials of the first kind. This is almost a **"factoid"** if the idea of a Chebyshev polynomial is not understood.

A compact way to introduce the concept of Chebyshev polynomials is with an example and its interpretation. As is well known, the Runge function is a difficult function for polynomial representations to capture. If the coefficients of a generic nth order polynomial, that is, $y = c_0 + c_1 x + c_2 x^2 + \cdots + c_n x^n$, are found using a collocation procedure. That is to say, the coefficients are found by forcing the polynomial to fit the data at n number of points. It is known that the procedure will capture the function at the sampling points but it can have large errors between the sampling points [A2]. That is to say, the errors are forced to zero at the sampling points.

However, if Chebyshev polynomials are used as the basis functions instead of the simple polynomials, 1, x, x^2, x^3, ..., the errors are minimized over the domain of the problem instead of at a discrete set of points **even though a collocation procedure is used.** The error is distributed as it is in a Fourier series representation. In fact, there is a close correspondence between Chebyshev polynomials and the cosine functions.

A recent book claims that Chebyshev polynomials can be found at the foundation of many numerical methods. One of the objectives of this Appendix is to make the reader aware of this book. As a preliminary remark in Chapter 1 of Reference [1], there is the following unattributed quote, "Chebyshev polynomials are everywhere dense in numerical analysis."

7A.4 REFERENCES

A1. Mason, J. C. and Handscomb, D. C. *Chebyshev Polynomials,* Boca Raton, FL: A CRC Press Company, 2003.
  Although it is not often mentioned, the Gauss points are the zeroes of the Chebyshev polynomials (of the first order). As an aside, a recent book on Chebyshev polynomials pursues the idea that practically every numerical method can be developed in terms of Chebyshev polynomials. The most interesting concept in the book is the exploration of the relationship between the Chebyshev polynomials and Fourier series.

A2. Harman, T. L., Dabney, J. and Richert, N. *Advanced Engineering Mathematics Using MATLAB*, Boston, MA: PWS Publishing Company, 1997, p. 342.

An Unsuccessful Example of Adaptive Refinement

7B.1   INTRODUCTION

The locally normalized Z/Z error estimator defined by Eqs. 7.3 and 7.4 in the main text is used to guide the adaptive refinement of the five-element demonstration problem in this Appendix. The objective of this demonstration is twofold: (1) to show that the Runge function and its derivatives are, indeed, challenging functions to approximate and (2) to show that a usually successful error estimator can fail to produce a convergent result.

7B.2   EXAMPLE 1

Before presenting this example of a negative result, it should be noted that the error estimator defined by Eqs. 7.3 and 7.4 successfully guided the adaptive refinement procedure to an acceptable converged result for this demonstration problem when the refinement guide was banded by refinement guides between 10 and 21%. However, the strain distribution was either unacceptable or the adaptive refinement process did not converge when the refinement guide was outside this band of values.

For example, when the refinement threshold for this case is equal to **4%**, the adaptive refinement produced the divergent results shown in Table 7B.1. The modified version of the locally normalized Z/Z error estimator successfully guided the adaptive refinement for this problem to converge in four iterations. This result is presented in Section 7.8.

Table 7B.1. The maximum errors vs. refinement step

Iteration no.	No. of elements	Max. error (%)
1	5	91
2	10	25
3	17	30
4	21	10
5	23	7
6	26	11
7	27	13
8	29	15
9	31	19
10	33	25
11	35	33
12	37	45
13	39	58

As can be seen in Table 7B.1, the adaptive refinement process ultimately diverged after the error estimates started to converge. After the seventh iteration, the refinement process began to follow a pattern. In this pattern, two lightly strained elements that accurately represented the exact result that are adjacent to an element with a high level of error are subdivided and the maximum error increases in the next iteration.

The maximum error increases because the lightly strained elements are the only ones subdivided. As a result, the lightly strained elements adjacent to the elements with high error decrease in size and their strain energy content is reduced so the denominator given by Eq. 7.4 decreases. Since the element with high error was not subdivided, the numerator of the error estimator for the lightly strained element does not decrease significantly because of the influence of the high error element. Consequently, the error estimate for the low strain element increases with each subsequent refinement.

7B.3 EXAMPLE 2

In order to better show the form of this divergence visually, several iterations of the case with a refinement guide of 40% are presented. The use of this unnaturally large refinement guide eliminates the initial convergence seen in Table 7B.1. As a result, the divergent behavior starts with a smaller finite element model so the structure of the strain distribution that causes the divergence can be seen more clearly.

Figure 7B.1a presents the discontinuous finite element strain results for the five-element model superimposed on the exact strain distribution. Figure 7B.1b presents the subsequent error analysis. As can be seen in Fig. 7B.1a, three of the elements represent lightly strained regions in the exact solution reasonably well. In contrast, the two elements with relatively high magnitudes of strains do not closely represent the exact solution.

The error estimates presented in Fig. 7B.1b contradict the observations just made about the accuracy of the strain representations of the individual elements. The error estimates computed for the three lightly strained elements that are relatively close to the exact solution have high values. In fact, two of the elements that accurately represent the exact solution have error estimates that exceed 50%. These two elements will be subdivided because they exceed the threshold value of 40%.

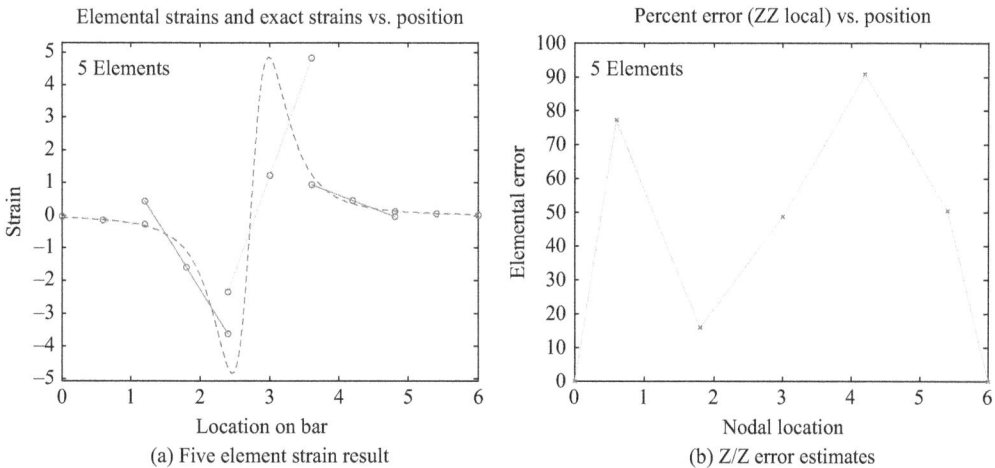

(a) Five element strain result   (b) Z/Z error estimates

Figure 7B.1. The initial five-element mesh.

The reason that these lightly strained elements that accurately represent the exact strain distribution have high error estimates is described in detail in the main text. In summary, these erroneous error estimates occur for two reasons: (1) the denominator given by Eq. 7.4 is small because these are low strain elements and (2) the numerators are overly large because the interelement jumps are relatively large as they contain contributions from adjacent elements with relatively higher errors.

In contrast, the error estimates for the two highly strained elements that are not good representations of the exact solution are lower than the error estimates for the relatively accurate elements. Only one of these high error elements is above the refinement guide. As a result, only one of these elements will be subdivided.

Before continuing with the adaptive refinement being performed, we will contrast the results presented in Fig. 7B.1b with the results produced by the modified version of the locally normalized Z/Z error estimator in Fig. 7.9b. As would be expected, the error estimates for the lightly strained elements are significantly higher in Fig. 7B.1b and the error estimates for the highly strained elements are slightly lower than they are in Fig. 7.9b.

When the five-element model is refined under the guidance of the error estimates contained in Fig. 7B.1b, a nine-element model is produced. The discontinuous finite element strain distribution is compared to the continuous exact result for the nine-element model in Fig. 7B.2a. The associated error estimates are shown in Fig. 7B.2b. As can be seen, only one element exceeds the refinement criterion of 40%. This element is a lightly strained, relatively accurate element and has a relatively large interelement jump at its right end. This is the only element that will be subdivided. We will see this pattern repeated in the subsequent refinements.

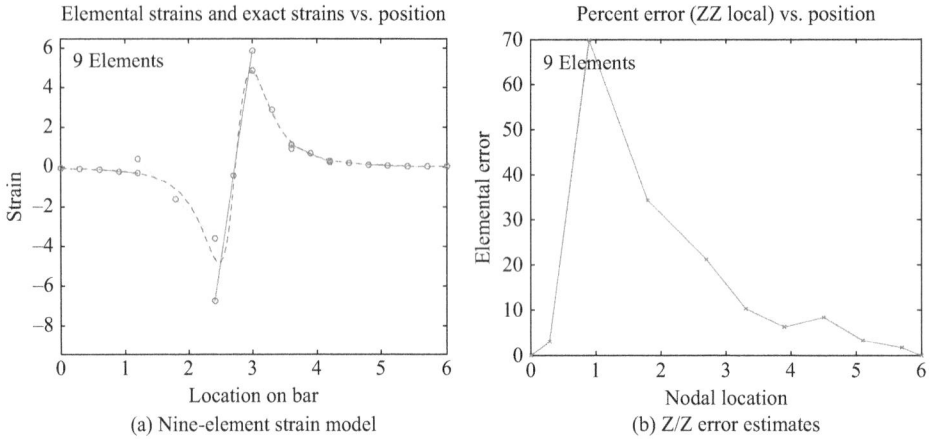

(a) Nine-element strain model (b) Z/Z error estimates

Figure 7B.2. The first adaptively refined model of nine elements.

Figures 7B.3–7B.5 contain the results of a further three iterations of the adaptive refinement process. In this series of refinements, we see that the pattern just described repeats itself. Only the low-strain element adjacent to the high-strain element produced in the previous refinement has an error estimate that exceeds the refinement criterion. Therefore, it is the only element refined and the maximum error has increased.

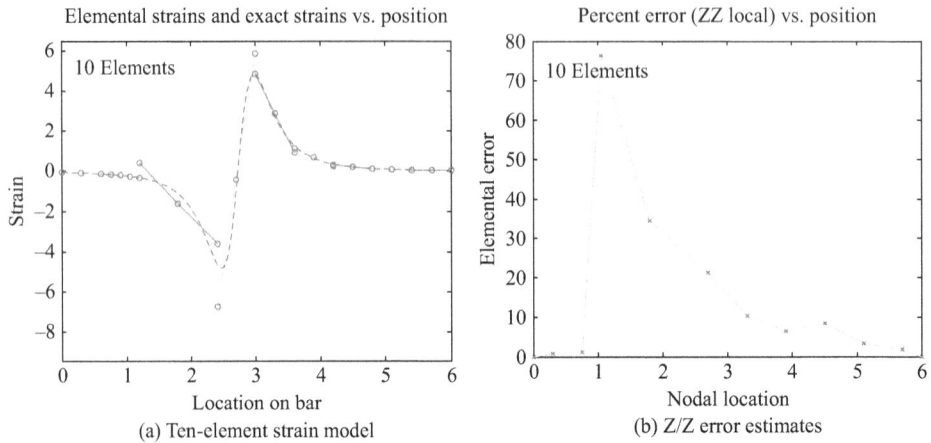

(a) Ten-element strain model (b) Z/Z error estimates

Figure 7B.3. The second adaptively refined model of 10 elements.

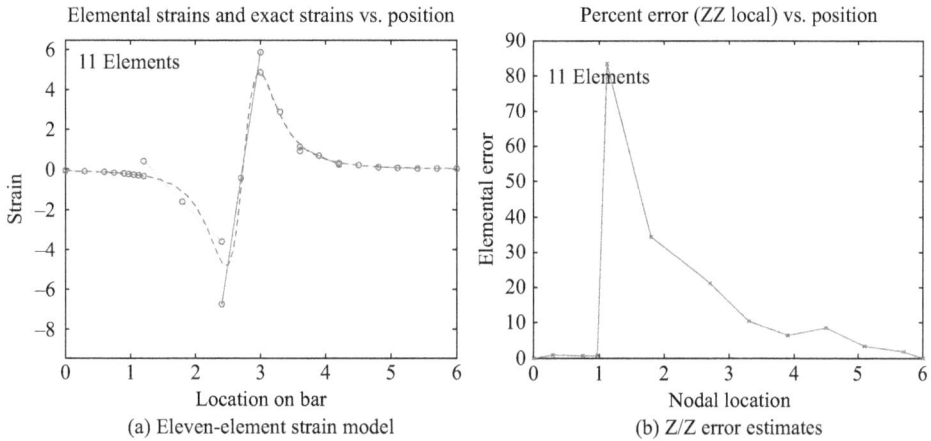

Figure 7B.4. The third adaptively refined model of 11 elements.

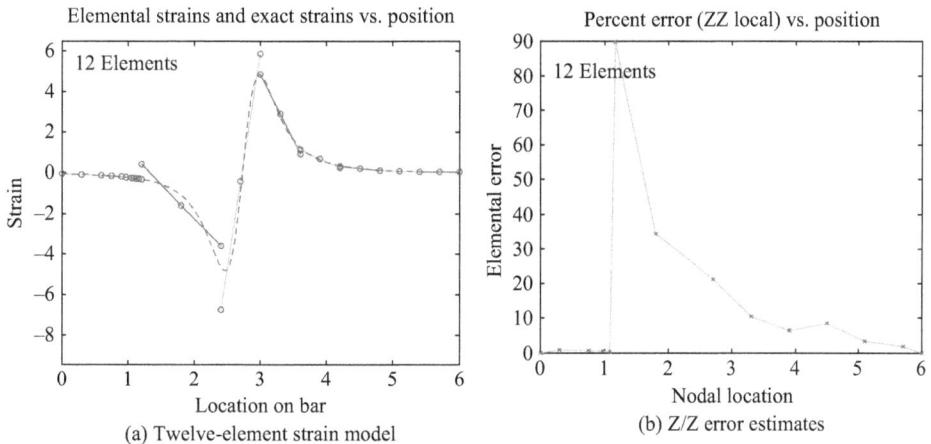

Figure 7B.5. The fourth adaptively refined model of 12 elements.

7B.4 SUMMARY

This Appendix has demonstrated a deficiency with the error estimator given by Eqs. 7.3 and 7.4. As described earlier, this problem was caused by the fact that the numerator contained contributions from adjacent elements and the denominator depended only on the behavior of the element being evaluated. This deficiency was corrected by introducing a contribution from the strain energy contained in the adjacent elements into the denominator of the error estimator. The form of this modification introduced into Eq. 7.5 was not very subtle, but it demonstrated that the deficiency could be corrected.

CHAPTER 8

A HIGH RESOLUTION POINT-WISE RESIDUAL ERROR ESTIMATOR

8.1 INTRODUCTION

In Chapter 6, an error estimator that quantifies the discretization errors in a finite element model in terms of the interelement jumps in the finite element strain representation was demonstrated. The interelement jumps in strains are shown to be an aggregation of the point-wise residuals that exist when a finite element solution does not satisfy the governing differential equations being solved. The most significant characteristic of this error estimator is the fact that the errors are estimated in terms of strains. As a consequence of this metric, a termination criterion can be formed that is directly related to quantities that are sought in continuum mechanics, namely, stresses and strains.

The point-wise error estimator just described has a secondary advantage. As is shown in the previous chapter, this point-wise error estimator has a higher resolution than the Zienkiewicz/Zhu (Z/Z) error estimator. This is only significant if the refinement guide being used is correlated with the level of the estimated error for an element. As noted in the previous chapter, a new type of refinement guide is developed and demonstrated in Chapter 11 that is not based on a correlation with the error estimator.

The new refinement guide compares the strain representation in high error elements with an estimate of the exact solution that is emerging from the finite element result. Since this refinement guide compares the finite element result to an improved solution, it can be classified as a recovery technique, as is the Z/Z error estimator.

In this chapter, a new type of **point-wise error estimator** is developed and demonstrated. This error estimator computes the residual that is produced when the finite element solution does **not satisfy the finite difference representation** of the governing differential equation at the interelement nodes.

In addition to highlighting the relationship between the finite element and the finite difference methods, the error estimator developed here has a higher resolution than the error estimators presented in Chapters 6 and 7. The resolution of the two residual-based error estimators is compared in Fig. 8.1 for two different levels of refinement of the control problem (fixed end bar loaded with the unsymmetrical Runge function) formed with three-node elements that has been used in the previous chapters.

The plots for the two error estimators for a five-element model are presented in Fig. 8.1a. As can be seen, the error estimator based on the finite difference residual produces a maximum estimate that is more than twice as large as the maximum produced by the error estimator based on interelement jumps. The maximum magnitudes are 808 units and 349 units, respectively. Both error estimators are quantified in Fig. 8.1 at the interelement nodes.

The results for the two error estimators are compared for the highly refined 40-element model in Fig. 8.1b. The maximum magnitudes for the two approaches are 68 units and 33 units, respectively. Both of the error estimates for this refined model are smaller by a factor of 10 in comparison with the smaller model. The point-wise error estimator is still more than twice as large as the interelement jump approach. In both cases, the elements that accurately represent the exact solution have error estimates that are near zero. That is to say, these figures demonstrate the higher resolution of the error estimator presented in this chapter.

The difference in the resolution for the two error estimators is easily explained. The interelement jumps are the result of an aggregation. The aggregation occurs because the residuals that form the interelement jumps are integrated over the whole domain of the elements. As a result, the plus and minus residuals are averaged. In contrast, the point-wise error estimator, by definition, is not an integrated value, so no amelioration of its magnitude occurs because of an averaging process.

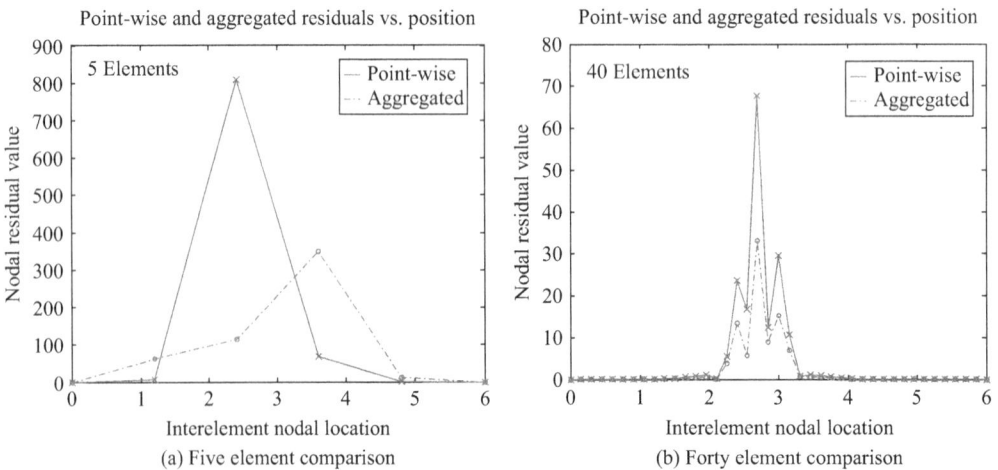

(a) Five element comparison

(b) Forty element comparison

Figure 8.1. Aggregated and point-wise residual comparisons.

However, the metric for the point-wise error estimator is not ideal. It is not expressed in terms that are as basic to solid mechanics as the strain-based error estimator presented in Chapter 6. This error estimator is expressed in terms of the magnitude of the distributed load at the point being evaluated. However, the error can be related to the level of the applied load.

The objectives of this chapter are the following: **(1)** to provide an overview of the point-wise error estimator developed here, **(2)** to present the theoretical basis for its development, and **(3)** to demonstrate its characteristics with examples of uniform and adaptive refinement.

Since the error estimator developed in this chapter uses the finite difference method to evaluate the accuracy of finite element results, the theoretical relationship between the two methods is presented in this chapter. This requires that an overview of the finite difference method be presented. On the operational level, the heart of the finite difference method is the estimation of the derivatives for a discrete set of dependent variables. This idea was introduced in Chapter 3. It is used to identify the modeling capabilities of individual elements.

In the next section, we will give an overview of the point-wise error estimator. Then, the theoretical and practical bases for the error estimator are presented. Finally, this error estimator is demonstrated with applications to uniformly and adaptively refined models.

8.2   AN OVERVIEW OF THE POINT-WISE RESIDUAL ERROR ESTIMATOR

The point-wise error estimator developed in this chapter is computed as follows. The finite element displacements for the problem being solved are substituted into a finite difference approximation of the governing differential equation at the point being evaluated. When an accurate finite element result is substituted, the finite difference approximation is nearly satisfied so the residual is small at the point being analyzed. Conversely, when an inaccurate finite element result is substituted, the residual is large.

The rationale behind this point-wise error estimator can be viewed as a variant of a maxim Sherlock Holmes gave to Watson when he said, [1]

"When you follow two separate chains of thought, Watson, you will find some point of intersection which should approximate the truth."

This maxim applies to this error analysis technique for two reasons. On one hand, the finite element and the finite difference methods use two significantly different approaches or chains of thought for finding approximate solutions. On the other hand, both solution techniques are attempting to solve the same problem so the two solutions will converge to the same result as the model is improved.

The point-wise error estimator developed here has another advantage over the error estimator based on interelement jumps in addition to its higher resolution. The error estimator developed here can estimate the residual errors at both the internal and interelement nodes of a finite element model. The error estimator based on interelement jumps does not apply to the interior nodes because the

---

[1] Sir Arthur Conan Doyle, *The Disappearance of Lady Francis Carfax.*

strain representations are continuous on the interior of an element. The value of being able to esti-mate the error at the interior points is shown in Fig. 8.2.

The strain distribution produced by the uniformly refined 40-element model is compared to the exact result in Fig. 8.2a. Figure 8.2b is a reproduction of Fig. 8.1b. This figure contains the error estimates at the interelement nodes for the estimators based on the point-wise residuals and the interelement jumps. As noted earlier, the point-wise residual approach has a higher resolution than the interelement jumps in the strains. The point-wise residual error estimates at the interior nodes of the individual elements are presented in Fig. 8.2c.

When Fig. 8.2b is compared to Fig. 8.2c, it can be seen that the error estimates at the interior nodes have a higher magnitude that the estimates at the interelement nodes, that is, 82 units

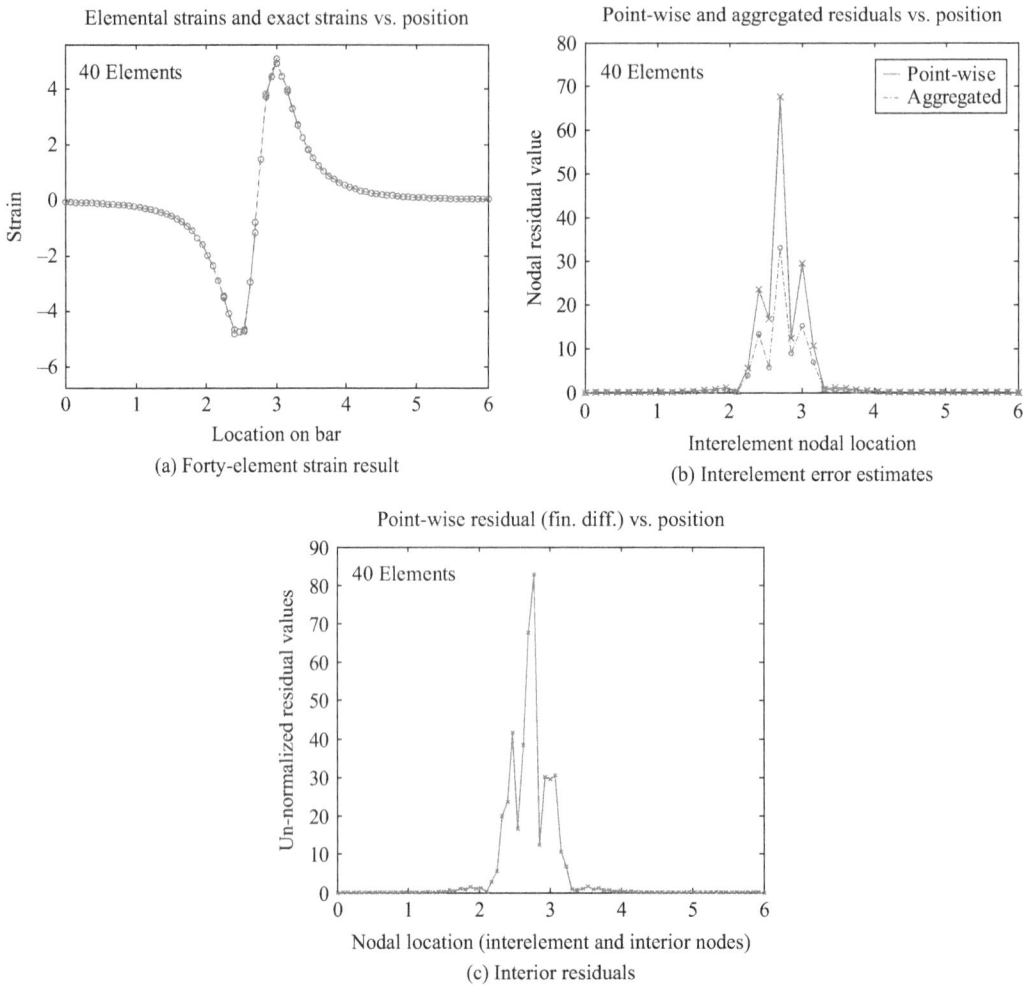

(a) Forty-element strain result

(b) Interelement error estimates

(c) Interior residuals

Figure 8.2. A comparison of errors in a 40-element model.

versus 68 units. Since the error estimates for the elements that represent the exact result well are nearly zero for both cases, the estimated errors at the interior nodes have a higher resolution than the estimates for the interelement nodes.

The higher resolution produced by the internal nodes can be explained in the following way. The error estimates at the interior nodes are solely due to the deficiencies in the element being analyzed. If the element does not represent the exact solution well, the error estimate will be high. On the other hand, if the element accurately represents the exact solution, the error estimate will be low. However, if a high-error element is adjacent to a low-error element, the error estimates at the interelement nodes contain contributions from both elements. Thus, the resulting error estimate is an average quantity. It will be lower than the estimate for the highly inaccurate element alone. Conversely, the estimated error will be higher than the error estimate for the accurate element alone. The comparison of Figs. 8.2b and c shows the value of evaluating an element on its own merits.

8.3 THE THEORETICAL BASIS FOR THE POINT-WISE RESIDUAL ERROR ESTIMATOR

As outlined in the previous section, the point-wise residual error estimator is based on the following two observations: (1) the finite element and finite difference methods are trying to solve the same problem and (2) the two methods approach this problem in significantly different ways. As a result, the failure of the finite element result to satisfy the finite difference approximation can be interpreted as a measure of the discretization error in the model. These observations are given substance in this section.

Both methods approximate the displacements that minimize the potential energy in a continuous solid mechanics equilibrium problem. The finite element method finds the nodal displacements that minimize the potential energy in a physical approximation of the continuous problem.

In contrast, the finite difference method finds the nodal displacements that satisfy an approximation of the necessary and sufficient conditions that guarantee that the potential energy is minimized. The necessary and sufficient conditions are the equilibrium equations and the boundary conditions for the continuous problem.

The relationship between the two approximate methods is shown schematically in Fig. 8.3. Both methods are based on the Principle of Minimum Potential Energy. This is shown by the two blocks common to both methods at the top of Fig. 8.3. They both approximate the displacements that minimize the potential energy in the continuum problem being analyzed. As outlined above, the two solution techniques find their approximate solutions in significantly different ways. The differences in the way the two methods approximate the displacements that minimize the potential energy in the continuous problem are described next.

As shown schematically in the right-hand branch of Fig. 8.3, **the finite element method** uses a direct approach for finding an approximate solution that minimizes the potential energy in the continuous problem. A representation of the continuous problem is formed by subdividing the domain of the problem into relatively simple regions, or finite elements. Then, a potential energy function is

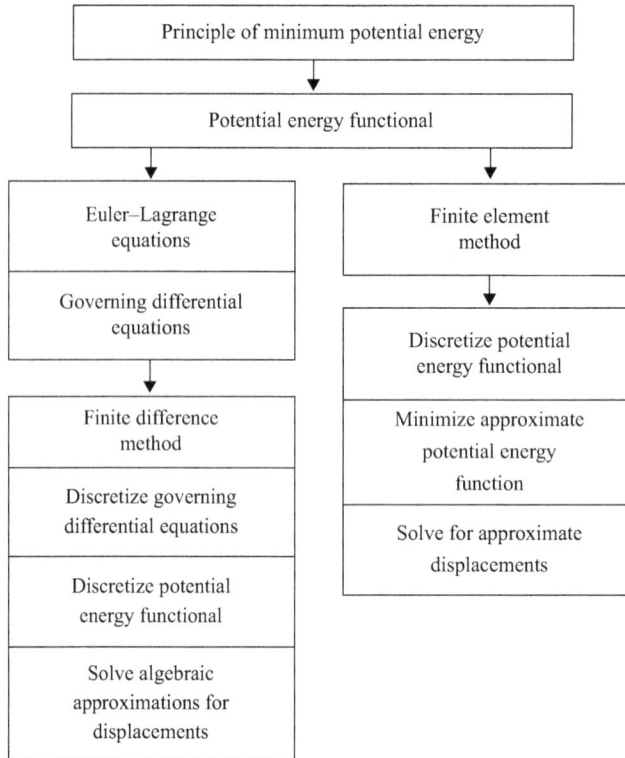

Figure 8.3. The relationship between the finite element and the finite difference methods.

formed for this discrete representation of the continuous problem. That is to say, the finite element method models the physical system being analyzed.

This potential energy function takes the form of a quadratic algebraic equation. When this quadratic function is minimized, a set of linear algebraic equilibrium equations is produced. These equations are then solved for the nodal displacements that approximate the exact solution of the problem.

The finite difference method minimizes the potential energy indirectly. That is accomplished in a two-step process. The first step is shown schematically by the two blocks at the top of the left-hand branch of Fig. 8.3. The Euler–Lagrange equations are applied to the potential energy functional to identify the conditions that must be satisfied in order to minimize the potential energy in the continuous problem. The conditions that must be satisfied are the equilibrium equations and the boundary conditions for the continuous problem.

The second step in the finite difference method is to approximate the equilibrium equations and the boundary conditions of the continuum problem. The derivative expressions in these equations are approximated with finite difference operators. The resulting set of algebraic equations is then solved to find the nodal displacements that approximate the displacements of the exact solution.

In other words, the finite difference method models the equations that are the necessary and sufficient conditions that must be satisfied to ensure that the potential energy of the system being analyzed is minimized.

In summary, both the finite element and the finite difference methods find approximations of the displacements that minimize the potential energy in the continuous physical system. In this way, the two methods are trying to solve the same problem. As a result, their solutions will converge in the limit.

However, the two approximate techniques find approximate solutions to the same continuum problem in two very different ways. The finite element method forms a discrete model of the continuous physical system. Then, it finds the displacements that minimize the strain energy in this discrete system. In contrast, the finite difference method forms a discrete model of the equilibrium equations and boundary conditions for the continuous system. Then, the displacements that satisfy these approximations are found.

The point-wise error estimator developed here computes the amount by which the displacements of the finite element result fails to satisfy the finite difference approximation of the conditions that minimize the potential energy in the continuous system. This residual quantity is taken as a measure of the discretization error in the finite element model.

8.4 COMPUTATION OF THE POINT-WISE RESIDUAL ERROR ESTIMATOR

Let us start with a disclaimer to assuage any concerns about the computational efficiency of this point-wise error estimator. The point-wise error estimation technique developed in this chapter **does not require that a finite difference problem be solved** in order to compute the error estimates at the nodal points. This is because **an approximate finite difference solution** is formed from the finite element solution of the problem being analyzed. This idea is shown schematically in Fig. 8.4.

The final entry on the left-hand column of Fig. 8.4 identifies the formulation of an augmented finite element solution. As we will see, this augmented finite element solution is evaluated to form the point-wise error estimators. This augmentation of the finite element solution consists of introducing the boundary conditions for the finite difference model into the finite element results. The procedure for representing the boundary conditions in a finite difference model is developed and demonstrated in Reference [1].

This augmented finite element solution is then interpreted as **an approximate finite difference solution.** This interpretation is shown in the final entry in the right-hand column of Fig. 8.4.

The right-hand column of Fig. 8.4 identifies the opportunity that is exploited in this chapter to form the point-wise error estimator. As shown in Fig. 8.4, the residuals for an actual finite difference solution are equal to zero. This indicates that the finite difference solution satisfies the finite difference approximation of the governing differential equations at the nodal points of the discrete model.

In contrast, when the augmented finite element solution is substituted into the finite difference approximations of the governing differential equations, the residuals are only equal to zero if the

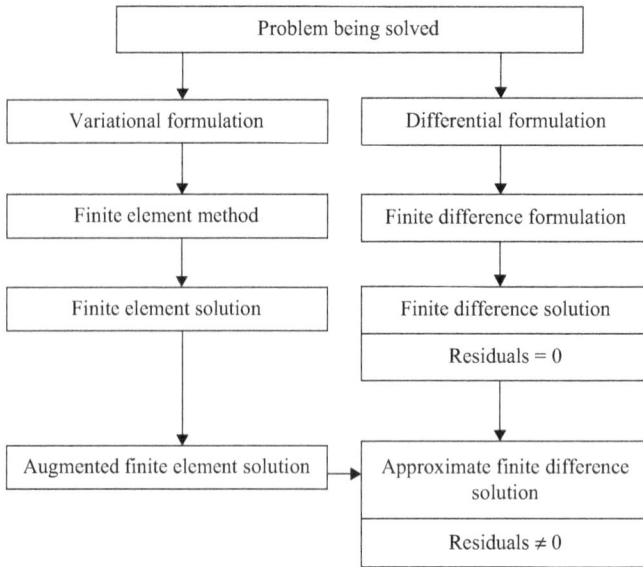

Figure 8.4. The relationship between the finite element and the finite difference solutions.

nodal displacements match the finite difference representation of the problem. When the residuals are not zero, the amount by which they fail to satisfy the approximate equilibrium model quantifies the discretization errors in the finite element model. This residual quantity has the same units as a distributed load that is applied to the finite difference model.

As we will see when the finite element model is uniformly refined in a later section, the residual quantities are reduced as the finite element model is improved. Then, we will demonstrate that this point-wise error estimator successfully guides the adaptive refinement process.

The practical application of this point-wise error estimator is made possible because the finite difference operators are formulated from a Taylor series expansion of the displacements [1]. As a result, finite difference operators can be formed for unevenly spaced meshes using the same rational process in one, two, and three dimensions. In practical terms, this means that the finite difference operators can be applied to any finite element model.

8.5   FORMULATION OF THE FINITE DIFFERENCE OPERATORS

In this section, a formal procedure for forming the finite difference operators for unevenly spaced nodes is presented for the one-dimensional case shown in Fig. 8.5. The objectives of this presentation are the following:

1. To present a capability that is needed to form an approximate finite difference solution by augmenting the finite element result.
2. To present a capability that is needed to make the adaptive refinement process possible.

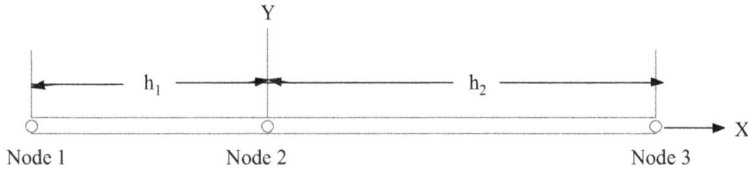

Figure 8.5. A bar with unevenly spaced nodes.

3. To show that this capability is extendable to two and three dimensions.
4. To make those who are unfamiliar with the finite difference method aware of its relative simplicity.

The first step in forming the finite difference operators for any nodal spacing is to represent the displacements with a Taylor series expansion. Since the highest order derivative present in the governing differential equation for a bar is of second order (see Eq. 4.1), a three-term Taylor series expansion is required in order to approximate this derivative. The required three-term Taylor series is the following:

$$u(x) = u_0 + \left(\frac{du}{dx}\right)_0 x + \frac{1}{2}\left(\frac{d^2u}{dx^2}\right)_0 x^2 \qquad \text{(Eq. 8.1)}$$

where the subscript zero indicates that the quantities are associated with the local origin.

The next step is to introduce the location of the three nodes and the displacements at the nodes of the template into the Taylor series expansion. As shown in Fig. 8.5, Node 1 is located a distance of h_1 in the negative x direction from the local origin which in this case is taken to be Node 2. Node 3 is located a distance h_2 in the positive x direction from Node 2. When these quantities are substituted one at a time into Eq. 8.1, the three equations produced by these substitutions have the following form:

$$\begin{Bmatrix} u_1 \\ u_2 \\ u_3 \end{Bmatrix} = \begin{bmatrix} 1 & -h_1 & (-h_1)^2/2 \\ 1 & 0 & 0 \\ 1 & h_2 & (h_2)^2/2 \end{bmatrix} \begin{Bmatrix} u \\ \dfrac{du}{dx} \\ \dfrac{d^2x}{dx^2} \end{Bmatrix} \qquad \text{(Eq. 8.2)}$$

Equation 8.2 consists of three simultaneous equations that must be solved in order to get the expressions that approximate the derivatives at the local origin in terms of the nodal displacements.

When the set of linear simultaneous algebraic equations given by Eq. 8.2 are solved, we have the following:

$$
\begin{Bmatrix} u \\ \dfrac{du}{dx} \\ \dfrac{d^2x}{dx^2} \end{Bmatrix} = \begin{bmatrix} 0 & 1 & 0 \\[2mm] \dfrac{-h_2}{h_1(h_1+h_2)} & \dfrac{(h_1-h_2)}{h_1h_2} & \dfrac{h_1}{h_2(h_1+h_2)} \\[2mm] \dfrac{2}{h_1(h_1+h_2)} & \dfrac{-2}{h_1h_2} & \dfrac{2}{h_2(h_1+h_2)} \end{bmatrix} \begin{bmatrix} u_1 \\ u_2 \\ u_3 \end{bmatrix}
\qquad \text{(Eq. 8.3)}
$$

The expressions on the right-hand side of Eq. 8.3 are the finite difference operators that approximate the quantities on the left-hand side evaluated at the local origin. This is seen definitively with the equation in the first row, namely, $u_0 = u_2$. This equation says that the displacement at the local origin is equal to the displacement u_2. That is to say, this equation is true by the definition of the local origin since we have taken Node 2 as the local origin.

The finite difference operator that approximates the second derivative at the location of the local origin, $x = x_2$, for an unevenly spaced mesh is given by the third row in Eq. 8.3. When the approximation of the second derivative is extracted, we have the following:

$$
\left(\dfrac{d^2u}{dx^2}\right)_0 = \begin{bmatrix} \dfrac{2}{h_1(h_1+h_2)} & \dfrac{-2}{h_1h_2} & \dfrac{2}{h_2(h_1+h_2)} \end{bmatrix} \begin{Bmatrix} u_1 \\ u_2 \\ u_3 \end{Bmatrix}
\qquad \text{(Eq. 8.4)}
$$

In order to put these results in a more recognizable form, we will reduce Eq. 8.3 to the case where $h_1 = h_2$. The finite difference operators for evenly spaced nodes are the following:

$$
\begin{Bmatrix} u \\ \dfrac{du}{dx} \\ \dfrac{d^2x}{dx^2} \end{Bmatrix}_0 = \begin{bmatrix} 0 & 1 & 0 \\[2mm] \dfrac{-1}{2h_1} & 0 & \dfrac{1}{2h} \\[2mm] \dfrac{1}{h^2} & \dfrac{-2}{h^2} & \dfrac{1}{h^2} \end{bmatrix} \begin{Bmatrix} u_1 \\ u_2 \\ u_3 \end{Bmatrix}
\qquad \text{(Eq. 8.5)}
$$

The finite difference operators for the partial derivatives that are present in the governing differential equations in multidimensions are found by starting with multidimensional Taylor series expansions. The multidimensional case is presented in detail in Reference [1].

8.6 THE FORMULATION OF THE POINT-WISE RESIDUAL ERROR ESTIMATOR

The point-wise residual error estimators are formed by substituting the nodal displacements for the augmented finite element solution into the finite difference representation of the governing differential equation. The governing differential equation for the one-dimensional bar is given in Eq. 4.1. It is repeated here for the convenience of the reader:

$$EA\frac{d^2u(x)}{dx^2} = f(x) \qquad \text{(Eq. 8.6)}$$

When Eq. 8.6 is put in the form for computing a residual, we have the following:

$$R(x) = EA\frac{d^2u(x)}{dx^2} - f(x) \qquad \text{(Eq. 8.7)}$$

where *R(x)* is the amount by which a given approximate solution fails to satisfy the governing differential equation at the origin of the local coordinate system. Note that the residual has the same units as the applied distributed load *f(x)*.

The specific expression of the residual that measures the amount that the finite element solution fails to satisfy the finite difference approximation of the equilibrium equation is found by substituting the nodal displacements from the augmented finite element solution into the finite difference approximation of Eq. 8.7 for computing the residual quantity. When this is done, we have the following:

$$R(x) = EA\left[\frac{2}{h_1(h_1+h_2)} \quad \frac{-2}{h_1h_2} \quad \frac{2}{h_2(h_1+h_2)}\right]\begin{Bmatrix} u_1 \\ u_2 \\ u_3 \end{Bmatrix} - f(x)$$

$$= EA\left[\left(\frac{2}{h_1(h_1+h_2)}\right)u_1 + \left(\frac{-2}{h_1h_2}\right)u_2 + \left(\frac{2}{h_2(h_1+h_2)}\right)u_3\right] - f(x) \qquad \text{(Eq. 8.8)}$$

This equation is the point-wise error estimator developed in this chapter. It represents the failure of the finite element solution to satisfy the finite difference approximation of the equilibrium equation for the problem being analyzed at a single point.

In contrast, the error estimator based on the interelement jumps in the strains is an aggregated version of the failure of the finite element solution to satisfy the governing differential equation. It is this aggregation of positive and negative residuals that reduces the resolution of the error estimator based on the interelement jumps in the strains.

8.7 A DEMONSTRATION OF THE POINT-WISE FINITE DIFFERENCE ERROR ESTIMATOR

In this section, the point-wise finite difference residual error estimator given by Eq. 8.8 is applied to a sequence of uniform refinements beginning with the five-element control problem. The results of analyzing the same problem with the integrated residual method using the interelement jumps in the strains developed in Chapter 6 are presented for comparison.

The results for the two types of error estimators contained in this section are presented for different points in Fig. 8.6. The point-wise residual error estimators are given for the internal nodes of the individual elements. This is done because it has been shown that the error estimates for these points have a higher resolution than when the same quantity is computed for the interelement nodes. The error estimates based on the interelement jumps in strain are presented at the interelement nodes because that is the only place where they can be computed.

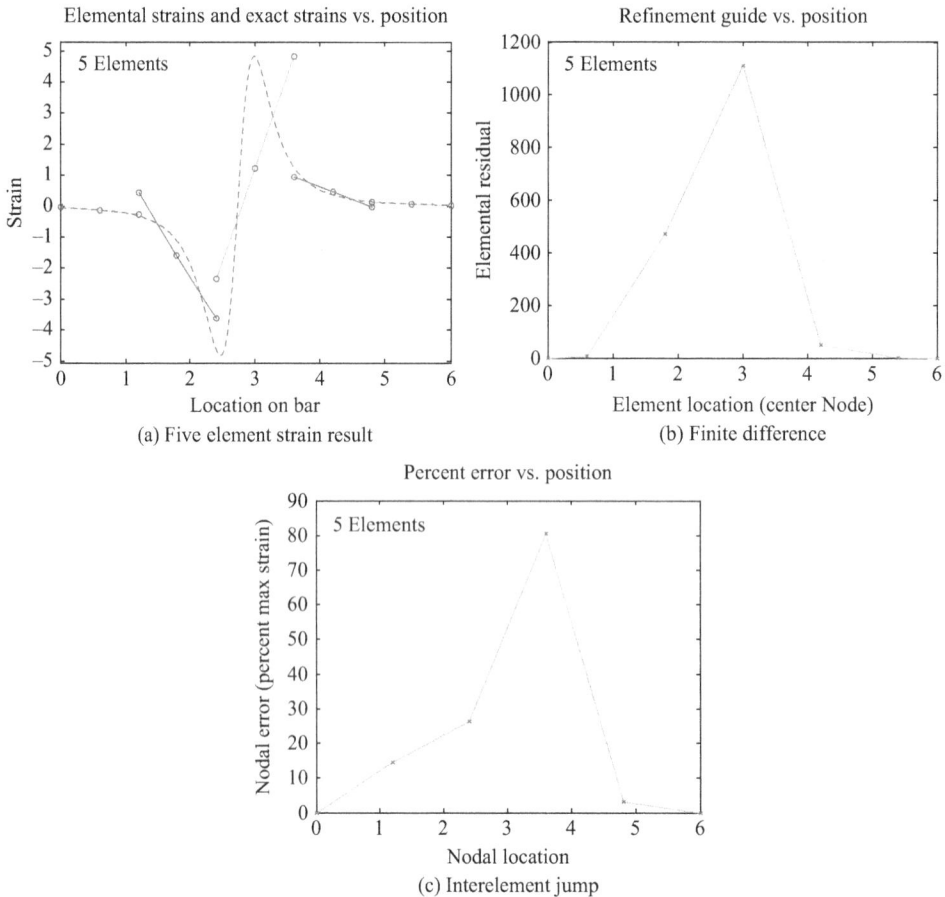

(a) Five element strain result

(b) Finite difference

(c) Interelement jump

Figure 8.6. A five-element model.

The results for the initial five-element model are shown in Fig. 8.6. Figure 8.6a compares the strains in the finite element model to the strains in the exact solution. Figure 8.6b contains the error estimates for the point-wise residual error measure at the interior node of the individual elements. Figure 8.6c contains the elemental error estimates produced by the integrated residual measures at the interelement nodes.

The higher resolution of the point-wise residual error estimator can be seen when Figs. 8.6b and 8.6c are compared. The first element on the left-hand end of the bar can be seen to be close to the exact result in Fig. 8.6a. The error estimate for this element is shown to be close to zero in Fig. 8.6b. As can be seen in Figs. 8.6a and 8.6c, the interelement jump and the error estimate produced by this interelement jump are not close to zero. A similar comparison can be presented for the two elements on the right-hand end of the bar to identify the higher resolution of the point-wise residual error estimator.

When the five-element model is uniformly refined two times, the results for the 10- and 20-element models are given in Figs. 8.7 and 8.8, respectively. Although the errors for these two refinements have the same approximate shapes, the distributions have significant differences.

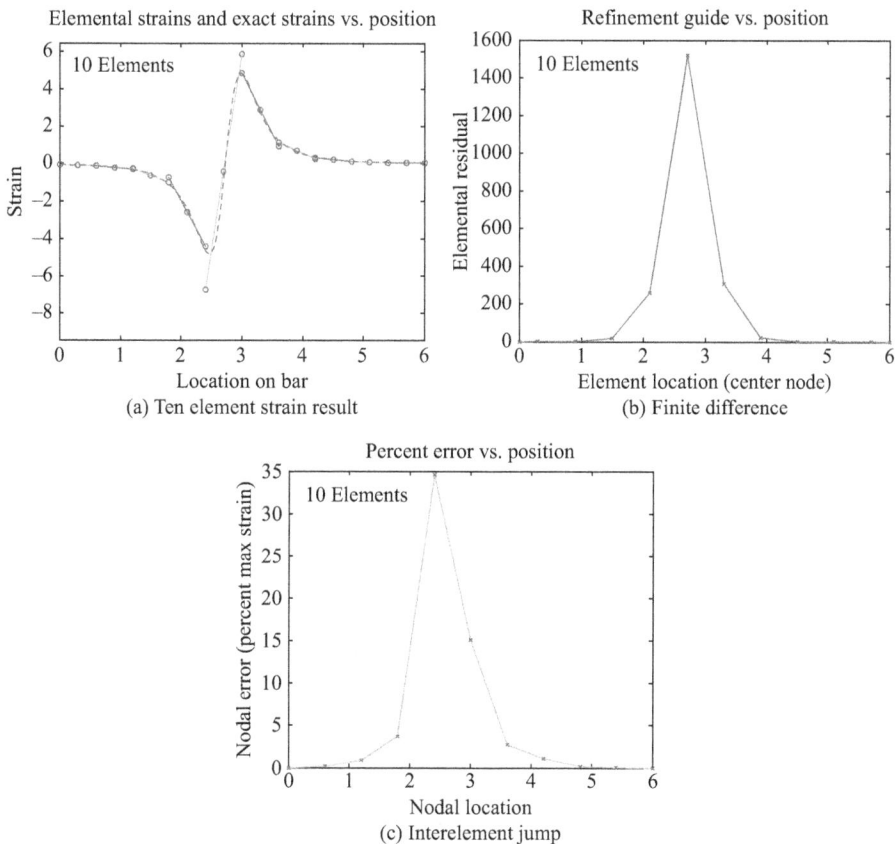

(a) Ten element strain result

(b) Finite difference

(c) Interelement jump

Figure 8.7. A 10-element model.

(a) Twenty element strain result

(b) Finite difference

(c) Interelement jump

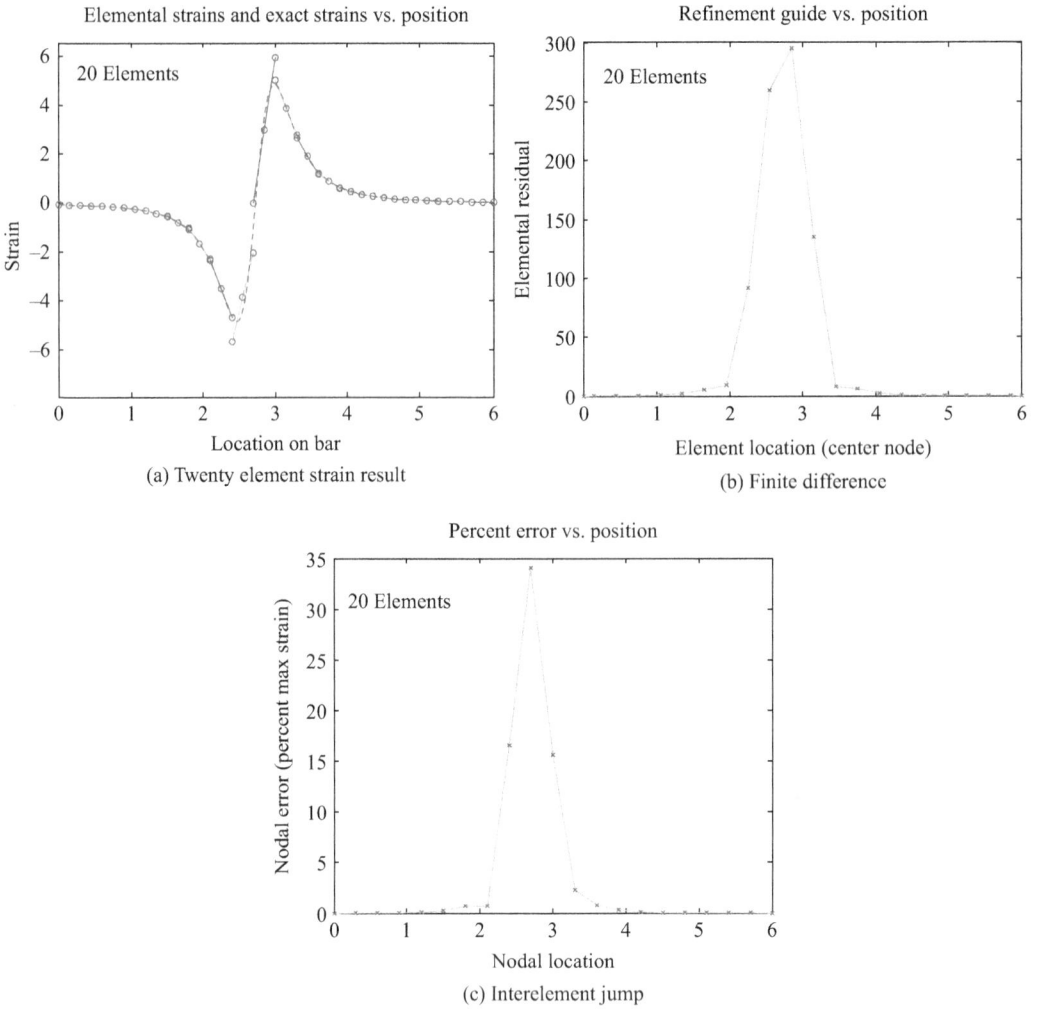

Figure 8.8. A 20-element model.

In the refinement to 10 elements, the maximum error estimate produced by the point-wise finite difference error estimator increased from approximately 1100 units to 1500 units. In contrast, the integrated residual measure decreased from approximately 80 units to 34 units. In the uniform refinement to 20 elements, the changes were different. The point-wise estimator dropped to approximately 300 units and the integrated residual stayed approximately constant at 34 units.

The results of uniformly refining the model two more times are shown in Figs. 8.9 and 8.10, respectively.

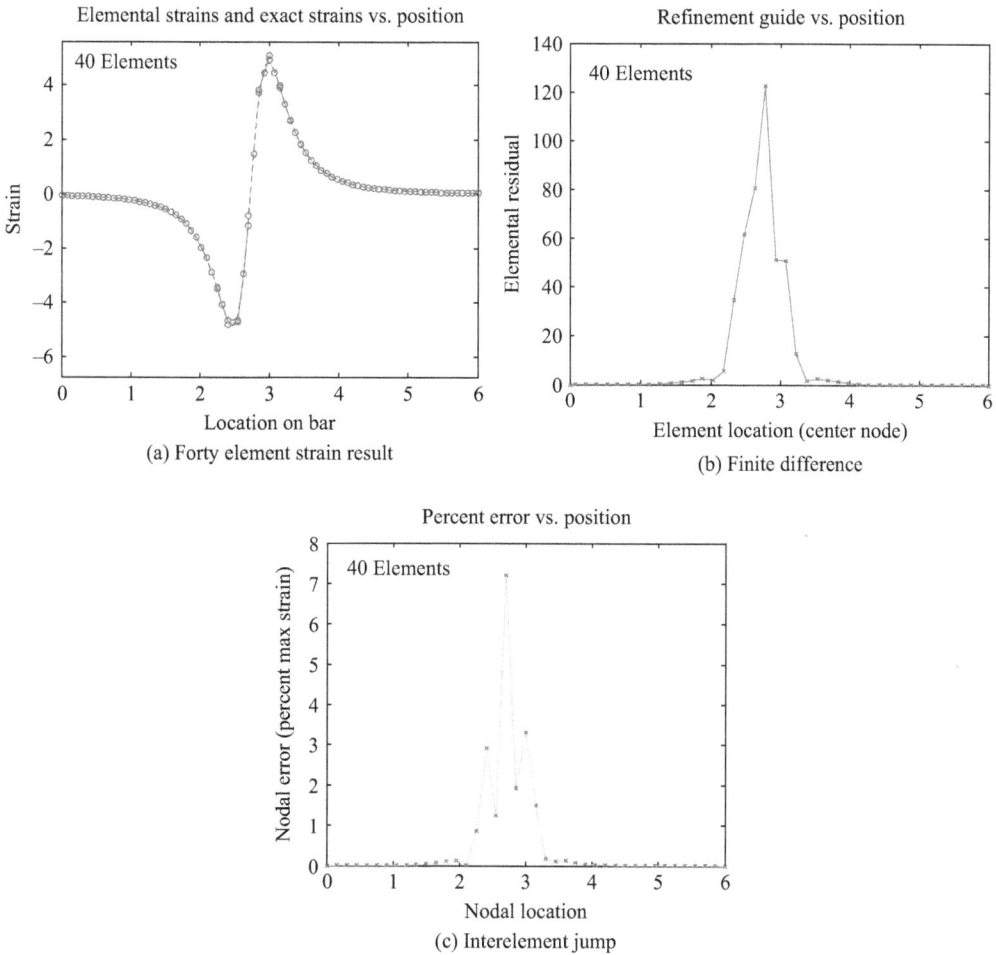

Elemental strains and exact strains vs. position

40 Elements

(a) Forty element strain result

Refinement guide vs. position

40 Elements

(b) Finite difference

Percent error vs. position

40 Elements

(c) Interelement jump

Figure 8.9. A 40-element model.

As can be seen, both refinements reduced the maximum magnitudes or estimated errors for both the point-wise and the integrated residual estimators. Furthermore, the shapes of the final error estimates are approximately the same.

In addition to showing the general behavior of the point-wise finite difference residual error estimator and comparing it to the integrated residual error estimator, this series of uniform refinements helped to identify the refinement threshold that will be used in the next section.

The demonstration of adaptive refinement presented in the next section will use a refinement threshold of 50 units. If the error estimate of an element exceeds this value it will be subdivided. This number was chosen by looking at the error estimates for the 80-element model. The maximum residual for this acceptable solution was approximately 40 units.

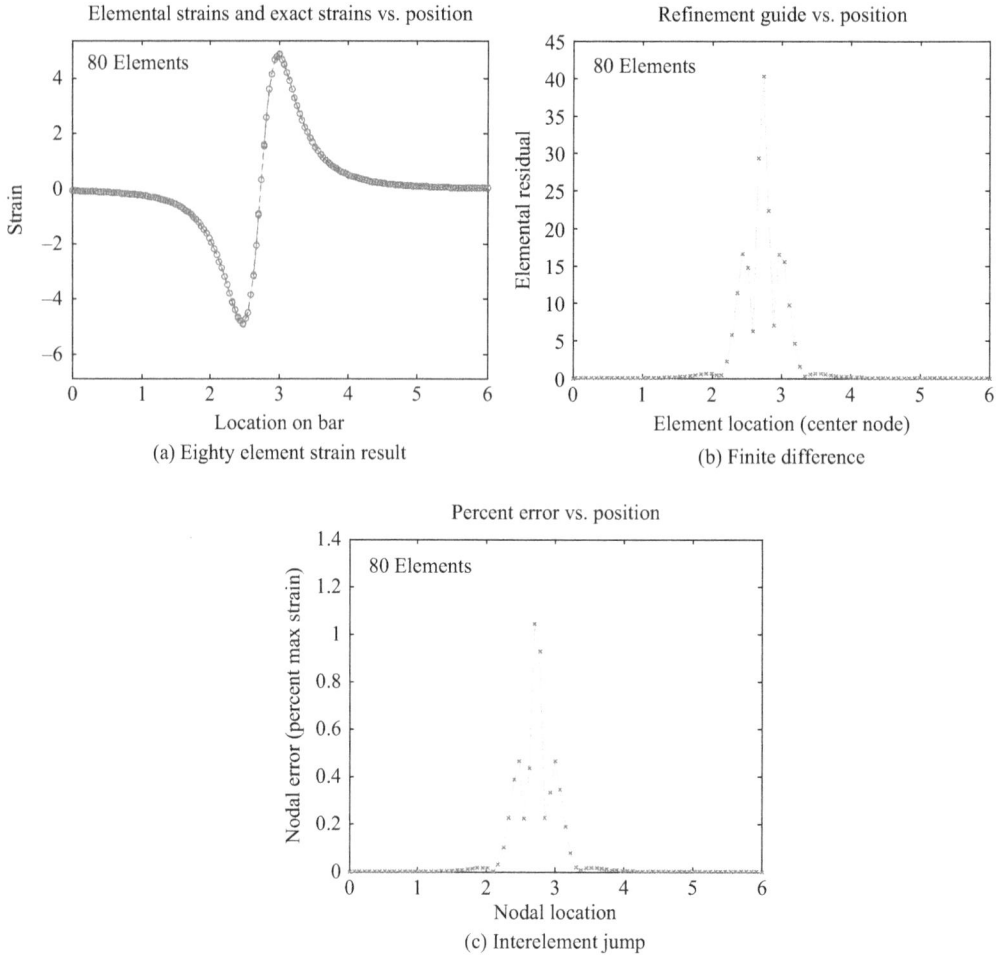

Figure 8.10. An 80-element model.

8.8 A DEMONSTRATION OF ADAPTIVE REFINEMENT

The point-wise finite difference error estimator is used to guide the adaptive refinement of the five-element demonstration problem in this section. In this demonstration, an element is subdivided into two equal length elements if the refinement guide is greater than or equal to a value of 50 units.

In the next chapter, the termination criteria will be considered in detail. As we will see, the choice is limited only by one's imagination and the objective of the analysis being performed. For example, if a brittle material is being considered, the termination criterion might be heavily weighted in favor of an accurate shear strain representation.

This refinement guide is simply the residual value produced by Eq. 8.8. The units are identical to the units on the right-hand side of the governing differential equation, that is, they are equal to the loading density. As a result, they can be interpreted physically in terms of the applied load. This refinement guide can be considered as being globally normalized by a value of one.

If the plot of the applied load shown in Fig. 4.4 is consulted, the maximum value of the applied load distribution is approximately 3000 units. This means that the termination criterion of 50 units is equivalent to stopping the analysis if the residual is equal to approximately 1.6% of the maximum value of the distributed load.

Figure 8.11 presents the results of solving and analyzing the point-wise residuals for the initial five-element model. Figures 8.12–8.15 present the results for the sequence of adaptive refinements. In each case, when the finite element strain distribution is compared to the exact result, the elements with the highest level of error also have high estimated errors. As a result, the elements that need to be subdivided to improve the finite element representation are, indeed, subdivided. That is to say, the point-wise finite difference residual error estimator is effective.

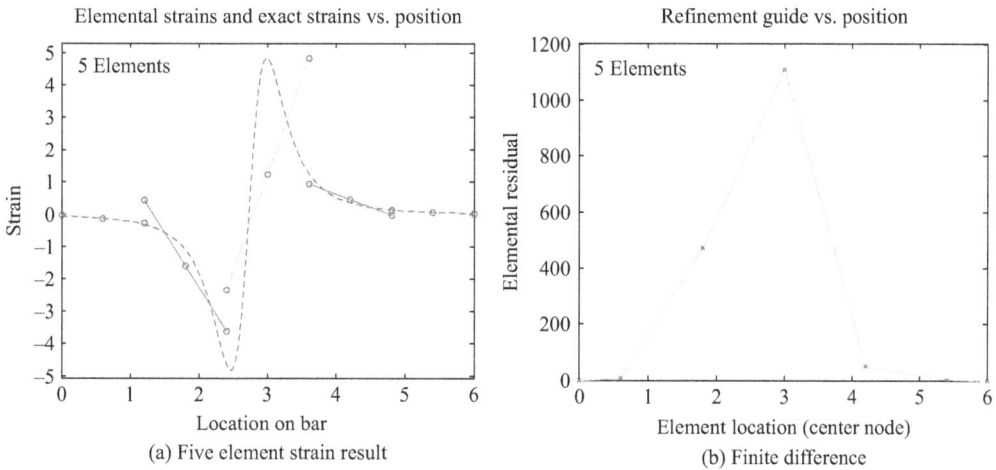

(a) Five element strain result   (b) Finite difference

Figure 8.11. The initial five-element mesh.

When the strain distribution for the final adaptive refinement shown in Fig. 8.15 is compared to the strain distribution for the uniformly refined model with 80 elements shown in Fig. 8.10, we see that the finite element representations in the critical areas in the neighborhoods of the two stress concentrations are similar. To better show this similarity, the stress concentration located at approximately 2.5 units along the bar for these two representations are shown in expanded form in Fig. 8.16.

When the right-hand sides of these two figures are compared, we see that the adaptive refinement process produced the same finite element representation as the uniformly refined model. The minimum point is represented identically and both distributions are represented with three elements. Furthermore, the jumps in the interelement strains look to be identical at this close resolution.

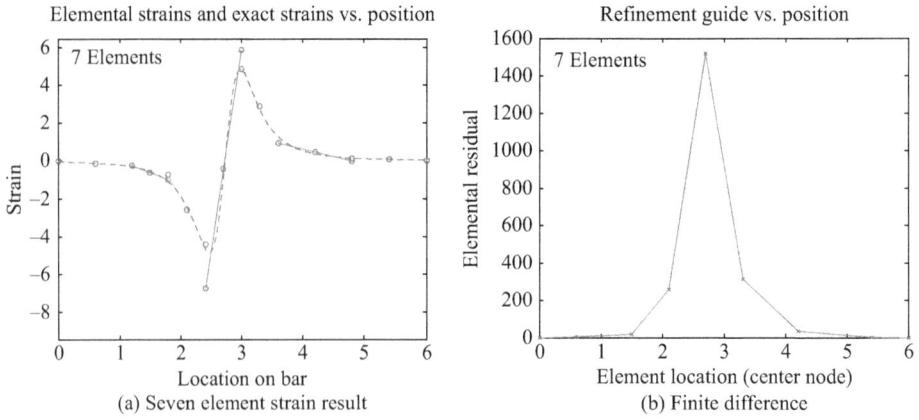

(a) Seven element strain result

(b) Finite difference

Figure 8.12. The first adaptively refined mesh of seven elements.

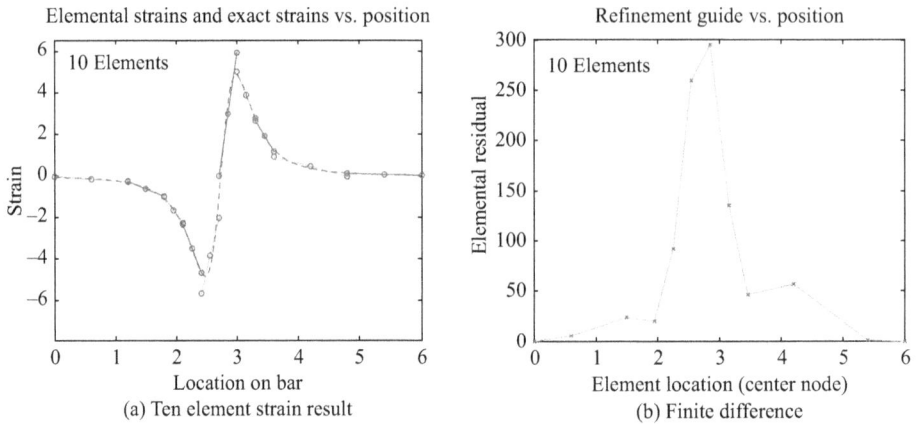

(a) Ten element strain result

(b) Finite difference

Figure 8.13. The second adaptively refined mesh of 10 elements.

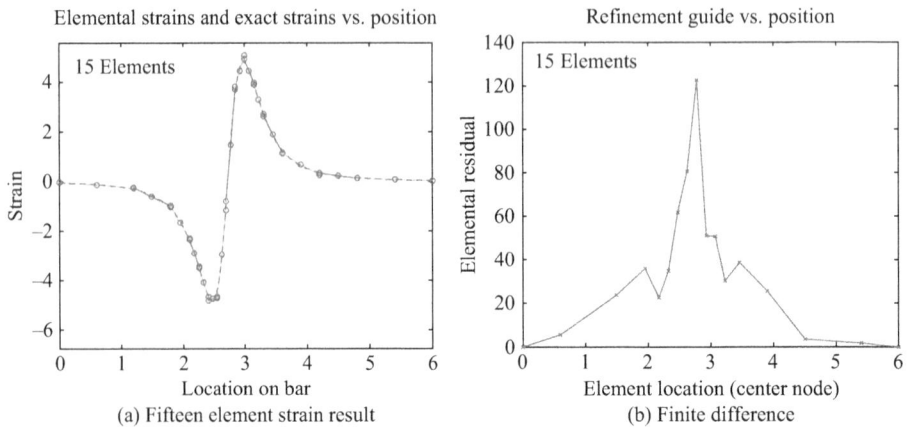

(a) Fifteen element strain result

(b) Finite difference

Figure 8.14. The third adaptively refined mesh of 15 elements.

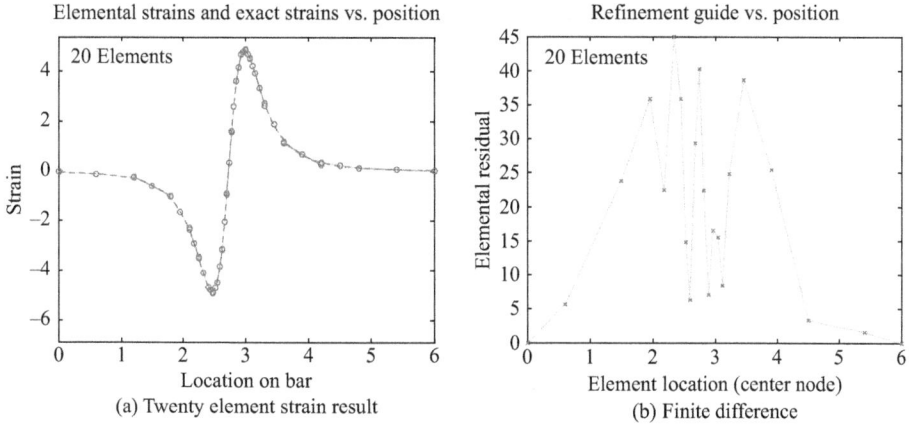

Elemental strains and exact strains vs. position Refinement guide vs. position

(a) Twenty element strain result

(b) Finite difference

Figure 8.15. The fourth adaptively refined mesh of 20 elements.

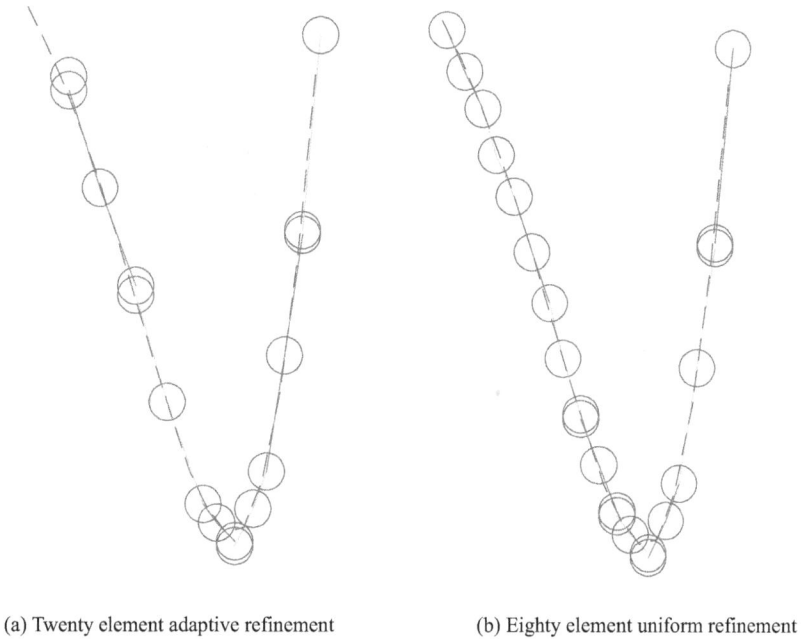

(a) Twenty element adaptive refinement

(b) Eighty element uniform refinement

Figure 8.16. Close-ups of a strain concentration.

The left-hand sides of the finite element representations contain different numbers of elements. The adaptively refined model has three full elements showing and one partial element. The uniformly refined model has five full elements and one partial element representing the same approximate portion of the bar. Furthermore, the jumps in the interelement strains are more pronounced in the adaptively refinement model. However, both finite element models represent the critical point at the strain concentration identically.

This result demonstrates the efficacy of the point-wise finite difference error estimator. Its advantage is largely due to the fact that it evaluates the modeling errors on the domain of a single element. There is no aggregating of the errors from more than one element.

8.9   A TEMPTATION TO AVOID AND A REASON FOR USING CHILD MESHES

One might be tempted to apply a loading to the finite element model based on the point-wise residuals as a way of getting a set of strains to improve the original finite element results. This strategy will not be successful because the strains created by the residual loads would be composed from the same polynomial components as the original finite element strain representation. That is to say, if the original model could have done a better job of representing the strains, it would have.

The fact that this strategy would not succeed is stated more formally by saying the error in the strain results for the finite model is orthogonal to the finite element result. The orthogonality of the error is not mentioned as an isolated fact. The idea of the orthogonality of the error implies something about mesh refinement.

When refining a mesh, it is wise to refine the existing mesh instead of constructing a new mesh. In the usual terminology, the refined mesh should be a child mesh. The reason for this is the fact that the child mesh can represent any strain distribution that the parent mesh can represent.

If a child mesh is not used, the new mesh might not be able to represent the strain distribution as well as the original mesh. That is to say, new sources of error could be introduced.

8.10   SUMMARY AND CONCLUSION

This chapter exploits the similarities and the differences between the finite element and the finite difference methods. The two approximate solution techniques have the following similarities: (1) both of these approximate solution techniques attempt to minimize the potential energy contained in the continuous problems and (2) both approximate solution techniques can be formulated from the same Taylor series basis.

These two similarities have the following implications: (1) the results for both approaches will produce the same result when they converge since they are attempting to solve the same problem and (2) the finite difference method can represent practically any problem that can be represented by the finite element method because both methods can be formulated from the same Taylor series basis.

The difference between the two methods is what makes the point-wise error estimator possible. Although both methods attempt to find an approximate solution that minimizes the potential energy, they attempt this minimization in significantly different ways. The finite element method attempts to minimize the potential energy by directly minimizing an approximation of the potential energy.

The finite difference method attempts to minimize the potential energy by satisfying the conditions that must be satisfied to minimize the potential energy. The conditions that must be satisfied to minimize the potential energy are the governing differential equations and the associated boundary condition.

The point-wise error estimates are found by quantifying the amount by which the finite element solution fails to satisfy the finite difference representation of the governing differential equations and the boundary conditions. This difference is quantified as a residual quantity that can be interpreted physically as a load intensity.

As was seen, the residual served as a high-resolution error estimator. This high resolution exists because the point-wise residual approach does not have the deficiencies of either the Kelly approach or the Z/Z approach. That is to say, it does not aggregate the error estimates in an integral and/or as a summation.

The point-wise error estimator was shown to successfully guide the adaptive refinement process. The termination criterion used was determined empirically by comparing the results for a succession of uniformly refined models. The fact that the termination criterion was determined empirically exposes a deficiency in this error estimator. The units of the error estimator are equivalent to the magnitude of a distributed load. This metric is not a quantity that is directly involved in the failure analysis in solid mechanics problems.

Even though the error estimator based on interelement jumps in strains does not have as high a resolution as the error estimator developed in this chapter, it is more useful as a termination criterion. The units of the error estimator based on interelement jumps in strain, that is, strains, are of primary interest in solid mechanics. Thus, they can directly identify the level of error in the finite element model because the strains are of primary interest in solid mechanics.

In Chapter 11, a refinement guide that compares the modeling capabilities of the individual finite elements to an estimate of the exact solution that is emerging from the finite element result is developed. This means that the level of refinement needed for rapid convergence does not depend on the resolution of the error estimator. Thus, the error estimator based on the interelement jumps would be a better pairing with the new refinement guide because it allows the accuracy of the result to be related to a quantity of primary interest in solid mechanics.

8.11   REFERENCE

1. Dow, J. O. *A Unified Approach to the Finite Element Method and Error Analysis Procedures*, New York: Academic Press, 1999.
   This book puts the finite element method, the finite difference method, and error analysis on a common Taylor series basis. It develops rational procedures for treating a variety of complex boundary conditions encountered in solid mechanics problems. In addition, finite difference models for multi-material models are developed.

CHAPTER 9

MODELING CHARACTERISTICS AND EFFICIENCIES OF HIGHER ORDER ELEMENTS

9.1 INTRODUCTION

In the previous chapters, the errors in finite element models are shown to result when the basis set of an element is unable to capture the complexity of the exact solution that exists on the domain of the element. This means that a higher order element can represent a larger portion of a complex solution with the same accuracy as a lower order element. As a result, we will see that the higher order elements can produce more efficient strain representations for a given level of accuracy than lower order elements.

The ability of higher order elements to better represent a critical portion of a complex strain distribution is shown in Fig. 9.1. A typical three-node element representation of a maximum point in a strain distribution is shown superimposed on the exact solution in Fig. 9.1a. When this model is uniformly refined by subdividing each element into three elements, the improved result is shown in Fig. 9.1b. The improvements to the strain representations are seen in two ways: (1) The finite element strain distribution is closer to the exact result and (2) The interelement jumps are reduced.

The same region is represented in Fig. 9.1c with a single four-node element. As can be seen, the single four-node element provides a better representation of the maximum point than do the three three-node elements that cover the same portion of the model. The strain representation produced by the four-node model is closer to the exact result and the interelement jumps are reduced. Furthermore, the four-node representation of the strain in the neighborhood of the maximum point is more efficient. It, by definition, contains four nodes and the three three-node elements contain seven nodes.

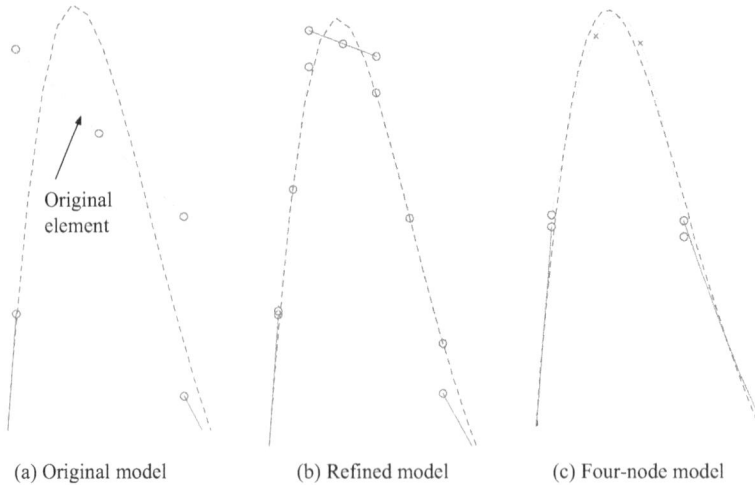

(a) Original model   (b) Refined model   (c) Four-node model

Figure 9.1. Three models of a point of maximum strain.

This chapter has two objectives. The first is to demonstrate and identify the advantages of using higher order elements in finite element models. The second is to identify the way that lower order elements compensate for their inability to represent a higher order strain state. These results provide insights into forming the high performance refinement guides that are developed in chapter 11.

The ability of the higher order elements to better represent the complexity of the exact solution is easily explained. The three-node element cannot capture the maximum point on its interior because it can, at most, represent a linear strain distribution. In contrast, a four-node element can capture a maximum point on its interior because a four-node element can represent curvature. Similarly, as we will see in later sections, a five-node element can also capture the **change in curvature** so it can represent every type of extreme point on its domain, that is, maximum, minimum, and inflection points can be captured on its interior.

The advantages of using higher order elements are presented with two demonstrations in this Chapter. In the first demonstration, initial models of the control problem that produce results with high levels of error are formed from three-, four- and five-node elements. These initial models are then adaptively refined with termination criteria that require different levels of accuracy, namely 4.0 and 0.4% of the maximum absolute strain, respectively.

The adaptive refinement process subdivides elements with high levels of error until the specified level of accuracy is reached everywhere in the model. The number of elements contained in the final refinement depends on the complexity of the exact strain distribution being represented in a given region of the model and the stringency of the termination criterion. We will see that the higher order elements produce results that are as efficient as the lower order elements when the lower level of accuracy is specified. In contrast, when a higher level of accuracy is specified, the results for the higher order elements are significantly more efficient. In summary, the higher order elements are at least as efficient as the lower order elements.

In the second demonstration of the advantages of using higher order elements, a model formed from five-node elements with a small number of elements that accurately represents the control problem is solved. Then, the individual five-node elements are replaced with increasing numbers of four-node elements until a result with approximately the same level of accuracy is attained. The participation levels of the different strain states that the individual elements are representing are identified for each refinement.

The comparisons contained in the second demonstration explain the greater efficiency of the higher order elements when the termination criterion is stringent. In addition, this comparison identifies how lower order elements approximate the additional strain modeling capability of higher order elements. As we will see, these results provide guidance in developing improved refinement guides.

9.2   ADAPTIVE REFINEMENT EXAMPLES (4.0% TERMINATION CRITERION)

The objective of the next two sections is to compare the efficiency of models formed from three-, four- and five-node elements. The modeling efficiency is measured in terms of the number of degrees of freedom that it takes to satisfy a given termination criterion.

In the models adaptively refined in this section, an element is subdivided if the interelement jump at one node of an element is larger than 4.0% of the maximum absolute strain in the finite element model. In the next section, the termination criterion is specified as 0.4%.

The initial finite element models used in the adaptive refinement process for the three- and four-node elements contain five evenly spaced elements. The initial model for the five-element representation contains four evenly spaced elements. Four elements are used in the five-node model so the initial models formed with the four- and five-node elements will have approximately the same number of degrees of freedom. The strain representations for the three initial models are compared to the exact solution in Fig. 9.2. This sequence of figures demonstrates the modeling capabilities of the different order elements.

Figure 9.2a shows that the most complex strain distribution that a three-node element is capable of representing is a linear strain distribution. This capability can be seen by the fact that each element is representing a linear variation in the strain distribution even when the actual strain distribution is more complex than a linear function.

Figure 9.2b shows that a four-node element is capable of representing a more complex strain distribution than a linear strain element. In this figure, the three four-node elements in the center of the model are clearly representing curvature, that is, the rate of change of slope. Another significant capability of a four-node element is shown in this figure. The maximum point is captured on the domain of the element. This cannot be done with a three-node element.

The full modeling capability of a five-node element can be seen in the strain distribution produced by the four element model shown Fig. 9.2c. The ability of a five-node element to represent a change in curvature on its domain is demonstrated by the element that contains the minimum point of the strain distribution on its domain. The strain distribution presented by this element contains

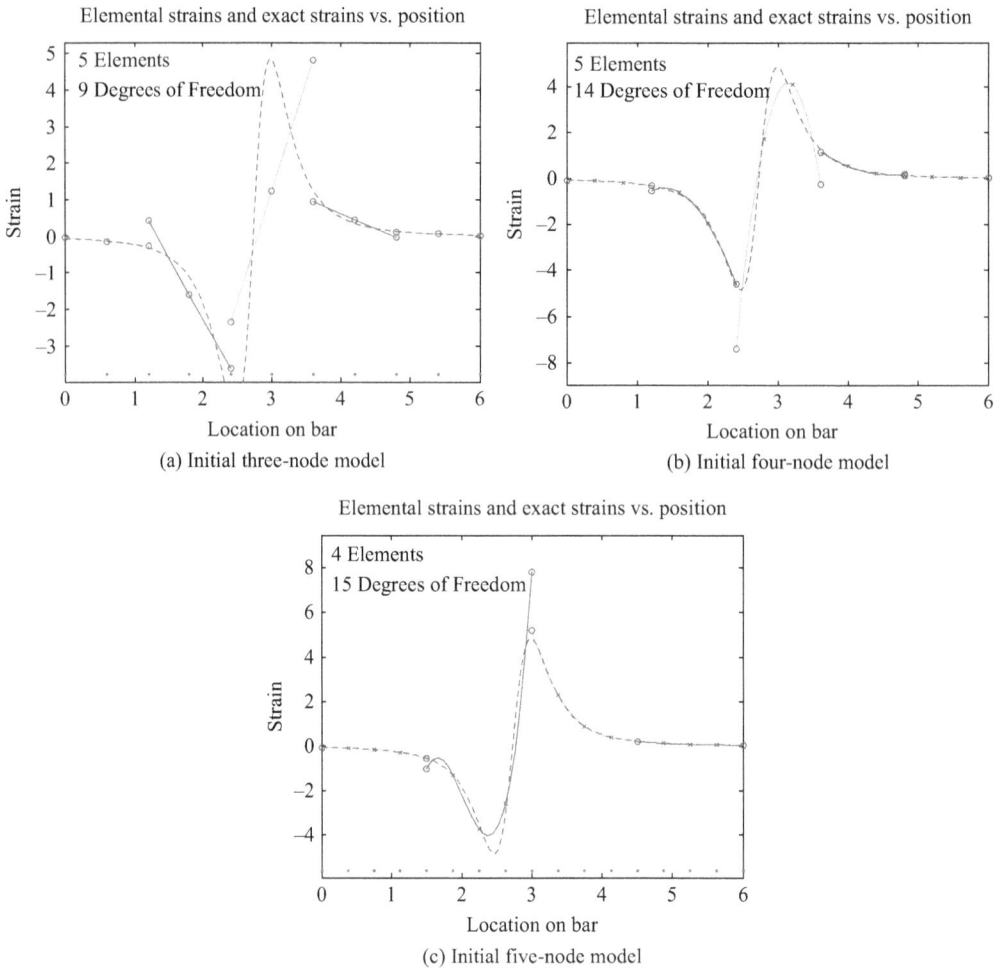

(a) Initial three-node model

(b) Initial four-node model

(c) Initial five-node model

Figure 9.2. Initial models.

both a maximum and a minimum. That is to say, the strain distribution in this element contains both a positive and a negative curvature. As we will see in a later section, this modeling capability means that a five-node element can accurately capture the strain distribution in the neighborhood of an inflection point.

The three initial models represented in Fig. 9.2 will now be adaptively refined with a termination criterion of 4.0%. We will compare the number of degrees of freedom in the final result for these three cases. Since the termination criteria for the three results are specified to be the same, the number of nodes contained in each model is used as a measure of the modeling efficiency of the different order elements.

When the initial model formed with five three-node elements shown in Fig. 9.2a is adaptively refined, it takes four refinements to achieve the desired level of accuracy of 4.0%. The models produced by the successive application of the adaptive refinement process contained the following number of elements: 12, 16, 18, and 20.

The final result is the 20-element model with 39 degrees of freedom shown in Fig. 9.3a. The points along the bottom of the plot indicate the locations of the nodes. Notice that the additional nodes are concentrated in the center of the model where the actual strain distribution is the most complex. The interelement jumps in the strain associated with this model are shown in Fig. 9.3b. As can be seen, the maximum error of approximately 3.2% is below the termination criterion of 4.0%.

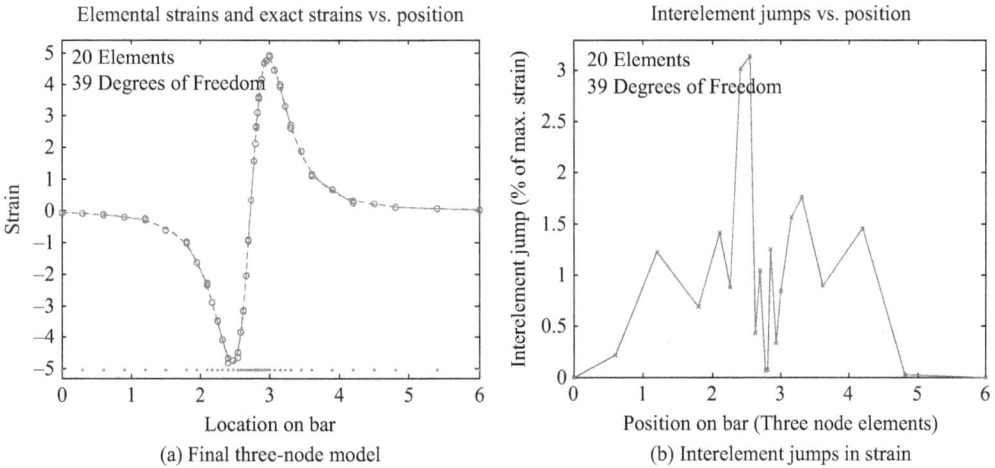

Figure 9.3. Adaptively refined three-node model.

As a note on modeling, consider the representations of the maximum and the minimum points shown in Fig. 9.3a. At the maximum point, an attempt is made to capture the curvature with two elements because there is an interior node near the maximum point. The contour of the strain distribution at the minimum point is represented by three elements. The minimum point is represented by an element with the minimum point near the interior node and two flanking elements.

Next, a representation of the control problem formed with four-node elements is adaptively refined starting with the initial model shown in Fig. 9.2b. The initial model has five elements and 14 degrees of freedom. In order to achieve a model with a maximum error of less than 4%, two refinements were required with the following number of elements: 8 and 11.

The final result of the adaptive refinement process is the 11-element model with 32 degrees of freedom shown in Fig. 9.4a. The interelement jumps in the strain associated with this model are shown in Fig. 9.4b. As can be seen, the maximum error of approximately 3.5% is below the termination criterion of 4.0%.

When the final results for the three- and four-node models are compared, the maximum error in both models is approximately the same, 3.2 versus 3.5%. The maximum error is located in the same

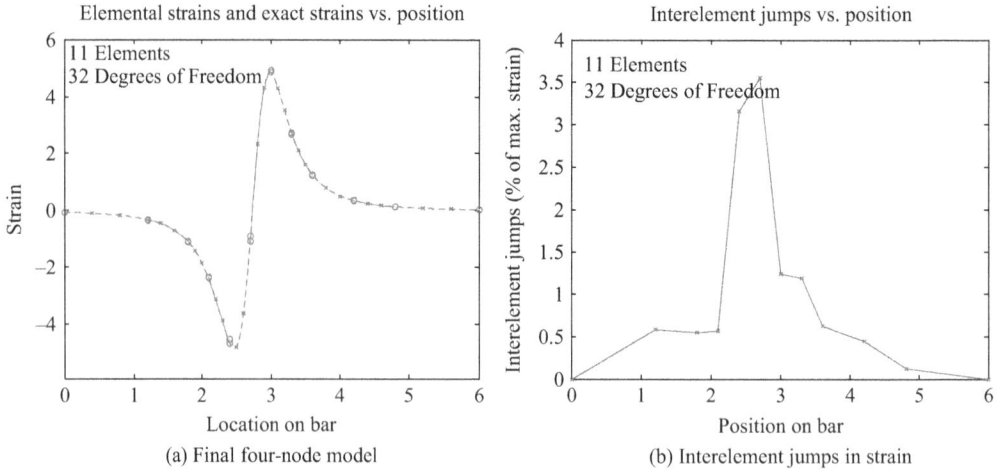

Figure 9.4. Adaptively refined four-node model.

region in both models, that is, near the point of minimum strain. The errors on the other regions of the problem are somewhat smaller for the four-node model. In summary, we can conclude that the three- and four-node element models represent the strain distribution equally well for this case.

Finally, the control problem that is initially modeled with four five-node elements and represented in Fig. 9.2.c is adaptively refined. The final result is the nine-element model with 35 degrees of freedom shown in Fig. 9.5a. The interelement jumps in the strain associated with this model are shown in Fig. 9.5b. As can be seen, the maximum error of approximately 3.9% is below the termination criterion of 4.0%.

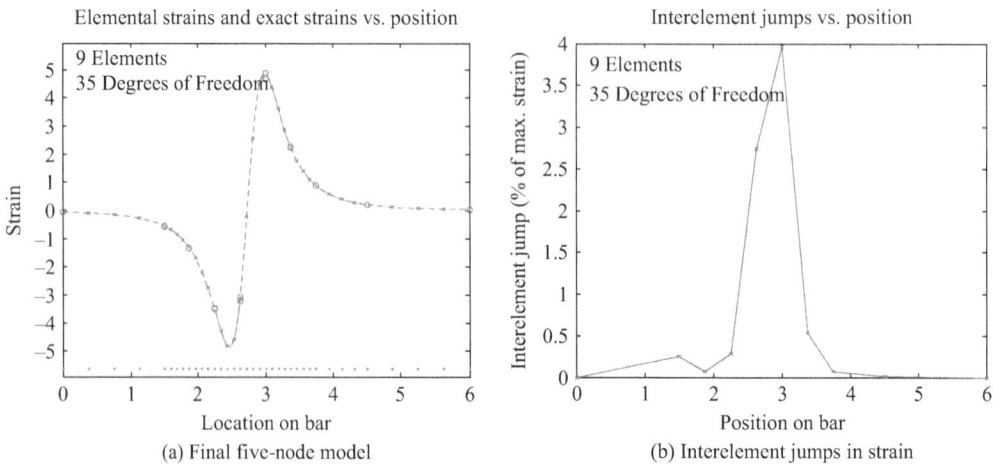

Figure 9.5. Adaptively refined five-node model.

The final model shown in Fig. 9.5a was produced after two refinements of six and nine elements. The maximum error in this model is located near the point of maximum strain. This differs from the three- and four-node cases where the maximum error is located near the point of minimum strain. Note that the levels of error on the other portions of the problem are lower for the five-node representation than for the three- or four-node representations.

When the sizes of the final models formed from the three-, four-, and five-node elements are compared, we see that they have 39, 32, and 35 degrees of freedom, respectively. That is to say, all three models have approximately the same number of degrees of freedom when the termination criterion requires an interelement jump of less than 4% of the maximum strain in the finite element model. Thus, it can be concluded that all three orders of elements represent this problem with approximately the same efficiency when the termination criterion is relatively loose.

9.3 ADAPTIVE REFINEMENT EXAMPLES (0.4% TERMINATION CRITERION)

In this section, the termination criterion is tightened by a factor of 10. The termination criterion and the refinement guide for the examples presented in this section are reduced to 0.4% of the maximum absolute strain in the finite element result. These quantities were specified as 4.0% in the previous section.

In the previous section, there was not a significant difference in the number of degrees of freedom required to produce the specified level of accuracy for the different order elements. However, it was seen that fewer iterations were required to achieve the desired level of accuracy with the higher order elements. That is to say, there was no increase in the modeling efficiency with an increased order of element. However, the adaptive refinement process with a basic refinement guide required fewer iterations.

The three initial models shown in Fig. 9.2 will now be adaptively refined with a termination criterion of 0.4%. The results for the final iterations for the three models formed with three-, four-, and five-node elements with the more restrictive termination criterion are presented in Figs. 9.6–9.8.

When the initial model formed with five three-node elements shown in Fig. 9.1a is refined with the more restrictive termination criterion, the resulting strain distribution for the final iteration is presented in Fig. 9.6a. The interelement jumps in the strain for this model are shown in Fig. 9.6b.

As can be seen, the locations of the nodes in the final model are concentrated in the center of the model where the strain distribution is the most complex. This indicates that the refinement guide modified the model so that the elements attempting to represent the most complex portions of the strain distribution were subdivided more often.

When the initial model formed with five four-node elements shown in Fig. 9.1b is refined with the more restrictive termination criterion, the resulting strain distribution for the final iteration is presented in Fig. 9.7a. The interelement jumps in the strain for this model are shown in Fig. 9.7b.

Elemental strains and exact strains vs. position

Interelement jumps vs. position

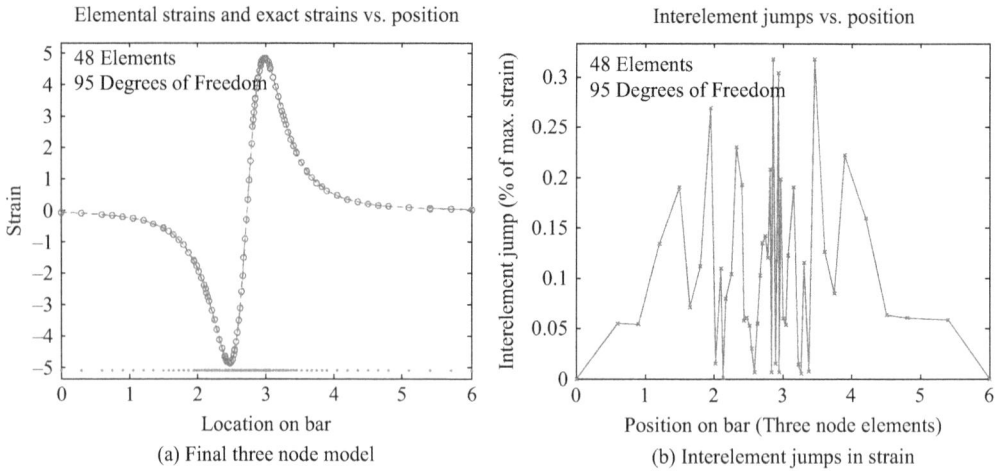

(a) Final three node model

(b) Interelement jumps in strain

Figure 9.6. Adaptively refined three-node model.

Elemental strains and exact strains vs. position

Interelement jumps vs. position

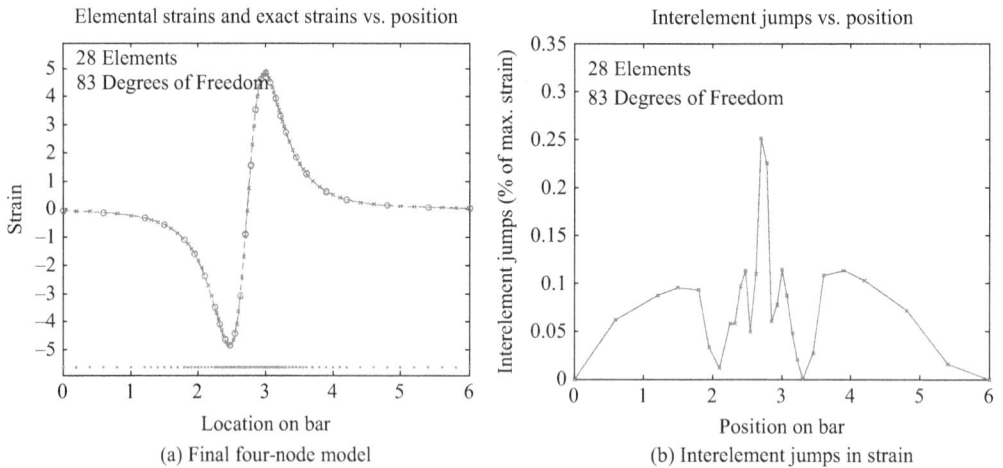

(a) Final four-node model

(b) Interelement jumps in strain

Figure 9.7. Adaptively refined four-node model.

As was the case for the model formed from the three-node elements, the adaptive refinement process concentrated the additional four-node elements in the model in the center where the exact strain distribution is the most complex. This is shown by the concentration of the nodes in the center of the model as indicated by the locations of the nodes at the bottom of Fig. 9.7a.

The adaptive refinement of the initial five-node model containing four elements shown in Fig. 9.1c produces the final strain distribution shown in Fig. 9.8a. The interelement jumps in the strain representation for this model are shown in Fig. 9.8b.

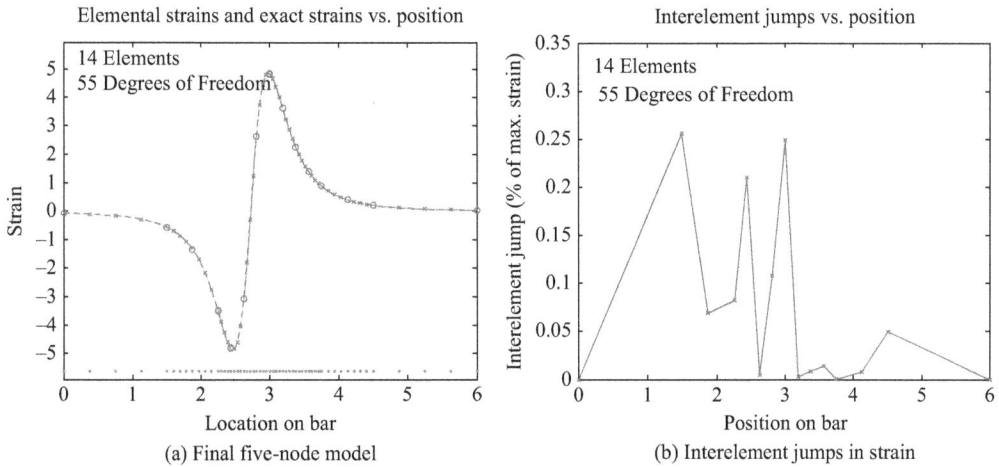

Elemental strains and exact strains vs. position

Interelement jumps vs. position

(a) Final five-node model

(b) Interelement jumps in strain

Figure 9.8. Adaptively refined five-node model.

As can be seen in Fig. 9.8a, the additional five-node elements added to the model by the adaptive refinement process were introduced in the center of the model as was the case for the models formed with the three- and four-node elements. However, the number of nodes contained in this model is significantly smaller than was the case for the three- and four-node models.

The conclusion produced by the results contained in Figs. 9.6–9.8 is simple. The higher order elements outperformed the lower order elements with the more restrictive termination criterion. The number of degrees of freedom for the models formed with three-, four-, and five node elements that satisfied the termination criterion were 95, 83, and 55, respectively. The numbers of elements in these models were 48, 28, and 14, respectively. That is to say, the higher order elements produced models that are more efficient than the lower order elements with this restrictive termination criterion.

9.4   IN-SITU IDENTIFICATION OF THE FIVE-NODE ELEMENT MODELING BEHAVIOR

In the previous two sections, we saw that the modeling efficiency of a finite element model depends on the order of the finite element used to represent the problem being solved and the restrictiveness of the termination criterion.

The objective of this section is to provide the basis for identifying the reason for the higher modeling efficiency of the higher order elements. This objective is accomplished by forming a carefully constructed model of the control problem with five-node elements. This model is constructed so that each of the five extreme points (three inflection points, a maximum point, and a minimum point) presented in Fig. 9.9 is contained on the domain of a single five-node element. As a result of this construction, the modeling capabilities of these five five-node elements are fully engaged by the complexity of the strain distribution that exists in the neighborhood of these critical points.

Strain distribution with max/min and inflection points

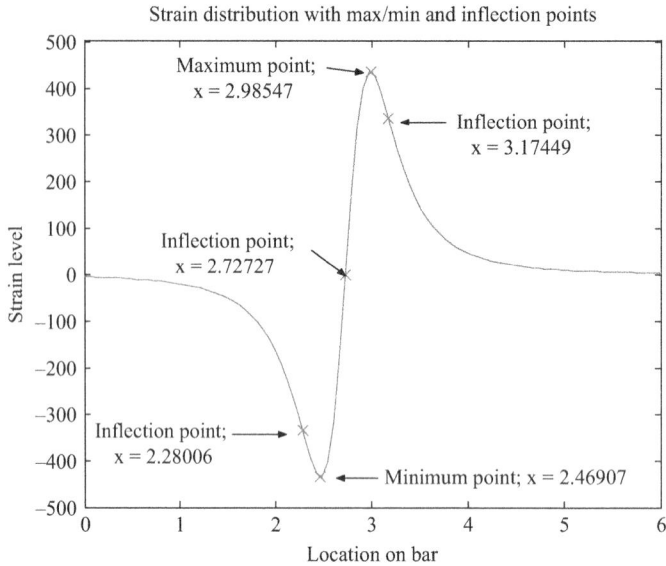

Figure 9.9 Locations of critical points.

After solving this problem, the level of participation of the individual strain gradient quantities contained in the basis set for the individual five-node elements is identified. As we will see, this 16-element model produces a result with very little error.

The locations of the interelement nodes are shown by the circles in Fig. 9.10a. These nodes are located at the following points: [0.0, 0.45, 0.9, 1.41, 1.6818, 2.1, 2.3116, 2.6, 2.85, 3.143, 3.3127, 3.5845, 3.8564, 4.1282, 4.4, 5.3, 6.0].

This model of the control problem contains three categories of elements. The first five elements consist of a transition from the left end of the bar to the interior five elements that each contains a critical point on their domain. The final six elements form a transition from the fully engaged elements in the center of the model to the boundary on the right end of the control problem. This transition section has six elements instead of five elements because it is longer due to the nonsymmetrical load on the problem. The function of the transition elements is to smooth out any anomalies in the strain distribution produced by the boundaries. The transition sections serve as a practical application of St. Venant's principle.

The three inflection points shown in Fig. 9.9 are located at $x = 2.28006$, $x = 2.72727$, and $x = 3.17449$, respectively. The three interior elements that contain an inflection point have the following interelement nodal locations: [2.1, 2.3116], [2.6, 2.85], and [3.143, 3.3127]. As can be seen when these coordinate locations are compared, the three inflection points are on the interior of these three elements.

The minimum and the maximum points are located at $x = 2.46907$ and $x = 2.98547$, respectively. The two five-node elements on the interior portion of the model that contain these two points

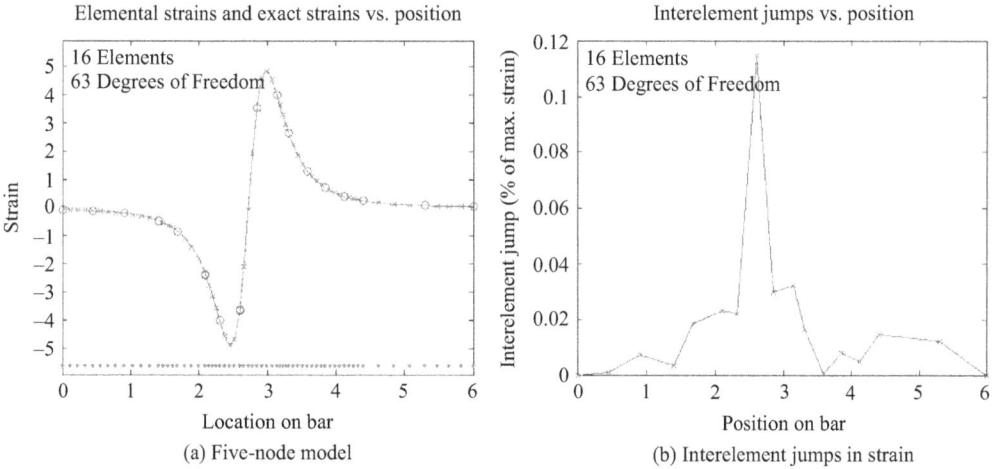

Figure 9.10. Specialized five-node model.

have the following interelement nodal locations: [2.3116, 2.6] and [2.85, 3.143]. As can be seen, these two critical points are located on the domain of these two elements.

The elements containing an inflection point engage the full modeling capability of a five-node element by definition. An inflection point has a curvature of zero and the sign of the curvature changes across the inflection point. That is to say, there is a change in curvature across the inflection point so the strain gradient term $(\varepsilon_{x,xxx})_0$ is engaged.

The fact that the elements with the maximum and minimum points on their domain engage the full modeling capacity can be seen in Fig. 9.10a. In these two elements, the curvature does not change sign. Instead, it changes magnitude on the domain of the elements. The magnitudes of the curvatures get smaller away from the critical points near the center of the elements. As a result of the change in the magnitudes of the curvature, the full capability of these two five-node elements is fully engaged.

Figure 9.10b presents a plot of the interelement jumps as a percentage of the maximum absolute strain in the finite element model. When these error measures are compared to those produced by any of the previous models solved in this chapter, it is seen that these error estimates are very low. This means that the individual elements provide good representations of the exact solution on their domains.

The participation of the four components of the basis set that produce strains in the five-node elements are presented in Table 9.1. Column 1 identifies the number of the element contained in the 16-element control problem. The elements contained in Fig. 9.10 are numbered from left to right.

The contents of Column 2 identify the level of participation of the constant strain terms, $(\varepsilon_x)_0$, at the center of the individual elements. Note that Rows 7 and 9 contain the strains with the largest magnitudes. As would be expected, these high magnitude quantities identify strains at points near the locations of the minimum and the maximum strains.

Table 9.1. Five-node strain gradient components

El. no.	$(\varepsilon_x)_0$	$(\varepsilon_{x,x})_0$	$(\varepsilon_{x,xx})_0$	$(\varepsilon_{x,xxx})_0$
1	−0.0901	−0.0918	−0.1413	−0.2648
2	−0.1507	−0.1931	−0.3550	−0.7855
3	−0.3036	−0.5040	−1.1656	−3.1308
4	−0.6286	−1.3084	−3.5344	−10.9268
5	−1.3877	−3.4802	−10.3260	−28.2096
6	−3.1314	−7.8150	−10.8585	100.6357
7	−4.8398	−1.3036	98.4584	744.5053
8	−0.0805	33.2002	3.8676	−1762.8378
9	4.8317	−1.0753	−100.0525	740.9016
10	3.2864	−8.0435	8.8713	133.3141
11	1.8534	−4.8569	13.5955	−26.6515
12	0.9409	−2.2030	6.3871	−20.0527
13	0.5208	−1.0513	2.7344	−8.2533
14	0.3128	−0.5460	1.2481	−3.4010
15	0.1283	−0.1700	0.3230	−0.6975
16	0.0517	−0.0507	0.0687	−0.1122

Column 3 identifies the level of the participation of the slope of the strain distribution, $(\varepsilon_{x,x})_0$, represented by the individual elements. Note that Row 8 contains the highest value for the slope of any of the elements in this model. This relatively steep slope can be clearly seen near the center in the strain distribution shown in Fig. 9.10.

Column 4 quantifies the contribution of the rate of change of the slope to the strain distribution in an individual element, that is, $[(\varepsilon_{x,x})_x]_0$ or $(\varepsilon_{x,xx})_0$. This is the curvature of the strain distribution. Note that the curvatures for Elements 7 and 9 have the highest magnitudes. These results correspond to the fact that the points of maximum and minimum strains are on the domain of these two elements.

Finally, Column 5 quantifies the rate of change of curvature, $(\varepsilon_{x,xxx})_0$, in the individual five-node elements. The fact that Elements 6–10 have relatively large magnitudes of rates of change of curvature indicates that the modeling capability of these five-node elements is fully engaged.

For example, note that Element 8, the element with the steepest slope, has the largest magnitude for the rate of change of curvature. Furthermore, this element has a relatively small curvature (at its center node). These two characteristics are the signature for an element that is representing an inflection point.

9.5   STRAIN CONTRIBUTIONS OF THE BASIS SET COMPONENTS

The fact that the modeling capabilities of the five elements containing the five critical points are fully engaged was shown in the previous section. The five rows associated with these elements are separated from the transition elements by the heavy lines before Element 6 and after Element 10 in Table 9.1.

As can be seen, the rate of change of curvature, $(\varepsilon_{x,xxx})_0$, presented in Column 5 for these five elements is significantly larger than this quantity for the elements in the transition from the boundary, that is, (Elements 1–5 and Elements 11–16). The large magnitude of these quantities indicates that the change in curvature for these elements plays a more significant role in representing the strain distribution in these elements than it does in the transition elements. The strains produced by the components of the basis set for the individual elements are presented in this Section.

The procedure for identifying the participation factors of the individual components of the basis set presented in Table 9.1 was discussed at length in Chapter 3. In brief, a set of simultaneous linear algebraic equations is formed by inserting the nodal displacements and locations for each of the nodes of the individual elements into the Taylor series representation of the element displacements. Then, this set of equations is inverted to identify the strain gradient participation factors as a function of the nodal displacements. As noted earlier, these gradient terms are identical in concept to the finite difference approximations of derivatives. The Taylor series representation of the displacements for a five-node element with physically interpretable coefficients is given as:

$$u(x) = (u_{\mathrm{rb}})_0 + (\varepsilon_x)_0\, x + (\varepsilon_{x,x})_0\, \frac{x^2}{2} + (\varepsilon_{x,xx})_0\, \frac{x^3}{6} + (\varepsilon_{x,xxx})_0\, \frac{x^4}{24} \qquad \text{(Eq. 9.1)}$$

The subscript 0 indicates that the strain gradient quantities are evaluated at the local origin, which is located at the center of the individual elements in this work.

The strains produced by the participation factors presented in Table 9.1 are computed by first taking the derivative of Eq. 9.1 to form an expression for the strain in a longitudinal bar, that is, $\varepsilon_x = du/dx$. When this operation is performed, we have the following:

$$\varepsilon(x) = (\varepsilon_x)_0 + (\varepsilon_{x,x})_0\, x + (\varepsilon_{x,xx})_0\, \frac{x^2}{2} + (\varepsilon_{x,xxx})_0\, \frac{x^3}{6} \qquad \text{(Eq. 9.2)}$$

This equation indicates that the strains in a five-node element are due to four strain gradient components. The values of these participation factors for the 16-element model shown in Fig. 9.10 are presented for the individual elements in Table 9.1, for example, $(\varepsilon_{x,xx})_0$.

The components of Eq. 9.2 are used to compute the maximum absolute contributions of the four strain gradient terms for the individual elements. These interelement strain quantities are presented in Table 9.2 for the individual elements. These quantities are computed by introducing the maximum distance from the local origin for each of the elements into the individual terms of Eq. 9.2.

In this case, the maximum distance from the origin is equal to one-half of the length of an element, that is, $h = L/2$. The first row of Table 9.2 identifies the Taylor series term contained in the various columns of this table.

Although it was seen in Table 9.1 that the five elements containing the five critical points contained significant participation factors for the rate of change of curvature, Table 9.2 shows that only three of these elements contain significant strains at the interelement nodes due to the change in curvature. These elements are Elements 7, 8, and 9.

In the next section, each of the five-node elements in this model is replaced with an increasing number of four-node elements until a model with an equivalent accuracy is produced. That is to say, each five-node element is first replaced with a single four-node element. Then, each five-node element is replaced by two four-node elements, etc., until the interelement jumps are all lower than those contained in Fig. 9.10b. As we will see, this occurs when each five-node element is replaced by five four-node elements.

Then, by analyzing the participation of the individual strain gradient components in the four-node elements, the way in which the increasing number of four-node elements compensate for not being able to represent the full modeling capability of a five-node element is identified.

Table 9.2. Five-node strain components

El. no.	$(\varepsilon_x)_0$	$(\varepsilon_{x,x})_0 h$	$(\varepsilon_{x,xx})_0 h^2/2$	$(\varepsilon_{x,xxx})_0 h^3/6$
1	0.0901	0.0207	0.0036	0.0005
2	0.1507	0.0434	0.0090	0.0015
3	0.3036	0.1285	0.0379	0.0087
4	0.6286	0.1778	0.0326	0.0046
5	1.3877	0.7277	0.2257	0.0430
6	3.1314	0.8268	0.0608	0.0199
7	4.8398	0.1880	1.0237	0.3721
8	0.0805	4.1500	0.0302	0.5738
9	4.8317	0.1575	1.0737	0.3883
10	3.2864	0.6825	0.0319	0.0136
11	1.8534	0.6601	0.1255	0.0111
12	0.9409	0.2995	0.0590	0.0084
13	0.5208	0.1429	0.0253	0.0035
14	0.3128	0.0742	0.0115	0.0014
15	0.1283	0.0765	0.0327	0.0106
16	0.0517	0.0178	0.0042	0.0008

As mentioned earlier, this knowledge will be used in Chapter 11 to develop a new approach for forming refinement guides.

As can be seen in the fifth column of Table 9.2 for the two groups of transition elements at the ends of the bar being analyzed, the contribution to the strain distribution of the rate of change of curvature in these elements is very small. For example, Element 5 makes the largest contribution, namely, 0.0430 strain units. This is approximately equal to 0.9% of the maximum absolute strain in the finite element model. This means that the replacement of this five-node element by a single four-node element will not significantly affect the strain results for the model.

9.6 COMPARATIVE MODELING BEHAVIOR OF FOUR-NODE ELEMENTS

In this section, the individual five-node elements contained in the control problem shown in Fig. 9.10 are replaced by increasing numbers of four-node elements. This process is continued until a model with at least as good accuracy as the model formed with five-node elements is produced. As we will see, a result with an equivalent accuracy is not produced until each of the individual five-node elements is replaced with five four-node elements.

One Four-Node Element: When each of the five-node elements in the control problem is replaced with a single four-node element, the result is presented in Fig. 9.11. The strain distribution produced by this four-node model presented in Fig. 9.11a is not as good a representation of the exact result as that produced by the five-node model shown in Fig. 9.10a.

There are visible jumps in the interelement strains in the four-node model that do not exist in the strain distribution produced by the five-node model. The differences in these jumps are quantified when the plots of the interelement jumps presented in Figs. 9.10b and 9.11b for the two cases are

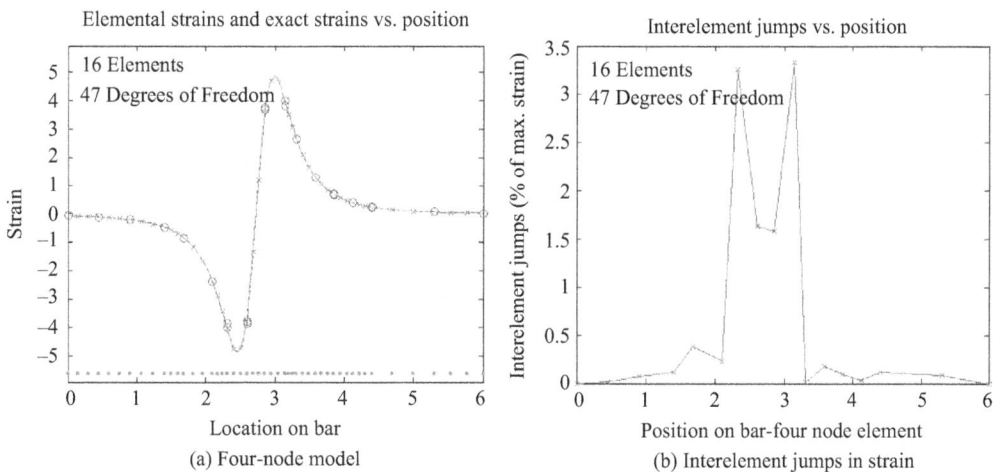

Figure 9.11. Initial four-node model.

compared. The maximum error in the four-node model is approximately 3.3%. This compares to a maximum error of approximately 0.11% for the five-node model.

It should be noted that the errors in the transition elements at both ends of the bar are small in both the four- and the five-node models. This is as could have been anticipated because of the low values of the interelement strains shown in Table 9.2 for the transition elements.

In addition, the maximum errors in the two models occur in different locations. In the five-node model, the maximum error occurs at the node connecting the element that includes the minimum point and the element containing the inflection point between the minimum and maximum points. However, the level of error in the two models is approximately the same at this point in the two models. The maximum errors in the four-node model occur at the nodes connecting the minimum and maximum points to the boundaries on each end of the elements. These are the errors that are approximately 30 times the maximum error in the five-node model.

The reason that the model formed from the five-node elements produces better results than the model formed from the four-node elements is obvious. The basis set for the five-node element contains one more term than does the basis set for the four-node element. As a result, an individual five-node element can represent a more complex strain distribution than can a four-node element.

This difference in the strain representation between the two models can be quantified by comparing the levels of participation of the strain gradient components in the basis sets for the individual elements in the two models. The strain gradient components for all 16 of the elements in this four-node model are presented in Table 9.3.

When Table 9.3 is compared to Table 9.1, the most interesting observation is that the participation of the curvatures, $(\varepsilon_{x,xx})_0$, is essentially identical for the two models. This means that the inability of a four-node element to represent the rate of change of curvature, $(\varepsilon_{x,xxx})_0$, is largely responsible for the difference in accuracy of the four- and five-node models. The other components are slightly different between the two representations because the finite element method is minimizing the strain energy with the modeling capabilities that it has available in a given model.

Table 9.3. Four-node strain gradient components

El. no.	$(\varepsilon_x)_0$	$(\varepsilon_{x,x})_0$	$(\varepsilon_{x,xx})_0$
1	−0.0843	−0.0932	−0.1413
2	−0.1449	−0.1970	−0.3550
3	−0.2978	−0.5244	−1.1656
4	−0.6230	−1.3286	−3.5344
5	−1.3824	−3.6036	−10.3260
6	−3.1269	−7.7025	−10.8585
7	−4.8367	0.2445	98.4585

(Continued on following page)

Table 9.3. (*Continued*)

El. no.	$(\varepsilon_x)_0$	$(\varepsilon_{x,x})_0$	$(\varepsilon_{x,xx})_0$
8	−0.0796	30.4462	3.8676
9	4.8305	0.5149	−100.0527
10	3.2835	−7.9476	8.8713
11	1.8494	−4.9062	13.5955
12	0.9364	−2.2401	6.3872
13	0.5160	−1.0665	2.7344
14	0.3078	−0.5523	1.2481
15	0.1232	−0.1841	0.3230
16	0.0466	−0.0521	0.0687

In this case, the strain energy for the model formed from the five-node elements is 878.90388 units. The strain energy for the model formed from the four-node elements is 877.78554 units for a 0.127243% difference. Thus, the five-node model is the better model from the strain energy point of view according to the Rayleigh–Ritz criterion because it contains more strain energy.

A truncated version of Table 9.3 that contains only the strain gradient quantities for the element containing the five critical points is presented as Table 9.4 so the strain gradient results for this case are presented in a format similar to that used for the later models formed with four-node elements. A table identifying the results for a smaller number of elements is used because the number of elements in the later models makes the full version of the table unwieldy without presenting any more useful information.

Table 9.4. Four-node strain gradient components

El. no.	$(\varepsilon_x)_0$	$(\varepsilon_{x,x})_0$	$(\varepsilon_{x,xx})_0$
6	−3.1269	−7.7025	−10.8585
7	−4.8367	0.2445	98.4585
8	−0.0796	30.4462	3.8676
9	4.8305	0.5149	−100.0527
10	3.2835	−7.9476	8.8713

Two Four-Node Elements: The result for the case where the individual five-node elements in the control problem are each replaced with two four-node elements is presented in Fig. 9.12. When the strain distribution produced by the four-node model presented in Fig. 9.12a is compared to the analogous results for a four-node model formed with one four-node element in place of each five-node element shown in Fig. 9.11a, there is a significant improvement in the results. The interelement jumps that are clearly visible in Fig. 9.11a are reduced in size and barely visible in Fig. 9.12a.

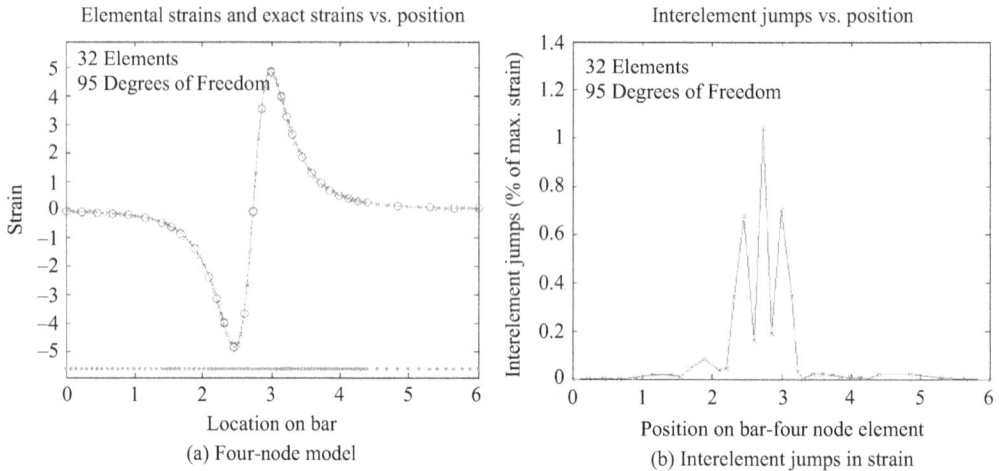

Figure 9.12. Second four-node model.

These improvements are quantified in the plots of the absolute values of the interelement jumps. The maximum error in this four-node model is approximately 1.05%. This compares to a maximum error of approximately 3.3% in the previous four-node model. However, this result has significantly larger errors than the results produced by the five-node model.

In this four-node model, the maximum error is located in the same position as the maximum error in the five-node model. However, the magnitude of this error is approximately 10 times larger than the error in the five-node model. This four-node model has peaks in the errors at the same locations as the peaks in the four-node model in the previous Section. However, these errors are approximately four times smaller than the errors in the previous four-node model where only one four-node element replaced each of the five-node elements in the original form of the control problem.

As would be expected, the strain energy for this four-node model (877.95619) is closer to the strain energy for the five-node model (878.90388) than that for the previous four-node model (877.78554) [0.107826 versus 0.127243 %]. Thus, this model is a better model of the control problem than the previous four-node model. However, the five-node model is still the better representation of the control problem.

This result validates the observation made in Section 9.3 that models formed with higher order elements are more efficient on a per degree-of-freedom basis than models formed with lower order elements when the termination criterion is restrictive. As we saw when Fig. 9.12b was compared to Fig 9.10b, the model formed from the five-node elements is more accurate than the model formed when each of the five-node elements is replaced by two four-node elements. That is to say, the five degrees of freedom in the higher order elements produce a better result than the seven degrees of freedom in the four-node model.

The following paragraphs explain why this four-node model of the control problem produces better results than the previous four-node model of this problem. This explanation is based on the result of extracting the participation factors for each of the four-node elements that replace the five-node elements containing the five critical points. These participation factors are then used to evaluate the modeling characteristics of this model.

The participation factors for the four-node elements that are replacing the individual five-node elements containing the five critical points are presented in Table 9.5. In this table, the results for the two four-node elements that are replacing a single five-node element containing a critical point are separated by **bold lines.** The five-node elements that the two four-node elements replace are identified in the final column.

Table 9.5. Two, four node elements represent 1 five node element.

El. no.	$(\varepsilon_x)_0$	$(\varepsilon_{x,x})_0$	$(\varepsilon_{x,xx})_0$	
11	−2.7317	−7.0681	−14.9625	6
12	−3.5536	−8.2257	−4.2525	
13	−4.5336	−6.0657	45.4651	7
14	−4.6284	8.1300	151.8777	
15	−2.0748	28.7995	110.3839	8
16	1.9315	29.2883	−105.9249	
17	4.5921	8.6543	−153.7103	9
18	4.5318	−6.0045	−46.1893	
19	3.6313	−8.2791	2.2383	10
20	2.9522	−7.5209	13.5831	

The ability of this four-node model to produce results that are more accurate than the previous four-node model can be attributed to the fact that the two four-node elements are attempting to approximate the change in curvature represented by the five-node element with a discrete representation. The single four-node element was incapable of making any contribution to an attempt to represent this higher order strain state.

This discrete approximation of the rate of change of the curvature by the two four-node elements can be seen qualitatively by the fact that the curvatures contained in the two elements that replace each of the individual five-node elements have different values. That is to say, there is a changing value for the curvature in these two elements that can be interpreted as a discrete representation of the rate of change in the curvature. The attempt by the two four-node elements to represent the additional modeling capability of the five-node element is quantified by forming a finite difference approximation of this gradient quantity.

The process by which the two four-node elements attempt to represent the additional modeling capability of a five-node element will be shown quantitatively with an example. The curvature values for Elements 11 and 12 contained in Table 9.5 are –14.9625 and –4.2525, respectively. These two elements replace the five-node element identified as Element 6 in Table 9.1. That is to say, these two elements cover the same portion of the domain of the problem as that covered by Element 6.

The interelement nodal locations for this five-node element are 2.1 and 2.3116. This gives a length for Element 6 of 0.2116. The change of curvature for Element 6 is given in Table 9.4 as 100.6357.

The discrete approximation of the rate of change in curvature produced by the two four-node elements is computed using a two-node central difference template. Since the local origins of the four-node elements are located at the center of the elements, the curvature values for the individual elements are taken to be acting at these local origins. Thus, the distance separating the two curvatures for these elements is equal to one-half the length of one element and one-half the length of the second element. Since the elements are of the same length, the total difference separating their local origins is one-half of the total length of the five-node element that the two four-node elements replace (the length of the two four-node elements). The distance separating the two nodes is 0.2116/2 or 0.1058.

The finite difference approximation of the rate of change of curvature produced by Elements 11 and 12 gives the following result:

$$(\varepsilon_{x,xxx})_0 \approx [(\varepsilon_{x,xx})_{12} - (\varepsilon_{x,xx})_{11}]/h$$

$$\approx (-4.2525) - (-14.9625)/0.1058$$

$$\approx 10.710/0.1058$$

$$\approx 101.2287$$

When this result is compared to the rate of change in curvature that exists in the five-node element that these two four-node elements replace, the percentage error is found to be $((101.2287 - 100.6357)/100.6357) * 100 = 0.59\%$. This result shows that the two four-node elements approximate the rate of change of curvature that exists in the five-node model with a high level of accuracy.

Although the two-element representation of the rate of change of curvature produces a good approximation of this higher order strain state, this four-node model does not represent the strain distribution as well as the five-node model. In the next model, the individual five-node elements are replaced by three four-node elements.

Three Four-Node Elements: When the five-node elements in the control problem are each replaced with three four-node elements, the result is presented in Fig. 9.13. The strain distribution

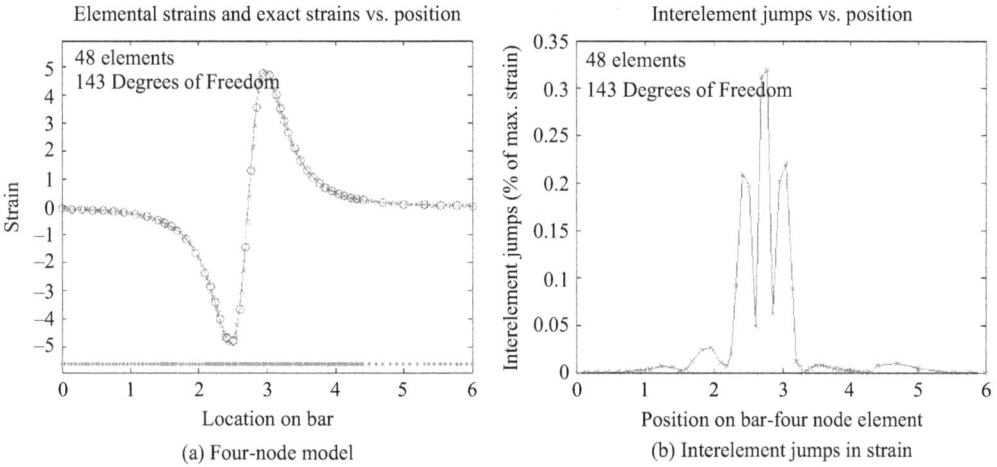

Figure 9.13. Third four-node model.

produced by this four-node model is contained in Fig. 9.13a. A comparison with the analogous four-node model shown in Fig. 9.12a shows that there is little visual improvement in the results. The primary difference between Figs. 9.12a and 9.13a is a slightly different shape in the region of the maximum and minimum strains.

The improvements in the strain distribution are quantified in the plots of the absolute values of the interelement jumps. The maximum error in this four-node model is approximately 0.32%. This compares to a maximum error of approximately 1.05% in the previous four-node model. However, this result has larger errors than the results produced by the five-node model.

It should be noted that each of the five-node elements is replaced with three four-node elements that contain eleven nodes, that is, five degrees of freedom versus 11 degrees of freedom. This result validates the conclusion of Section 9.3 that models formed with higher order elements are more efficient on a per degree of freedom basis than models formed with lower order elements when the termination criterion is restrictive.

As would be expected, the strain energy for this four-node model is larger than that for the previous four-node model, that is, 877.95782 units versus 877.95619 units (0.107641% versus 0.107826% of the five-node model strain energy). The strain energy content for the five-node model is still larger than the strain energy content for this four-node model. Thus, the five-element model is the better model from the strain energy point of view.

The participation factors for the four-node elements that are replacing the individual five-node elements containing the five critical points are presented in Table 9.6. In this table, the results for the three four-node elements that are replacing a single five-node element containing a critical point are separated by **bold lines.** The five-node elements that the three four-node elements replace are identified in the final column.

The more accurate results produced by this model can be attributed to the fact that the three four-node elements approximate the change in curvature represented by the five-node element better

Table 9.6. Three, four node elements represent 1 five node element.

El. no.	$(\varepsilon_x)_0$	$(\varepsilon_{x,x})_0$	$(\varepsilon_{x,xx})_0$	
16	−2.6093	−6.8039	−15.4580	
17	−3.1262	−7.8000	−11.9635	6
18	−3.7001	−8.3017	−0.6701	
19	−4.3684	−7.1543	30.8940	
20	−4.8363	−1.1800	97.9393	7
21	−4.3951	11.7878	165.6257	
22	−2.6595	26.4479	140.5937	
23	−0.0801	32.9843	4.4615	8
24	2.5283	27.0717	−135.5863	
25	4.3409	10.4159	−166.9332	
26	4.8304	−0.9518	−99.7937	9
27	4.3649	−7.1281	−31.2908	
28	3.7491	−8.3113	−0.9783	
29	3.2832	−8.0317	9.7226	10
30	2.8471	−7.3285	14.4406	

than the two four-node elements in the previous example. This can be seen by the fact that the curvatures contained in the three elements that replace each of the individual five-node elements have values that change monotonically. That is to say, the changing values of the curvature in the three elements have a higher resolution than when the single five-node element is replaced with two four-node elements.

Although this attempt by the three four-node elements to represent the strain distribution as well as the single five-node element is more successful than was the case when two four-node elements replaced the single five-node element, the result was not as good as the representation when only one five node element was used. In the next model the individual five-node elements are replaced by four four-node elements.

Four Four-Node Elements: The result for the case where the individual five-node elements in the control problem are each replaced with four four-node elements is presented in Fig. 9.14.

When the strain distribution produced by the four-node model presented in Fig. 9.14a is compared to the analogous results for a four-node model formed with three four-node elements in place of each five-node element shown in Fig. 9.13a, there is no significant visible improvement in the results.

These improvements are quantified in the plots of the absolute values of the interelement jumps presented in Fig. 9.14b. The maximum error in this four-node model

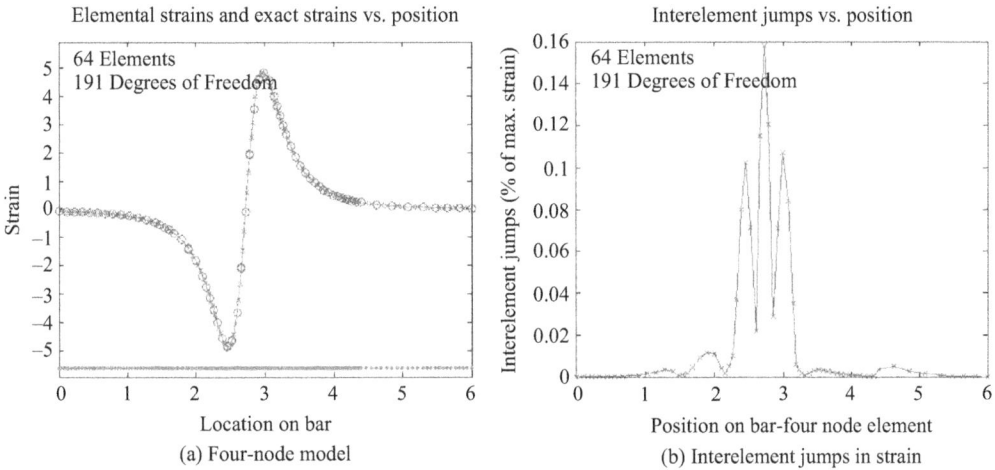

Figure 9.14. Fourth four-node model.

is approximately 0.16%. This compares to a maximum error of approximately 0.32% in the previous four-node model. However, this result has larger errors than the results produced by the five-node model.

It should be noted that each of the five-node elements is replaced with four four-node elements that contain eleven nodes, that is, five degrees of freedom versus 15 degrees of freedom. This result validates the conclusion of Section 9.3 that models formed with higher order elements are more efficient on a per degree of freedom basis than models formed with lower order elements when the termination criterion is restrictive.

As would be expected, the strain energy for this four-node model is larger than that for the previous four-node model, that is, 877.95795 units versus 877.95782 units (0.107626% versus 0.107641% of the five-node model strain energy). The strain energy content for the five-node model is still larger than the strain energy content for this four-node model. Thus, the five-element model is the better model from the strain energy point of view.

The participation factors for the four-node elements that are replacing the individual five-node elements containing the five critical points are presented in Table 9.7. In this table, the results for the four four-node elements that are replacing a single five-node element containing a critical point are separated by **bold lines.** The five-node elements that the four four-node elements replace are identified in the final column.

The more accurate results produced by this model can be attributed to the fact that the four four-node elements are attempting to approximate the change in curvature represented by the five-node element with a discrete representation. This can be seen by the fact that the curvatures contained in the four elements that replace each of the individual five-node elements have a different value. That is to say, the changing values of the curvature are approximating the change in curvature contained in the five-node element that the four-node elements replace.

Table 9.7. Four, four node elements represent 1 five node element.

El. no.	$(\varepsilon_x)_0$	$(\varepsilon_{x,x})_0$	$(\varepsilon_{x,xx})_0$	
21	−2.5499	−6.6678	−15.5933	
22	−2.9241	−7.4599	−13.9741	6
23	−3.3367	−8.0848	−8.9979	
24	−3.7736	−8.3115	1.4538	
25	−4.2789	−7.5403	24.5674	
26	−4.7298	−4.2690	69.1902	7
27	−4.8157	2.8041	127.7086	
28	−4.2425	13.7682	170.3348	
29	−2.9292	25.0049	151.6778	
30	−1.1093	32.0468	64.5304	8
31	0.9535	32.3200	−56.1734	
32	2.8048	25.6827	−147.5076	
33	4.1779	14.4451	−171.1735	
34	4.7988	3.1705	−129.9600	9
35	4.7283	−4.1490	−70.4562	
36	4.2741	−7.5253	−24.8291	
37	3.8080	−8.3070	−2.8154	
38	3.4558	−8.2111	6.6255	10
39	3.1150	−7.8036	10.0794	
40	2.7956	−7.2269	14.7673	

Although this attempt by the four four-node elements to represent the strain distribution as well as the single five-node element is more successful than was the case when three four-node elements replaced the single five-node element, the result was not as good as the representation when only one five-node element was used. In the next model the individual five-node elements are replaced by five four-node elements.

Five Four-Node Elements: When the five-node elements in the control problem are each replaced with five four-node elements, the maximum error is smaller than it is in the five-node model. As shown in Fig. 9.15b, the maximum error in this four-node model is approximately 0.08%. This compares to the maximum error in the five-node model of 0.11%, which is shown in Fig. 9.10b.

The strain distribution produced by this four-node model is presented in Fig. 9.15a. A comparison with the strain distribution for the previous four-node model presented in Fig. 9.14a shows that there is little visual improvement in the results. The primary difference between Figs. 9.14a and 9.15a is seen in the region of the inflection point between the maximum and the minimum points. The two interelement nodes on either side of the inflection point in Fig. 9.14a are darker than they are in Fig. 9.15a. This means that the size of the interelement jump is reduced in the refined model.

The changes in the levels of the error are seen in the differences in the plots of the absolute values of the interelement jumps. The maximum error in this four-node model is approximately 0.08%. This compares to a maximum error of approximately 0.16% for the previous four-node model. When Fig. 9.15b is superimposed on Fig. 9.14b, the errors are smaller by a factor of two everywhere on the domain of the problem.

The replacement of the individual five-node elements by an increasing number of four-node elements was continued until the maximum jump in the interelement strains was smaller than the jumps in the five-element model. We could have stopped the process earlier because we had already made the point that the higher order elements are more efficient than the lower order elements when the termination criterion is restrictive. However, this four-node model provides the most significant example of this difference in modeling efficiency. In this case, it took 16 degrees of freedom in the four-node model to better the results produced by five degrees of freedom in the five-node model from the interelement jump in strain point of view.

As would be expected, the strain energy for this four-node model is larger than that for the previous four-node model, that is, 877.95797 units versus 877.95795 units (0.107624% versus 0.107826% of the five-node model strain energy). The strain energy content for the five-node model

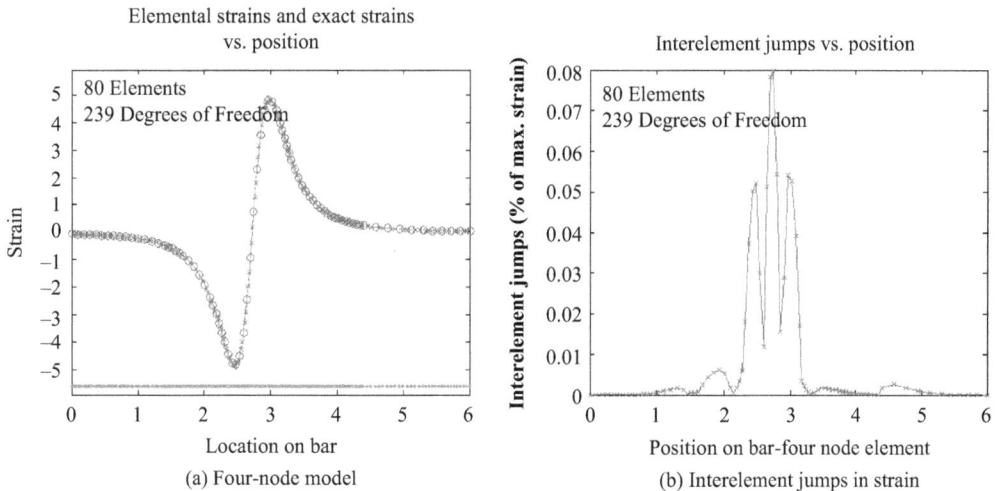

Figure 9.15. Fifth four-node model.

is still larger than the strain energy content for this four-node model. Thus, the five-element model is the better model from the strain energy point of view.

The fact that the two measures of accuracy are different for the two models deserves comment. The higher level of strain energy in the five-node model indicates that it has a smaller mean square error in the strain representation over the domain of the whole problem than does the four-node model. The higher maximum inner-element jump in strain for the five-node model means that the five-node model has a higher aggregated error in the point-wise satisfaction of the governing differential equations in two adjacent elements. As we have seen, the interelement jumps have the advantage over the differences in the overall strain energy as an error measure because the interelement jumps identify the locations of the discretization errors in the finite element model.

The participation factors for the four-node elements that are replacing the individual five-node elements containing the five critical points are presented in Table 9.8. In this table, the results for the five four-node elements that are replacing a single five-node element containing a critical point are separated by **bold lines.** The five-node elements that the five four-node elements replace are identified in the final column.

The more accurate results produced by this model can be attributed to the fact that the five four-node elements that approximate the change in curvature represented by the five-node element better

Table 9.8. Five, four node elements represent 1 five node element.

El. no.	$(\varepsilon_x)_0$	$(\varepsilon_{x,x})_0$	$(\varepsilon_{x,xx})_0$	
26	−2.5148	−6.5854	−15.6438	
27	−2.8074	−7.2331	−14.7510	
28	−3.1262	−7.8074	−10.0505	6
29	−3.4662	−8.2126	−6.5897	
30	−3.8176	−8.3068	2.8449	
31	−4.2235	−7.7273	21.0822	
32	−4.6204	−5.6301	53.6253	
33	−4.8363	−1.3023	97.8296	7
34	−4.7260	5.7123	144.4827	
35	−4.1390	14.9911	172.2478	
36	−3.0830	24.0706	157.2576	
37	−1.7001	30.5912	97.0559	
38	−0.0801	33.2036	4.5146	8
39	1.5509	31.0145	−89.6159	
40	2.9630	24.7757	−153.6088	

(*Continued on following page*)

Table 9.8. (*Continued*)

El. no.	$(\varepsilon_x)_0$	$(\varepsilon_{x,x})_0$	$(\varepsilon_{x,xx})_0$	
41	4.0676	15.6948	−172.7289	
42	4.6987	6.1747	−146.7328	
43	4.8304	−1.0783	−99.7004	9
44	4.6192	−5.5565	−54.5320	
45	4.2179	−7.7175	−21.2729	
46	3.8433	−8.2971	−3.9962	
47	3.5608	−8.2837	4.3014	
48	3.2832	−8.0382	9.7902	10
49	3.0166	−7.6443	13.1428	
50	2.7651	−7.1645	14.9342	

than the other four-node models. This can be seen by the fact that the curvatures contained in the five elements that replace each of the individual five-node elements have a different value. That is to say, the changing values of the curvature are approximating the change in curvature contained in the five-node element that the four-node elements replace.

9.7 SUMMARY, CONCLUSION, AND RECOMMENDATIONS FOR FUTURE WORK

This chapter has two objectives: (1) to identify the advantages of higher order finite elements and (2) to identify the way in which lower order elements approximate the modeling capabilities of higher order elements. These objectives have been satisfied by identifying and quantifying the way in which lower order elements compensate for their inability to represent the higher order strain states that are part of the basis function of higher order elements.

The manner in which a lower order element represents a higher order strain state is obvious from a **qualitative** point of view. A group of lower order elements will approximate a higher order strain state with a discrete representation. For example, a group of four-node elements will approximate the ability of a five-node element to represent the rate of change in curvature in the following way. Since a four-node element can represent curvature in the strain distribution, a group of four-node elements can approximate a change in curvature when the four-node elements represent monotonically changing values of curvature.

The open question is **quantitative.** How many four-node elements does it take to adequately represent the rate of change in curvature? The answer obviously depends on the magnitude of the change in curvature and on the degree of accuracy that is desired. If the termination criterion is measured in terms of interelement jumps in the strain, the number of lower order elements that must replace a higher order element can be approximated because of the Taylor series nature of the basis functions of the element.

For example, if the rate of change of curvature in a five-node element is known, the error that will exist in a four-node element of a given length can be estimated with the appropriate term in a Taylor series expansion. That means that the length of a four-node element that will not exceed the termination criterion can be estimated. This estimated length can be used to estimate the number of four-node elements needed to adequately approximate the rate of change in curvature that exists in the five-node element. Thus, the number of lower order elements needed to replace a higher order element is dependent on the size of the termination criterion.

The examples presented in the main text have shown that higher order elements are more efficient than lower order elements for representing complex strain distributions when the termination criterion is stringent. In Section 9.2, models of the control problem were adaptively refined with a termination criterion of 4.0% of the maximum absolute strain in the exact result. The number of degrees of freedom contained in the models formed from three-, four-, and five-node elements is approximately the same.

When the adaptive refinement procedure was repeated with the more stringent termination criterion of 0.4% in Section 9.3, the model formed with five-node elements had fewer degrees of freedom than the model formed with four-node elements. Similarly, the four-node elements formed a model that was more efficient than the model formed with three-node elements.

The higher efficiency of the five-node element was demonstrated in a different way in Section 9.5. In this demonstration, a highly efficient model of the control problem is formed that utilizes the full modeling capability of individual five-node elements in critical locations. Five-node elements are located so that a single element contains each of the five critical points that exist in the control problem. These critical points are shown in Fig. 9.9.

Each of these critical points is located on the domain of an individual five-node element. As a consequence, each of the five-node elements containing a critical point is required to represent the rate of change of curvature in the strain distribution to accurately represent the exact solution to this problem.

The modeling capabilities of four-node elements are compared to the five-node element by replacing the individual five-node elements with an increasing number of four-node elements. This procedure is continued until the four-node model has approximately the same accuracy as the five-node model. A model with approximately the same accuracy as the five-node model is produced when the individual five-node elements are replaced by five four-node elements.

That is to say, it takes a group of four-node elements with 16 nodes to replace a single higher order element with five nodes. This result demonstrates that higher order elements are more efficient than lower order elements in terms of degrees of freedom when the modeling capability of the higher order element is fully engaged.

An explanation for this difference in efficiency is easily constructed. Since the strain energy content for each of the four-node models is close to the five-node model (0.127% to 0.107%) while the maximum interelement jumps vary from approximately 3.3% to 0.08%, a different mechanism in estimating the error is implied when the exact solution is complex and the termination criterion is stringent. Instead of a least square error criterion (strain energy), the error estimator in this situation can be interpreted as measuring the aggregated point-wise in the strain distribution.

The procedures for comparing the modeling characteristics of the different order elements suggest an approach for forming a rational refinement guide. If an estimate of the participation of the higher order strain gradient components contained in an estimate of the exact solution can be extracted from the finite element solution, an estimate of the number of lower order elements needed to reduce the interelement jumps to an acceptable level can be made. That is to say, a refinement guide that is developed from first principles can be developed. Such a refinement guide is developed in Chapter 11.

CHAPTER 10

FORMULATION OF A **10-NODE** QUADRATIC STRAIN ELEMENT

10.1   INTRODUCTION

The stiffness matrix for a 10-node quadratic strain element, such as the one shown in Fig. 10.1, is formed using the physically interpretable strain gradient notation in this chapter. The objective of this chapter is to present the theoretical and computational advantages that accrue from introducing knowledge of continuum mechanics into the notation used to derive the stiffness matrices. The element formulation process is further explicated with two Appendices. The first Appendix presents the numerical results for the individual steps for an example 10-node finite element. The second Appendix presents annotated versions of the Matlab m-files that create the numerical results.

The development of a 10-node element is presented for three reasons: (1) A 10-node element is more efficient than lower order elements in representing complex strain distributions and stress concentrations in finite element models, (2) The development of this higher order element highlights

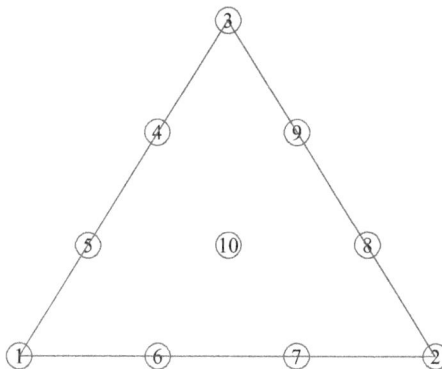

Figure 10.1. Element nodal numbering.

the advantages of using the physically interpretable strain gradient notation, and (3) The development provides the capabilities needed to extend all of the procedures presented here in one dimension to two dimensions.

Ten-node elements produce more efficient models than lower order elements because each element can represent strain distributions that are more complex than can be represented by lower order elements. The ability of a 10-node element to represent quadratic strain distributions means that a single element can capture maximum, minimum, and inflection points on its domain. In contrast, it takes several lower order elements to represent these extreme points.

The development of the 10-node element further demonstrates the fact that strain gradient notation embeds knowledge of continuum mechanics in the equations that surface during the development of an element. It is this notation that allows the modeling capabilities of an individual element to be identified by visual inspection during the formulation process. In contrast to the four-node element that contains several strain modeling errors, the 10-node element is seen to be capable of representing complete quadratic strain distributions, a capability that allows the 10-node element to represent the critical points mentioned earlier.

Furthermore, the role of the compatibility equations is clearly identified in the derivation of the strain models for the 10-node element. In the case of a 10-node planar element, there are 20 degrees of freedom. As we will see in a later Section, there are 21 terms in the complete strain representations for the three strain components. The fact that there are 21 strain states to be represented and only 20 degrees of freedom available in a 10-node element seems to contradict the idea that a 10-node element does not contain any strain modeling errors.

This difference in the number of degrees of freedom and the number of strain states required for complete quadratic strain representations does not identify a modeling deficiency. This difference highlights the role of the compatibility equations in continuum mechanics.

The compatibility equation for a planar continuum problem is given as the following, $\gamma_{xy,xy} = \varepsilon_{x,yy} + \varepsilon_{y,xx}$. That is to say, one of the quadratic variations in the shear strain, $\gamma_{xy,xy}$, is equal to the sum of two of the quadratic variations in the normal strains, $\varepsilon_{x,yy}$ and $\varepsilon_{y,xx}$. The compatibility equation is a constraint equation that exists because there are three strain components in the plane that are derived from only two independent displacement components.[1] The compatibility equation is utilized in Section 10.3 with Fig. 10.5 and Eq. 10.11 to form the strain representations for the 10-node element.

The developmental steps are presented in the following order:

1. The strain gradient displacement interpolation functions for a 10-node quadratic strain element are identified in terms of strain gradient quantities. (Eqs. 10.1 and 10.2)
2. The truncated Taylor series expansions for complete quadratic strain representations are presented. (Eqs. 10.6 through 10.8)
3. The differences between the strain gradient components of the quadratic strain representation and the required displacement interpolation functions are reconciled by identifying the role of the compatibility equation. (Eq. 10.11 and Fig. 10.5)

---

[1] This development also identifies a role for the higher order compatibility equations identified in Reference [1]. They would enter the development if a 15-node element were constructed.

4. The strain energy expression in strain gradient coordinates is formed. (Eq. 10.14 through 10.16)
5. The 15 integrals needed to form the strain energy expression for the element are identified and contrasted to the 210 integrals required for an equivalent isoparametric element.[2] (Section 10.5)
6. The strain energy expression is expanded and simplified. (Section 10.6 and Table 10.1)
7. The transformation from the strain gradient coordinates to the nodal displacement coordinates is formed. (Eq. 10.17 through 10.21)
8. The element stiffness matrix is formed. (Eqs. 10.22 through 10.24)

10.2   IDENTIFICATION OF THE LINEARLY INDEPENDENT STRAIN GRADIENT QUANTITIES

The individual terms that make up the two displacement interpolation polynomials for a 10-node quadratic strain element are identified with the artifact of a Pascal's triangle. These terms are identified at the nodes of a 10-node triangle shown in Fig. 10.2. As can be seen, the resulting terms form a complete third-order polynomial.[3]

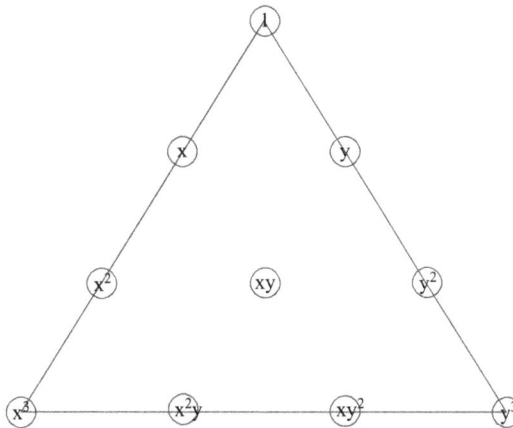

Figure 10.2. Polynomial terms for a 10-node triangle.

---

[2] The number of integrals for an isoparametric element is identified in the following way. The 20×20 stiffness matrix is symmetric. So there are a total of 190 off-diagonal terms, $(400 - 20)/2 = 380/2 = 190$. There are also 20 diagonal terms, so a total of $190 + 20 = 210$ integrals must be evaluated.

[3] It should be noted that the nodes of the 10-node finite element need not be located so they represent straight lines. The edge nodes can be located so they conform to a quadratic curve. The artifact of a Pascal's triangle is used because it produces polynomials that form coordinate transformations that are guaranteed to invert. See Reference [1] for further examples that demonstrate the value of this process.

The displacement interpolation polynomials for a 10-node triangle expressed in strain gradient components are the following [1]:

$$u(x,y) = (u_{rb})_0 + (\varepsilon_x)_0 x + [(\gamma_{xy}/2) - r_{rb}]_0 y + (\varepsilon_{x,x})_0 \frac{x^2}{2} + (\varepsilon_{x,y})_0 xy + (\gamma_{xy,y} - \varepsilon_{y,x})_0 \frac{y^2}{2}$$

$$+ (\varepsilon_{x,xx})_0 \frac{x^3}{6} + (\varepsilon_{x,xy})_0 \frac{x^2 y}{2} + (\varepsilon_{x,yy})_0 \frac{x y^2}{2} + (\gamma_{xy,yy} - \varepsilon_{y,xy})_0 \frac{y^3}{6}$$

(Eq. 10.1)

$$v(x,y) = (v_{rb})_0 + [(\gamma_{xy}/2) + r_{rb}]_0 x + (\varepsilon_y)_0 y + (\gamma_{xy,x} - \varepsilon_{x,y})_0 \frac{x^2}{2} + (\varepsilon_{x,y})_0 xy + (\varepsilon_{y,y})_0 \frac{y^2}{2}$$

$$+ (\gamma_{xy,xx} - \varepsilon_{x,xy})_0 \frac{x^3}{6} + (\varepsilon_{y,xx})_0 \frac{x^2 y}{2} + (\varepsilon_{y,xy})_0 \frac{x y^2}{2} + (\varepsilon_{y,yy})_0 \frac{y^3}{6}$$

The subscript zero indicates that the coefficients are evaluated at the local origin of the Taylor series expansion.

The 20 linearly independent strain gradient quantities contained in Eq. 10.1 correspond to the number of degrees of freedom contained in a 10-node element. These coefficients are the following:

$$
\begin{array}{cccccc}
 & & u_{rb} & v_{rb} & r_{rb} & \\
 & & \varepsilon_x & \varepsilon_y & \gamma_{xy} & \\
 \varepsilon_{x,x} & \varepsilon_{x,y} & \varepsilon_{y,x} & \varepsilon_{y,y} & \gamma_{xy,x} & \gamma_{xy,y} \\
 \varepsilon_{x,xx} & \varepsilon_{x,xy} & \varepsilon_{x,yy} & \varepsilon_{y,xx} & \varepsilon_{y,xy} & \varepsilon_{y,yy} & \gamma_{xy,xx} & \gamma_{xy,yy}
\end{array}
$$

(Eq. 10.2)

The three elements in the first row of Eq. 10.2 represent the three rigid body motions that a planar element can experience. The second row identifies the fact that a 10-node element can represent the three constant strain states. The ability of the 10-node element to represent these six quantities means that the element meets the convergence criteria for a finite element. In other words, if an element is reduced to an infinitesimal size, it can represent the rigid body motions and the strains at the individual point it is representing in the continuum.

The strain gradient terms contained in the third row indicate that the element can represent linear variations in the three strain components in the x and y directions. In other words, the contents of the first three rows indicate that a 10-node triangle can represent any strain distribution that a six-node element can model. The eight terms contained in the fourth row identify the additional strain modeling capability of the 10-node element. These capabilities will be discussed in detail in the next Section.

When the displacement interpolation functions given by Eqs. 10.1 are expressed in matrix form, we have the following:

$$\begin{Bmatrix} u(x,y) \\ v(x,y) \end{Bmatrix} = \begin{bmatrix} \left[T_{Disp} \right]_{RB} & \left[T_{Disp} \right]_6 & \left[T_{Disp} \right]_{10} \end{bmatrix} \begin{Bmatrix} \{\varepsilon,\}_{RB} \\ \{\varepsilon,\}_6 \\ \{\varepsilon,\}_{10} \end{Bmatrix}$$

(Eq. 10.3)

where

$$\{\varepsilon,\}_{RB} = [u_{RB} \quad v_{RB} \quad r_{RB}]^T$$

$$\{\varepsilon,\}_6 = [\varepsilon_x \quad \varepsilon_y \quad \gamma_{xy} \quad \varepsilon_{x,x} \quad \varepsilon_{y,x} \quad \gamma_{xy,x} \quad \varepsilon_{x,y} \quad \varepsilon_{y,y} \quad \gamma_{xy,y}]^T \qquad \text{(Eq. 10.4)}$$

$$\{\varepsilon,\}_{10} = [\varepsilon_{x,xx} \quad \varepsilon_{x,xy} \quad \varepsilon_{x,yy} \quad \varepsilon_{y,xx} \quad \varepsilon_{y,xy} \quad \varepsilon_{y,yy} \quad \gamma_{xy,xx} \quad \gamma_{xy,yy}]^T$$

and

$$[T_{Disp}]_{RB} = \begin{bmatrix} 1 & 0 & -y \\ 0 & 1 & x \end{bmatrix}$$

$$[T_{Disp}]_6 = \begin{bmatrix} x & 0 & y/2 & x^2/2 & -y^2/2 & 0 & xy & 0 & y^2/2 \\ 0 & y & x/2 & 0 & xy & x^2/2 & -x^2/2 & y^2/2 & 0 \end{bmatrix} \qquad \text{(Eq. 10.5)}$$

$$[T_{Disp}]_{10} = \begin{bmatrix} x^3/6 & x^2y/2 & xy^2/2 & 0 & -y^3/6 & 0 & 0 & y^3/6 \\ 0 & -x^3/6 & 0 & x^2y/2 & x^2y/2 & y^3/6 & x^3/6 & 0 \end{bmatrix}$$

The subscript "RB" indicates quantities associated with the rigid body displacements and rigid body rotation for a planar element. The subscript "6" indicates the strain states that produce deformations in a six-node linear strain element. The subscript "10" indicates the additional strain states that a 10-node finite element can represent.

10.3 IDENTIFICATION OF THE ELEMENTAL STRAIN MODELING CHARACTERISTICS

We will now finish identifying the strain modeling capabilities of the 10-node finite element. This will be accomplished by comparing the strain representation extracted from the displacement polynomials for the 10-node element to the equivalent Taylor series strain representation for these three components expressed in strain gradient notation. As we will see, the 10-node element represents the strains in the continuum with complete quadratic models of the three strain components.

The Taylor series expansions for the quadratic representations for the three strain components are the following:

$$\varepsilon_x(x,y) = (\varepsilon_x)_0 + (\varepsilon_{x,x})_0\, x + (\varepsilon_{x,y})_0\, y + (\varepsilon_{x,xx})_0\, x^2/2 + (\varepsilon_{x,xy})_0\, x\,y + (\varepsilon_{x,yy})_0\, y^2/2 \qquad \text{(Eq. 10.6)}$$

$$\varepsilon_y(x,y) = (\varepsilon_y)_0 + (\varepsilon_{y,x})_0\, x + (\varepsilon_{y,y})_0\, y + (\varepsilon_{y,xx})_0\, x^2/2 + (\varepsilon_{y,xy})_0\, x\,y + (\varepsilon_{y,yy})_0\, y^2/2 \qquad \text{(Eq. 10.7)}$$

$$\gamma_{xy}(x,y) = (\gamma_{xy})_0 + (\gamma_{xy,x})_0\, x + (\gamma_{xy,y})_0\, y + (\gamma_{xy,xx})_0\, x^2/2 + (\gamma_{xy,xy})_0\, x\,y + (\gamma_{xy,yy})_0\, y^2/2 \qquad \text{(Eq. 10.8)}$$

When the definition of the strain displacement relation for the normal strain in the x direction, $\varepsilon_x = du/dx$, is applied to the displacement interpolation polynomial in the x direction, the resulting strain representation is the following:

$$\varepsilon_x(x, y) = (\varepsilon_x)_0 + (\varepsilon_{x,x})_0\, x + (\varepsilon_{x,y})_0\, y + (\varepsilon_{x,xx})_0\, x^2/2 + (\varepsilon_{x,xy})_0\, x\, y + (\varepsilon_{x,yy})_0\, y^2/2 \quad \text{(Eq. 10.9)}$$

As can be seen, the strain representation for the normal strain in the x direction extracted from the displacement polynomial for the 10-node finite element matches the expected strain representation given by the Taylor series expansion shown in Eq. 10.6.

When the individual strain gradient terms are extracted from Eq. 10.9, they are as given in Fig. 10.3. As can be seen, these are the coefficients for the constant term, the two linear terms, and the coefficients for the three quadratic terms. That is to say, the 10-node element is capable of representing a complete quadratic model of the normal strains in the x direction.

$$\varepsilon_x$$
$$\varepsilon_{x,x} \quad \varepsilon_{x,y}$$
$$\varepsilon_{x,xx} \quad \varepsilon_{x,xy} \quad \varepsilon_{x,yy}$$

Figure 10.3. Strain gradient terms.

When the same process is applied to form the normal strain representation in the y direction, $\varepsilon_y = dv/dy$, the resulting strain representation is the following:

$$\varepsilon_y(x, y) = (\varepsilon_y)_0 + (\varepsilon_{y,x})_0\, x + (\varepsilon_{y,y})_0\, y + (\varepsilon_{y,xx})_0\, x^2/2 + (\varepsilon_{y,xy})_0\, x\, y + (\varepsilon_{y,yy})_0\, y^2/2 \quad \text{(Eq. 10.10)}$$

As can be seen, the strain representation for the normal strain in the y direction extracted from the displacement polynomial for the finite element matches the expected strain representation given by Eq. 10.7.

When the individual strain gradient terms are extracted from Eq. 10.10, they are as given in Fig. 10.4. As can be seen, these are the coefficients for the constant, the linear terms, and the quadratic terms. That is to say, the 10-node element can represent quadratic normal strains in the y direction.

We will now form the shear strain expression for the 10-node element from the displacement representations given by the components of Eq. 10.1. As we will see, the shear strain expression will **seem** to be incorrect until we apply knowledge from continuum mechanics. When we introduce the

$$\varepsilon_y$$
$$\varepsilon_{y,x} \quad \varepsilon_{y,y}$$
$$\varepsilon_{y,xx} \quad \varepsilon_{y,xy} \quad \varepsilon_{y,yy}$$

Figure 10.4. Strain gradient terms.

displacement interpolation functions into the definition of shear strain, $\gamma_{xy} = dv/dx + du/dy$, we get the following result:

$$\gamma_{xy}(x,y) = (\gamma_{xy})_0 + (\gamma_{xy,x})_0\, x + (\gamma_{xy,y})_0\, y + (\gamma_{xy,xx})_0\, x^2/2 + (\varepsilon_{x,yy} + \varepsilon_{y,xx})_0\, xy + (\gamma_{xy,yy})_0\, y^2/2 \quad \text{(Eq. 10.11)}$$

The strain gradient terms contained in Eq. 10.11 are presented in Fig. 10.5. When the terms contained in this shear strain expression for the 10-node element are compared to terms contained in the ideal Taylor series expression given by Eq. 10.8, we see that the coefficients of the xy terms in the two expressions do not match. As shown in Eq. 10.8 and Fig. 10.5, the expected term is $\gamma_{xy,xy}$. The xy term contained in Eq. 10.11 and shown in Fig. 10.5 is $\varepsilon_{x,yy} + \varepsilon_{y,xx}$.

$$\gamma_{xy}$$
$$\gamma_{xy,x} \quad \gamma_{xy,y}$$
$$\gamma_{xy,xx} \qquad \gamma_{xy,yy}$$
$$(\varepsilon_{x,yy} + \varepsilon_{y,xx})$$

Figure 10.5. Strain gradient terms.

This difference **seems** to exhibit both an error of omission and an error of commission. The absence of $\gamma_{xy,xy}$ in Eq. 10.11 **seems** to indicate an error of omission. The existence, in its place, of two normal shear terms **seems** to be an error of commission because the shear strain and the normal strains are not coupled in a typical material.

However, these two anomalies in the shear strain expression can be explained using knowledge of continuum mechanics. If the shear term, $\gamma_{xy,xy}$, was present in the shear strain expression, there would have been 21 independent variables in the basis set for the 10-node element. Since a 10-node element has only 20 degrees of freedom, this would be a problem. However, the seeming anomalies are easily explained.

When the coefficients for the xy term in the two shear expressions given by Eqs. 10.8 and 10.11 are equated, the result is the following: $\gamma_{xy,xy} = \varepsilon_{x,yy} + \varepsilon_{y,xx}$. This equation that relates the shear strain and the two normal strain components can be recognized as the compatibility equation for two-dimensional linear elasticity. This equation is a constraint equation. It exists because there are three strain components and only two displacement dimensions in planar problems.

Now that the existence of the normal strain terms in the shear strain expression has been explained, we can recognize that the shear strain in a 10-node element is a complete quadratic representation. The identification of this result is due to the knowledge of continuum mechanics that is embedded in the physically interpretable strain gradient notation.

The compatibility equations for two- and three-dimensional elasticity are developed in Reference [1] when the strain gradient coefficients are identified with a more formal process. In addition, it should be noted, because it is not commonly discussed, that higher order compatibility equations exist.

In two dimensions, the higher order compatibility equations consist of the two equations formed by taking the x and y derivatives of the compatibility equation just discussed. These higher order compatibility equations are constraint equations needed because there are more strain gradient terms in the complete strain representations than there are nodal degrees of freedom in a 15-node element. For example, they would surface if a 15-node planar finite element were developed. These higher order compatibility equations are discussed in Reference [1].

In order to prepare for forming the stiffness matrix for the 10-node element, we will assemble the three strain components given by Eqs. 10.9–10.11 in matrix form as:

$$
\begin{Bmatrix} \varepsilon_x \ (x,y) \\ \varepsilon_y \ (x,y) \\ \gamma_{xy} \ (x,y) \end{Bmatrix} = \begin{bmatrix} [T_{Strain}]_{RB} & [T_{Strain}]_6 & [T_{Strain}]_{10} \end{bmatrix} \begin{Bmatrix} \{\varepsilon, \}_{RB} \\ \{\varepsilon, \}_6 \\ \{\varepsilon, \}_{10} \end{Bmatrix} \qquad \text{(Eq. 10.12)}
$$

where

$$
[T_{Strain}]_{RB} = \begin{bmatrix} 0 & 0 & 0 \\ 0 & 0 & 0 \\ 0 & 0 & 0 \end{bmatrix}
$$

$$
[T_{Strain}]_6 = \begin{bmatrix} 1 & 0 & 0 & x & 0 & 0 & y & 0 & 0 \\ 0 & 1 & 0 & 0 & x & 0 & 0 & y & 0 \\ 0 & 0 & 1 & 0 & 0 & x & 0 & 0 & y \end{bmatrix}
$$

$$
[T_{Strain}]_{10} = \begin{bmatrix} x^2/2 & xy & y^2/2 & 0 & 0 & 0 & 0 & 0 \\ 0 & 0 & 0 & x^2/2 & xy & y^2/2 & 0 & 0 \\ 0 & 0 & xy & xy & -y^2/2 & 0 & x^2/2 & y^2/2 \end{bmatrix}
$$

The rigid body motions do not produce any strains by their very definitions. The rigid body motions are included in Eq. 10.12 in preparation for forming the element stiffness matrix in a later Section. We will see that the introduction of the null matrix associated with the rigid body motions in the strain representation given by Eq. 10.12 is instrumental in reducing the number of integrations that must be evaluated during the formulation of the stiffness matrix.

As we will see, any term that is related to a rigid body motion will produce no strain energy. This produces several terms that are equal to zero. These zero terms are identified as a result of the use of the physically interpretable notation.

10.4   FORMULATION OF THE STRAIN ENERGY EXPRESSION

The strain energy for a planar region experiencing plane stress in the continuum is given by the following integral expression:

$$
SE = 1/2 \int_\Omega \begin{Bmatrix} \varepsilon_x \ (x,y) \\ \varepsilon_y \ (x,y) \\ \gamma_{xy} \ (x,y) \end{Bmatrix}^T [E] \begin{Bmatrix} \varepsilon_x \ (x,y) \\ \varepsilon_y \ (x,y) \\ \gamma_{xy} \ (x,y) \end{Bmatrix} d\Omega \qquad \text{(Eq. 10.13)}
$$

where

$$[E] = \frac{E}{(1-v^2)} \begin{bmatrix} 1 & v & 0 \\ v & 1 & 0 \\ 0 & 0 & a \end{bmatrix}$$

$$a = \frac{1-v}{2}$$

The parameters E and v are Young's modulus and Poisson's ratio, respectively.

The strain energy expression for the 10-node element is formed by substituting Eq. 10.12 into Eq. 10.13. When this is done, the result is the following:

$$SE = 1/2 \begin{Bmatrix} \{\varepsilon,\}_{RB} \\ \{\varepsilon,\}_6 \\ \{\varepsilon,\}_{10} \end{Bmatrix}^T \int_\Omega \begin{Bmatrix} [T_{Strain}]_{RB} \\ [T_{Strain}]_6 \\ [T_{Strain}]_{10} \end{Bmatrix} [E] \begin{Bmatrix} [T_{Strain}]_{RB} \\ [T_{Strain}]_6 \\ [T_{Strain}]_{10} \end{Bmatrix}^T d\Omega \begin{Bmatrix} \{\varepsilon,\}_{RB} \\ \{\varepsilon,\}_6 \\ \{\varepsilon,\}_{10} \end{Bmatrix} \qquad \text{(Eq. 10.14)}$$

The strain gradient quantities are not included under the integral sign because they are not functions of x or y. They are the independent variables for the strain energy expression. It is in this step that the discrete approximation of the continuous problem is introduced.

When the integrand is expanded, the integrals in the strain energy expression become:

$$\bar{U} = \int_\Omega \begin{bmatrix} [T_{Strain}]_{RB}^T [E][T_{Strain}]_{RB} & [T_{Strain}]_{RB}^T [E][T_{Strain}]_6 & [T_{Strain}]_{RB}^T [E][T_{Strain}]_{10} \\ [T_{Strain}]_6^T [E][T_{Strain}]_{RB} & [T_{Strain}]_6^T [E][T_{Strain}]_6 & [T_{Strain}]_6^T [E][T_{Strain}]_{10} \\ [T_{Strain}]_{10}^T [E][T_{Strain}]_{RB} & [T_{Strain}]_{10}^T [E][T_{Strain}]_6 & [T_{Strain}]_{10}^T [E][T_{Strain}]_{10} \end{bmatrix} d\Omega \quad \text{(Eq. 10.15)}$$

This expression is not as imposing as it appears because of the use of the physically interpretable strain gradient notation. As defined in Eq. 10.12, the matrix $[T_{Strain}]_{RB}$ introduces the strains due to the rigid body motions. Since the rigid body motions, by definition, do not produce any deformations, the matrix $[T_{Strain}]_{RB}$ is the null matrix. Consequently, the first row and the first column of Eq. 10.15 contain only zeros. Thus, Eq. 10.15 reduces to the following:

$$\bar{U} = \int_\Omega \begin{bmatrix} 0 & 0 \\ 0 & [\bar{U}_{22}] \end{bmatrix} d\Omega \qquad \text{(Eq. 10.16)}$$

where

$$[\bar{U}_{22}] = \int_\Omega \begin{bmatrix} [T_{Strain}]_6^T [E][T_{Strain}]_6 & [T_{Strain}]_6^T [E][T_{Strain}]_{10} \\ [T_{Strain}]_{10}^T [E][T_{Strain}]_6 & [T_{Strain}]_{10}^T [E][T_{Strain}]_{10} \end{bmatrix} d\Omega$$

10.5   IDENTIFICATION AND EVALUATION OF THE REQUIRED INTEGRALS

When Eq. 10.16 is expanded, there are only 15 integrals that must be evaluated. This is the expected number because, as discussed earlier in Section 10.3, the three strain approximations are quadratic functions. When a quadratic expression is squared, the highest-order term formed is a fourth-order term.

The specific integrals that must be evaluated are easily identified using Pascal's triangle. The 15 relatively simple integrals that must be evaluated are the following:

$$I_1 = \int_\Omega \, dA$$

$$I_2 = \int_\Omega x \, dA \quad I_3 = \int_\Omega y \, dA$$

$$I_4 = \int_\Omega x^2 \, dA \quad I_5 = \int_\Omega xy \, dA \quad I_6 = \int_\Omega y^2 \, dA$$

$$I_7 = \int_\Omega x^3 \, dA \quad I_8 = \int_\Omega x^2 y \, dA \quad I_9 = \int_\Omega xy^2 \, dA \quad I_{10} = \int_\Omega y^3 \, dA$$

$$I_{11} = \int_\Omega x^4 \, dA \quad I_{12} = \int_\Omega x^3 y \, dA \quad I_{13} = \int_\Omega x^2 y^2 \, dA \quad I_{14} = \int_\Omega xy^3 \, dA \quad I_{15} = \int_\Omega y^4 \, dA$$

The first six integrals are the area, the two first moments of the area, and the three second moments of area. These are the same integrals that must be evaluated during the formulation of a six-node linear strain element.

The additional nine integrals exist due to the presence of the three quadratic terms, x^2, xy, and y^2, in the representations of the three strain components. Thus, the stiffness matrix for a 10-node element contains 15 integrals. It should be noted that this relatively small number of integrals contrasts to the 210 integrals that must be evaluated for a 10-node quadratic strain element formed using the isoparametric approach.[4]

In the example that is presented in Appendix 10A, these integrals are evaluated using Green's theorem. An exact integration scheme is used to emphasize that with the reduced number of simple integrals that must be evaluated an approximate integration procedure need not be used. Other exact or approximate integration procedures can be used. Green's theorem is chosen because of its elegance. Its use is demonstrated in the Appendices.

The use of Gauss quadrature is avoided in this work because it muddies the waters for readers new to finite element analysis. The use of Gauss quadrature introduces two detours that get in the way of the orderly flow of formulating the finite element stiffness matrix.

Integration using Gauss quadrature is a two-step process. The first detour consists of a mapping that takes the actual domain of the finite element onto a standard shape. The fact that this mapping

---

[4] The number of integrals for an isoparametric element is identified in the following way. The 20 × 20 stiffness matrix is symmetric. So there are a total of $(400 - 20)/2 = 380/2 = 190$ off-diagonal terms. There are then 20 diagonal elements, so a total of $190 + 20 = 210$ integrals must be evaluated.

has the same structure as the interpolation functions can be confusing to someone new to the process. The second detour consists of the form of the integral approximation itself. The integrals are approximated by evaluating the integrands at a set of Gauss points and summing weighted values of these integrands.[5] The rationale behind this approximation is not obvious without knowledge of the derivation of this process.

In addition to adding detours to the process of forming the stiffness matrices, Gauss quadrature has a significant drawback. As demonstrated in Chapter 5, the isoparametric mapping can introduce errors into the strain representations for the higher order strain states.[6] The rigid body motions and the constant strain states are represented accurately. As a result, the elements will converge in the limit, albeit inefficiently, as the model is refined. The very fact that unneeded errors can exist in isoparametric elements should identify the need for their replacement as the standard finite element formulation procedure.

10.6   EXPANSION OF THE STRAIN ENERGY KERNEL

When the quantities in the nonzero quadrant of Eq. 10.16 are expanded, the result for this 17×17 matrix expressed in terms of the 15 integrals identified in the previous section is shown in Table 10.1.

10.7   FORMULATION OF THE STIFFNESS MATRIX

The objective of this section is to generate the final form for the finite element stiffness matrix for the 10-node element. Since the strain energy expression contained in Eq. 10.14 is expressed in terms of strain gradient quantities, it must be transformed so the independent variables are nodal displacements. The displacements are required to be the independent variables so the elements can be assembled using the usual procedures.

This transformation to nodal displacements is a three-step process. First, the coordinate transformation from strain gradient coordinates to nodal displacements must be formed. Then, the strain energy expression must be transformed to the nodal displacements. Finally, the stiffness matrix can be extracted from the strain energy expression.

The required transformation matrix is formed using the displacement interpolation functions given by Eq. 10.3. The first step in forming the transformation is to evaluate Eq. 10.3 at each of the nodes of the 10-node element. This process can be likened to introducing the boundary conditions into an equation.

---

[5] The Gauss points are the zeros of the Chebyshev polynomial of the same order as the Gauss quadrature used to evaluate an integral.

[6] If an element has a nonstandard shape, a nonconstant Jacobian appears in the denominator of the strain representation. This produces a strain representation that is a rational polynomial, which can be interpreted as a polynomial with an infinite number of terms. This produces an inaccurate strain representation in the higher order strain states that exist in the element.

Table 10.1 The fourth quadrant of eq. 10.16 designated as u22

I_1	vI_1	0	I_2	vI_2	0	I_3	vI_3	0	$I_4/2$	I_5	$I_6/2$	$vI_4/2$	I_5	$vI_6/2$	0	0
vI_1	I_1	0	vI_2	I_2	0	vI_3	I_3	0	$vI_4/2$	I_5	$vI_6/2$	$I_4/2$	I_5	$I_6/2$	0	0
0	0	aI_1	0	0	aI_2	0	0	aI_3	I_5	0	I_5	I_5	0	0	$aI_4/2$	$aI_6/2$
I_2	vI_2	0	I_4	vI_4	0	I_5	I_5	0	$I_7/2$	I_8	$I_9/2$	$vI_7/2$	I_8	$vI_9/2$	0	0
vI_2	I_2	0	I_4	I_4	0	I_5	I_5	0	$vI_7/2$	vI_8	$vI_9/2$	$I_7/2$	I_8	$I_9/2$	0	0
0	0	aI_2	0	0	aI_4	0	0	0	0	0	aI_8	aI_8	0	0	$aI_7/2$	$aI_9/2$
I_3	vI_3	0	I_5	I_5	0	I_6	vI_6	0	$I_8/2$	I_9	$I_{10}/2$	$vI_8/2$	vI_9	$vI_{10}/2$	0	0
vI_3	I_3	0	I_5	I_5	0	vI_6	I_6	0	$vI_8/2$	vI_9	$vI_{10}/2$	$I_8/2$	I_9	$I_{10}/2$	0	0
0	0	aI_3	0	0	0	0	0	aI_6	0	0	aI_9	aI_9	0	0	$aI_8/2$	$aI_{10}/2$
$I_4/2$	$vI_4/2$	I_5	$I_7/2$	$vI_7/2$	0	$I_8/2$	$vI_8/2$	0	$I_{11}/4$	$I_{12}/2$	$I_{13}/4$	$vI_{11}/4$	$vI_{12}/2$	$vI_{13}/4$	0	0
I_5	I_5	0	I_8	vI_8	0	I_9	vI_9	0	$I_{12}/2$	I_{13}	$I_{14}/2$	$vI_{12}/2$	vI_{13}	$vI_{14}/2$	0	0
$I_6/2$	$vI_6/2$	I_5	$I_9/2$	$vI_9/2$	aI_8	$I_{10}/2$	$vI_{10}/2$	aI_9	$I_{13}/4$	$I_{14}/2$	$I_{15}/4+aI_{13}$	$vI_{13}/4+aI_{13}$	$I_{14}/2$	$vI_{15}/4$	$aI_{12}/2$	$aI_{14}/2$
$vI_4/2$	$I_4/2$	I_5	$vI_7/2$	$I_7/2$	aI_8	$vI_8/2$	$I_8/2$	aI_9	$vI_{11}/4$	$vI_{12}/2$	$(v/4+a)I_{13}$	$I_{11}/4+aI_{13}$	$I_{12}/2$	$I_{13}/4$	$aI_{12}/2$	$aI_{14}/2$
I_5	I_5	0	I_8	I_8	0	vI_9	I_9	0	$vI_{12}/2$	vI_{13}	$vI_{14}/2$	$I_{12}/2$	I_{13}	$I_{14}/2$	0	0
$vI_6/2$	$I_6/2$	0	$vI_9/2$	$I_9/2$	0	$vI_{10}/2$	$I_{10}/2$	0	$vI_{13}/4$	$vI_{14}/2$	$vI_{15}/4$	$I_{13}/4$	$I_{14}/2$	$I_{15}/4$	0	0
0	0	$aI_4/2$	0	0	$aI_7/2$	0	0	$aI_8/2$	0	0	$aI_{12}/2$	$aI_{12}/2$	0	0	$aI_{11}/4$	$aI_{13}/4$
0	0	$aI_6/2$	0	0	$aI_9/2$	0	0	$aI_{10}/2$	0	0	$aI_{14}/2$	$aI_{14}/2$	0	0	$aI_{13}/4$	$aI_{15}/4$

In order to illustrate this process, the displacement interpolation polynomials given by Eq. 10.3 are evaluated at the ith node. When the nodal coordinates for the ith node are substituted into Eq 10.3, the displacements at the ith node are given as:

$$\begin{Bmatrix} u_i \\ v_i \end{Bmatrix} = \begin{bmatrix} \left[T_{Disp} \right]_{RB} & \left[T_{Disp} \right]_6 & \left[T_{Disp} \right]_{10} \end{bmatrix}_i \begin{Bmatrix} \{\varepsilon,\}_{RB} \\ \{\varepsilon,\}_6 \\ \{\varepsilon,\}_{10} \end{Bmatrix} \qquad \text{(Eq. 10.17)}$$

where

$$\left[T_{Disp} \right]_{RB_i} = \begin{bmatrix} 1 & 0 & -y_i \\ 0 & 1 & x_i \end{bmatrix}$$

$$\left[T_{Disp} \right]_{6_i} = \begin{bmatrix} x_i & 0 & y_i/2 & x_i^2/2 & -y_i^2/2 & 0 & x_i y_i & 0 & y_i^2/2 \\ 0 & y_i & x_i/2 & 0 & x_i y_i & x_i^2/2 & -x_i^2/2 & y_i^2/2 & 0 \end{bmatrix} \qquad \text{(Eq. 10.18)}$$

$$\left[T_{Disp} \right]_{10_i} = \begin{bmatrix} x_i^3/6 & x_i^2 y/2 & xy_i^2/2 & 0 & -y_i^3/6 & 0 & 0 & y_i^3/6 \\ 0 & -x_i^3/6 & 0 & x_i^2 y/2 & x_i^2 y/2 & y_i^3/6 & x_i^3/6 & 0 \end{bmatrix}$$

and

$$\{\varepsilon,\}_{RB} = \begin{bmatrix} u_{RB} & v_{RB} & r_{RB} \end{bmatrix}^T$$

$$\{\varepsilon,\}_6 = \begin{bmatrix} \varepsilon_x & \varepsilon_y & \gamma_{xy} & \varepsilon_{x,x} & \varepsilon_{y,x} & \gamma_{xy,x} & \varepsilon_{x,y} & \varepsilon_{y,y} & \gamma_{xy,y} \end{bmatrix}^T \qquad \text{(Eq. 10.19)}$$

$$\{\varepsilon,\}_{10} = \begin{bmatrix} \varepsilon_{x,xx} & \varepsilon_{x,xy} & \varepsilon_{x,yy} & \varepsilon_{y,xx} & \varepsilon_{y,xy} & \varepsilon_{y,yy} & \gamma_{xy,xx} & \gamma_{xy,yy} \end{bmatrix}^T$$

The subscript "i" indicates that the quantity is associated with the ith node. The subscript "RB" indicates quantities associated with the rigid body displacements and rigid body rotation for a planar element. The subscript "6" indicates the strain states that produce deformations in a six-node linear strain element. The subscript "10" indicates the additional strain states that a 10-node finite element can represent.

When the 10 pairs of equations that result from substituting the nodal locations for the 10 nodes into Eq. 10.3 are assembled into a matrix, the result is the following:

$$\{u\} = [\Phi]\{\varepsilon,\} \qquad \text{(Eq. 10.20)}$$

where

$$\{u\}^T = \begin{bmatrix} u_1 & v_1 & u_2 & v_2 & u_3 & v_3 & u_4 & v_4 & u_5 & v_5 & u_6 & v_6 & u_7 & v_7 & u_8 & v_8 & u_9 & v_9 & u_{10} & v_{10} \end{bmatrix}$$

$$[\Phi] = \begin{bmatrix} \left[[T_{Disp}]_{RB} \quad [T_{Disp}]_6 \quad [T_{Disp}]_{10} \right]_1 \\ \left[[T_{Disp}]_{RB} \quad [T_{Disp}]_6 \quad [T_{Disp}]_{10} \right]_2 \\ \left[[T_{Disp}]_{RB} \quad [T_{Disp}]_6 \quad [T_{Disp}]_{10} \right]_3 \\ \left[[T_{Disp}]_{RB} \quad [T_{Disp}]_6 \quad [T_{Disp}]_{10} \right]_4 \\ \left[[T_{Disp}]_{RB} \quad [T_{Disp}]_6 \quad [T_{Disp}]_{10} \right]_5 \\ \left[[T_{Disp}]_{RB} \quad [T_{Disp}]_6 \quad [T_{Disp}]_{10} \right]_6 \\ \left[[T_{Disp}]_{RB} \quad [T_{Disp}]_6 \quad [T_{Disp}]_{10} \right]_7 \\ \left[[T_{Disp}]_{RB} \quad [T_{Disp}]_6 \quad [T_{Disp}]_{10} \right]_8 \\ \left[[T_{Disp}]_{RB} \quad [T_{Disp}]_6 \quad [T_{Disp}]_{10} \right]_9 \\ \left[[T_{Disp}]_{RB} \quad [T_{Disp}]_6 \quad [T_{Disp}]_{10} \right]_{10} \end{bmatrix}$$

$$\{\varepsilon,\} = \begin{Bmatrix} \{\varepsilon,\}_{RB} \\ \{\varepsilon,\}_6 \\ \{\varepsilon,\}_{10} \end{Bmatrix}$$

This expression is the transformation from the nodal coordinates to the strain gradient coordinates. In order to get the desired coordinate transformation from strain gradient coordinates to nodal coordinates, Eq. 10.20 must be inverted. When this is done we have the following:

$$\{\varepsilon,\} = [\Phi]^{-1}\{u\} \qquad \text{(Eq. 10.21)}$$

We are now in a position to form the stiffness matrix for the 10-node element. When Eq. 10.16 is substituted into Eq. 10.14, the strain energy for the 10-node element expressed in terms of strain gradient coordinates is the following:

$$SE = 1/2\{\varepsilon,\}^T [\bar{U}]\{\varepsilon,\} \qquad \text{(Eq. 10.22)}$$

This strain energy expression is transformed to nodal coordinates with the application of Eq. 10.21. When this transformation is applied, the result is the following:

$$SE = 1/2\{u\}^T [\Phi]^{-T} [\bar{U}][\Phi]^{-1}\{u\} \qquad \text{(Eq. 10.23)}$$

When the principle of minimum potential energy is applied as done in Reference [1], the stiffness matrix for the 10-node finite element is found to be the following:

$$[K]_{10\,Node} = [\Phi]^{-T} [\bar{U}][\Phi]^{-1} \qquad \text{(Eq. 10.24)}$$

10.8   SUMMARY AND CONCLUSION

The development of the 10-node element highlights all of the advantages that are provided by the physically interpretable strain gradient notation. The evaluation of the strain representations by visual inspections validates that the element can model complete quadratic strain distributions. As expected, the number of integrals that must be evaluated during the formulation of the stiffness matrix is significantly reduced as compared to the isoparametric formulation procedure. Only 15 integrals must be evaluated instead of the 210 that surface in an isoparametric element.

In addition to the pedagogical advantages that accrue from the elimination of the isoparametric mapping and of the use of Gauss quadrature integration, the elimination of the isoparametric mapping means that no errors are induced by the mapping in the linear and quadratic strain states if an element contains a curved edge as is the case for an isoparametric 10-node element. In fact, the lack of elegance that exists because of the very possibility of these induced errors seems to disqualify the use of isoparametric elements since an improved formulation procedure exists.

Finally, the evaluation of the shear strain representation in the element provides a concrete embodiment of the role of the compatibility equation as a constraint equation. The fact that a 10-node strain gradient element provides a complete quadratic representation of the three strain components is not obvious at first glance. This is because of the presence of the normal strain components, $\varepsilon_{x,yy}$ and $\varepsilon_{y,xx}$ in the shear strain representation.

A question concerning the accuracy of the strain representations in the 10-node element might surface because of the presence of these normal strain components in the shear strain expression. One might think that the unexpected presence of these normal strain terms in the shear strain expression identifies a modeling error, as was the case for the four-node quadrilateral element. However, this supposition would be wrong.

In the case of the 10-node element, the existence of the normal strain terms in the shear strain expression is the functional equivalent of the compatibility equation for planar elasticity. The two normal strain components in the shear strain representation are $\varepsilon_{x,yy}$ and $\varepsilon_{y,xx}$. They replace the shear strain component, $\gamma_{xy,xy}$. That is to say, the presence of these two normal strain terms are the concrete embodiment of the compatibility equation, $\gamma_{xy,xy} = \varepsilon_{x,yy} + \varepsilon_{y,xx}$.

The development of the 10-node element is not presented as an academic exercise to showcase strain gradient notation. This element has significant practical applications. Since this element can represent extreme points on its domain, it takes fewer elements to represent critical points in a finite element model. That is to say, this quadratic strain element is more efficient than lower order elements in representing complex strain distributions. Furthermore, as was seen in Chapter 9, the use of a quadratic strain element in the one-dimensional case reduced the number of adaptive refinement iterations needed to achieve the level of accuracy specified in the adaptive refinement process.

In summary, the 10-node element is recommended for use for the following reasons:

1. The element is computationally efficient.
2. The element can represent complex strain distribution better than lower order elements.
3. The efficient modeling capability of this element can reduce the number of iterations of the adaptive refinement process to achieve the desired level of accuracy.
4. The element contains no induced errors if it has a curved edge.

APPENDIX 10A

A NUMERICAL EXAMPLE FOR A 10-NODE STIFFNESS MATRIX

10A.1   INTRODUCTION

The procedure for formulating the stiffness matrix for 10-node elements developed in the main text is demonstrated in this Appendix with a numerical example. This presentation couples the equations presented in the main body of Chapter 10 for computing the stiffness matrix with the **output** of the Matlab programs contained in Appendix 10B that perform the requisite computations.

The objectives of these two Appendices are threefold. Their prime goal is to provide a detailed presentation of the element formulation procedure and to demonstrate its straightforward nature. The second objective is to provide detailed results that will allow the computer implementation to be checked if anyone desires to do so. The third objective is to provide the capabilities needed to extend all of the procedures presented in this book in one dimension to two dimensions.

The stiffness matrix for the 10-node finite element is given in the main text as Eq. 10.24. This equation is reproduced here for the convenience of the reader as:

$$[K]_{10\text{Node}} = [\Phi]^{-T} [\bar{U}][\Phi]^{-1} \qquad \text{(Eq. 10A.1)}$$

After the geometry of the element being used in this example is presented, the individual components of Eq. 10A.1 are developed in the sections that follow. Finally, the component matrices will be combined according to Eq. 10A.1 to form the stiffness matrix for this example of a 10-node element.

10A.2   ELEMENT GEOMETRY AND NODAL NUMBERING

The geometry and the nodal numbering of the element used in this example are shown in Fig. 10A.1. The node in the lower left is identified as Node 1. Then, the nodes on the vertices or the corners of the element are numbered consecutively in the counter-clockwise direction as Nodes 2 and 3. Node number 10 is located at the centroid of the triangle and serves as the local origin for the element used in this example. The interior edge nodes are numbered as shown.

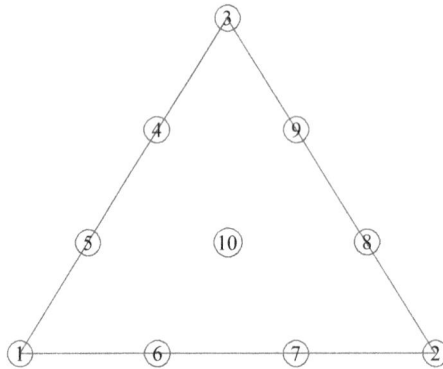

Figure 10A.1. Element nodal numbering.

The locations of the individual nodes are given in the local or elemental coordinate system in Table 10A.1. This isosceles triangle has a base of three units and a height of three units.

Table 10A.1. Element nodal locations

Node no.	1	2	3	4	5	6	7	8	9	10
X Coordinate	−1.5	1.5	0.0	−0.5	−1.0	−0.5	0.5	1.0	0.5	0.0
Y Coordinate	−1.0	−1.0	2.0	1.0	0.0	−1.0	−1.0	0.0	1.0	0.0

10A.3   FORMULATION OF THE TRANSFORMATION TO NODAL DISPLACEMENT COORDINATES

The heart of the finite element method lies in the ability to approximate the displacements on a domain in terms of a finite number of nodal displacements that serve as the independent variables for the element. The interpolation functions with physically interpretable coefficients used in this development serve two critical roles. In one role, these functions are used to form the transformation from strain gradient coordinates to nodal displacements. This transformation, designated as [Φ], is formed in this section.

In the second role, the interpolation functions provide the basis for forming the strain representations that approximate the exact solution in the finite element. The strain representations are then used to form a strain energy expression in terms of strain gradient coordinates. These roles are explicated in the next Section.

The displacement interpolation functions for the 10-node element are presented in the main text as Eqs. 10.3–10.5. These equations are reproduced here for the convenience of the reader as:

$$\begin{Bmatrix} u(x,y) \\ v(x,y) \end{Bmatrix} = \begin{bmatrix} \begin{bmatrix} T_{Disp} \end{bmatrix}_{RB} & \begin{bmatrix} T_{Disp} \end{bmatrix}_{6} & \begin{bmatrix} T_{Disp} \end{bmatrix}_{10} \end{bmatrix} \begin{Bmatrix} \{\varepsilon,\}_{RB} \\ \{\varepsilon,\}_{6} \\ \{\varepsilon,\}_{10} \end{Bmatrix} \qquad \text{(Eq. 10A.2)}$$

where

$$\begin{aligned} \{\varepsilon,\}_{RB} &= \begin{bmatrix} u_{RB} & v_{RB} & r_{RB} \end{bmatrix}^{T} \\ \{\varepsilon,\}_{6} &= \begin{bmatrix} \varepsilon_x & \varepsilon_y & \gamma_{xy} & \varepsilon_{x,x} & \varepsilon_{y,x} & \gamma_{xy,x} & \varepsilon_{x,y} & \varepsilon_{y,y} & \gamma_{xy,y} \end{bmatrix}^{T} \\ \{\varepsilon,\}_{10} &= \begin{bmatrix} \varepsilon_{x,xx} & \varepsilon_{x,xy} & \varepsilon_{x,yy} & \varepsilon_{y,xx} & \varepsilon_{y,xy} & \varepsilon_{y,yy} & \gamma_{xy,xx} & \gamma_{xy,yy} \end{bmatrix}^{T} \end{aligned} \qquad \text{(Eq. 10A.3)}$$

and

$$\begin{aligned} \begin{bmatrix} T_{Disp} \end{bmatrix}_{RB} &= \begin{bmatrix} 1 & 0 & -y \\ 0 & 1 & x \end{bmatrix} \\ \begin{bmatrix} T_{Disp} \end{bmatrix}_{6} &= \begin{bmatrix} x & 0 & y/2 & x^2/2 & -y^2/2 & 0 & xy & 0 & y^2/2 \\ 0 & y & x/2 & 0 & xy & x^2/2 & -x^2/2 & y^2/2 & 0 \end{bmatrix} \\ \begin{bmatrix} T_{Disp} \end{bmatrix}_{10} &= \begin{bmatrix} x^3/6 & x^2y/2 & xy^2/2 & 0 & -y^3/6 & 0 & 0 & y^3/6 \\ 0 & -x^3/6 & 0 & x^2y/2 & x^2y/2 & y^3/6 & x^3/6 & 0 \end{bmatrix} \end{aligned} \qquad \text{(Eq. 10A.4)}$$

The subscript "RB" indicates quantities associated with the rigid body displacements and rigid body rotation for a planar element. The subscript "6" indicates the strain states that produce deformations in a six-node linear strain element. The subscript "10" indicates the additional strain states that a 10-node finite element can represent.

The transformation from nodal displacements to strain gradient coordinates is formed by evaluating Eq. 10A.2 at each of the nodes of the 10-node element. This process can be likened to introducing the boundary conditions into an equation. Then, the 10 sets of equations are combined to form the transformation matrix.

In order to illustrate this process, the displacement interpolation polynomials given by Eq. 10A.2 are evaluated for Node 1. When the nodal coordinates of Node 1 are substituted into Eq. 10A.2, the displacements at Node 1 are given in terms of the strain gradient quantities as:

$$\left\{ \begin{matrix} u_1 \\ v_1 \end{matrix} \right\} = \left[\left[T_{Disp} \right]_{RB} \quad \left[T_{Disp} \right]_6 \quad \left[T_{Disp} \right]_{10} \right]_1 \left\{ \begin{matrix} \{\varepsilon,\}_{RB} \\ \{\varepsilon,\}_6 \\ \{\varepsilon,\}_{10} \end{matrix} \right\} \tag{Eq. 10A.5}$$

where

$$\left[T_{Disp} \right]_{RB1} = \begin{bmatrix} 1 & 0 & -y_1 \\ 0 & 1 & x_1 \end{bmatrix}$$

$$\left[T_{Disp} \right]_{61} = \begin{bmatrix} x_1 & 0 & y_1/2 & x_1^2/2 & -y_1^2/2 & 0 & x_1 y_1 & 0 & y_1^2/2 \\ 0 & y_1 & x_1/2 & 0 & x_1 y_1 & x_1^2/2 & -x_1^2/2 & y_1^2/2 & 0 \end{bmatrix}$$

$$\left[T_{Disp} \right]_{10_1} = \begin{bmatrix} x_1^3/6 & x_1^2 y_1/2 & x_1 y_1^2/2 & 0 & -y_1^3/6 & 0 & 0 & y_1^3/6 \\ 0 & -x_1^3/6 & 0 & x_1^2 y_1/2 & x_1^2 y_1/2 & y_1^3/6 & x_1^3/6 & 0 \end{bmatrix}$$

and

$$\left\{ \begin{matrix} x_1 \\ y_1 \end{matrix} \right\} = \left\{ \begin{matrix} -1.5 \\ -1.0 \end{matrix} \right\}$$

The subscript "1" on the displacement vector and the coefficient matrix indicates that these quantities are associated with the Node 1.

When the 10 pairs of equations that result from substituting the nodal locations for the 10 nodes into Eq. 10A.2 are assembled into a matrix, the result is the following:

$$\{u\} = [\Phi]\{\varepsilon,\} \tag{Eq. 10A.6}$$

where

$$\{u\}^T = [u_1 \ v_1 \ u_2 \ v_2 \ u_3 \ v_3 \ u_4 \ v_4 \ u_5 \ v_5 \ u_6 \ v_6 \ u_7 \ v_7 \ u_8 \ v_8 \ u_9 \ v_9 \ u_{10} \ v_{10}]$$

$$[\Phi] = \begin{bmatrix} \left[[T_{Disp}]_{RB} & [T_{Disp}]_6 & [T_{Disp}]_{10} \right]_1 \\ \left[[T_{Disp}]_{RB} & [T_{Disp}]_6 & [T_{Disp}]_{10} \right]_2 \\ \left[[T_{Disp}]_{RB} & [T_{Disp}]_6 & [T_{Disp}]_{10} \right]_3 \\ \left[[T_{Disp}]_{RB} & [T_{Disp}]_6 & [T_{Disp}]_{10} \right]_4 \\ \left[[T_{Disp}]_{RB} & [T_{Disp}]_6 & [T_{Disp}]_{10} \right]_5 \\ \left[[T_{Disp}]_{RB} & [T_{Disp}]_6 & [T_{Disp}]_{10} \right]_6 \\ \left[[T_{Disp}]_{RB} & [T_{Disp}]_6 & [T_{Disp}]_{10} \right]_7 \\ \left[[T_{Disp}]_{RB} & [T_{Disp}]_6 & [T_{Disp}]_{10} \right]_8 \\ \left[[T_{Disp}]_{RB} & [T_{Disp}]_6 & [T_{Disp}]_{10} \right]_9 \\ \left[[T_{Disp}]_{RB} & [T_{Disp}]_6 & [T_{Disp}]_{10} \right]_{10} \end{bmatrix}$$

$$\{\varepsilon,\} = \begin{Bmatrix} \{\varepsilon,\}_{RB} \\ \{\varepsilon,\}_6 \\ \{\varepsilon,\}_{10} \end{Bmatrix}$$

This expression is the transformation from the nodal coordinates to the strain gradient coordinates. The transformation matrix $[\Phi]$ given by Eq. 10A.6 for the example being used here is presented in Table 10A.2.

As shown in Eq. 10A.1 and as we will see in the next section, the transformation from strain gradient coordinates to nodal coordinates is needed to form the finite element stiffness matrix. This transformation is formed by inverting Eq. 10A.6 to give the following:

$$\{\varepsilon,\} = [\Phi]^{-1}\{u\} \tag{Eq. 10A.7}$$

When this matrix is formed for the example problem using Matlab, the result is presented in Table 10A.3.[7]

It should be noted that it is at this point that the finite element and the finite difference methods are unified. For example, the fourth row of the $[\Phi]^{-1}$ matrix expresses the strain in the x direction, ε_x.

---

[7] This matrix is formed in the m-file entitled FormPhi that is presented in Appendix 10B. The various matrices contained in this Appendix are presented so this result can be checked independently.

Table 10A.2. The phi matrix for the example problem

Row no.	u_{rb}	v_{rb}	r_{rb}	ε_x	ε_y	γ_{xy}	$\varepsilon_{x,x}$	$\varepsilon_{y,x}$	$\gamma_{xy,x}$	$\varepsilon_{x,y}$	$\varepsilon_{y,y}$	$\gamma_{xy,y}$	$\varepsilon_{x,xx}$	$\varepsilon_{x,xy}$	$\varepsilon_{x,yy}$	$\varepsilon_{y,xx}$	$\varepsilon_{y,xy}$	$\varepsilon_{y,yy}$	$\gamma_{xy,xx}$	$\gamma_{xy,yy}$
1	1.00	0.00	1.00	-1.50	0.00	-0.50	1.13	-0.50	0.00	1.50	0.00	0.50	-0.56	-1.13	-0.75	0.00	0.17	0.00	0.00	-0.17
2	0.00	1.00	-1.50	0.00	-1.00	-0.75	0.00	1.50	1.13	-1.13	0.50	0.00	0.00	0.56	0.00	-1.13	-0.75	-0.17	-0.56	0.00
3	1.00	0.00	1.00	1.50	0.00	-0.50	1.13	-0.50	0.00	-1.50	0.00	0.50	0.56	-1.13	0.75	0.00	0.17	0.00	0.00	-0.17
4	0.00	1.00	1.50	0.00	-1.00	0.75	0.00	-1.50	1.13	-1.13	0.50	0.00	0.00	-0.56	0.00	-1.13	0.75	-0.17	0.56	0.00
5	1.00	0.00	-2.00	0.00	0.00	1.00	0.00	-2.00	0.00	0.00	0.00	2.00	0.00	0.00	0.00	0.00	-1.33	0.00	0.00	1.33
6	0.00	1.00	0.00	0.00	2.00	0.00	0.00	0.00	0.00	0.00	2.00	0.00	0.00	0.00	0.00	0.00	0.00	1.33	0.00	0.00
7	1.00	0.00	-1.00	-0.50	0.00	0.50	0.13	-0.50	0.00	-0.50	0.00	0.50	-0.02	0.13	-0.25	0.00	-0.17	0.00	0.00	0.17
8	0.00	1.00	-0.50	0.00	1.00	-0.25	0.00	0.00	0.13	-0.13	0.50	0.00	0.00	0.02	0.00	0.13	-0.25	0.17	-0.02	0.00
9	1.00	0.00	0.00	-1.00	0.00	0.00	0.50	0.00	0.00	-0.50	0.00	0.00	-0.17	0.17	0.00	0.00	0.00	0.00	0.00	0.00
10	0.00	1.00	-1.00	0.00	1.00	-0.50	0.00	0.00	0.50	0.00	0.00	0.00	0.00	-0.13	0.00	0.00	0.00	0.00	-0.17	0.00
11	1.00	0.00	1.00	-0.50	0.00	-0.50	0.13	-0.50	0.00	-0.50	0.00	0.50	-0.02	0.02	-0.25	0.00	0.17	0.00	0.00	-0.17
12	0.00	1.00	-0.50	0.00	-1.00	-0.25	0.00	0.50	0.13	-0.13	0.50	0.00	0.00	-0.13	0.00	-0.13	-0.25	-0.17	-0.02	0.00
13	1.00	0.00	1.00	0.50	0.00	-0.50	0.13	-0.50	0.00	-0.50	0.00	0.50	0.02	-0.02	0.25	0.00	0.17	0.00	0.00	-0.17
14	0.00	1.00	0.50	0.00	-1.00	0.25	0.00	-0.50	0.13	-0.13	0.50	0.00	0.00	0.00	0.00	-0.13	0.25	-0.17	0.02	0.00
15	1.00	0.00	0.00	1.00	0.00	0.00	0.50	0.00	0.00	0.00	0.00	0.00	0.17	-0.17	0.00	0.00	0.00	0.00	0.00	0.00
16	0.00	1.00	1.00	0.00	0.00	0.50	0.00	0.00	0.50	0.00	0.00	0.00	0.00	0.13	0.00	0.00	0.00	0.00	0.17	0.00
17	1.00	0.00	-1.00	0.50	0.00	0.50	0.13	-0.50	0.00	-0.50	0.00	0.50	0.02	-0.17	0.25	0.00	-0.17	0.00	0.00	0.17
18	0.00	1.00	0.50	0.00	1.00	0.25	0.00	0.50	0.13	-0.13	0.50	0.00	0.00	0.13	0.00	0.13	0.25	0.17	0.02	0.00
19	1.00	0.00	0.00	0.00	0.00	0.00	0.00	0.00	0.00	0.00	0.00	0.00	0.00	0.00	0.00	0.00	0.00	0.00	0.00	0.00
20	0.00	1.00	0.00	0.00	0.00	0.00	0.00	0.00	0.00	0.00	0.00	0.00	0.00	0.00	0.00	0.00	0.00	0.00	0.00	0.00

Table 10A.3. The inverse of the phi matrix for the example problem

Row no.	u₁	v₁	u₂	v₂	u₃	v₃	u₄	v₄	u₅	v₅	u₆	v₆	u₇	v₇	u₈	v₈	u₉	v₉	u₁₀	v₁₀
1	0.00	0.00	0.00	0.00	0.00	0.00	0.00	0.00	0.00	0.00	0.00	0.00	0.00	0.00	0.00	0.00	0.00	0.00	1.00	0.00
2	0.00	0.00	0.00	0.00	0.00	0.00	0.00	0.00	0.00	0.00	0.00	0.00	0.00	0.00	0.00	0.00	0.00	0.00	0.00	1.00
3	-0.04	0.08	-0.04	-0.08	0.08	-0.00	-0.25	0.00	0.12	-0.25	0.12	-0.25	0.13	0.25	0.13	0.25	-0.25	0.00	-0.00	-0.00
4	0.17	0.00	-0.17	-0.00	0.00	0.00	0.00	-0.00	-0.50	0.00	-0.50	0.00	0.50	0.00	0.50	0.00	0.00	0.00	0.00	0.00
5	-0.00	0.08	0.00	0.08	0.00	-0.17	-0.00	0.50	0.00	-0.25	-0.00	-0.25	-0.00	-0.25	0.00	-0.25	-0.00	0.50	0.00	0.00
6	0.08	0.17	0.08	-0.17	-0.17	-0.00	0.50	0.00	-0.25	-0.50	-0.25	-0.50	-0.25	0.50	-0.25	0.50	0.50	0.00	0.00	-0.00
7	-0.00	-0.00	0.00	-0.00	-0.00	-0.00	0.00	0.00	1.00	0.00	-0.00	-0.00	0.00	0.00	1.00	0.00	0.00	0.00	-2.00	0.00
8	-0.00	-0.00	-0.00	0.00	-0.00	0.00	0.00	0.00	0.00	0.00	0.00	0.00	0.00	-0.50	0.00	0.00	0.00	0.00	0.00	0.00
9	0.00	0.00	0.00	0.00	-0.00	-0.00	-0.50	0.00	-0.00	1.00	0.50	0.50	-0.50	0.00	-0.00	1.00	0.50	0.00	0.00	-2.00
10	0.00	-0.00	0.00	0.00	0.00	0.00	-0.50	0.00	0.00	0.00	0.50	-0.00	-0.50	-0.00	0.00	0.00	0.50	0.00	-0.00	0.00
11	-0.00	-0.00	0.00	0.00	-0.00	0.00	-0.00	0.00	0.00	-0.25	0.00	0.00	0.00	0.50	0.00	-0.25	-0.00	0.50	-0.00	-1.50
12	0.00	0.00	0.00	-0.00	-0.00	-0.00	0.50	0.50	-0.25	-0.00	0.50	0.50	0.50	-0.50	-0.25	0.00	0.50	0.50	0.00	0.00
13	-1.00	0.00	1.00	0.00	0.00	0.00	0.00	0.00	0.00	0.00	3.00	0.00	-3.00	0.00	0.00	0.00	0.00	0.00	0.00	0.00
14	-0.50	-0.00	-0.50	-0.00	-0.00	0.00	0.00	0.00	1.00	0.00	0.50	-0.00	0.50	0.00	1.00	0.00	0.00	0.00	-2.00	0.00
15	-0.25	-0.00	0.25	0.00	-0.00	0.00	-1.00	0.00	1.00	0.00	-0.25	0.00	0.25	-0.00	-1.00	0.00	0.00	0.00	0.00	0.00
16	0.00	-0.50	0.00	-0.50	-0.00	-0.00	0.00	0.00	-0.00	1.00	0.00	0.50	0.00	0.50	-0.00	1.00	0.00	0.00	-0.00	-2.00
17	0.00	-0.25	0.00	0.25	-0.00	0.00	0.00	-1.00	-0.00	1.00	0.00	-0.25	0.00	0.25	-0.00	-1.00	0.00	1.00	-0.00	0.00
18	0.00	-0.12	0.00	-0.13	0.00	1.00	-0.00	-1.50	-0.00	0.75	0.00	-0.38	0.00	-0.37	-0.00	0.75	0.00	-1.50	-0.00	1.50
19	-0.50	-1.00	-0.50	1.00	-0.00	-0.00	0.00	0.00	1.00	0.00	0.50	3.00	0.50	-3.00	1.00	0.00	0.00	0.00	-2.00	0.00
20	-0.13	-0.25	-0.13	0.25	1.00	0.00	-1.50	-1.00	0.75	1.00	-0.37	-0.25	-0.37	0.25	0.75	-1.00	-1.50	1.00	1.50	0.00

Since strain is defined in linear elasticity as du/dx, this derivative of u is approximated by the following finite difference operator:

$$du/dx = 0.17\ u_1 - 0.17\ u_3 - 0.50\ u_5 - 0.50\ u6 + 0.50\ u_7 + 0.50\ u_8 u$$

The unification of the finite element and the finite difference method can be seen by the recognition that finite element stiffness matrices and finite difference derivative approximation are based on the $[\Phi]$ and the $[\Phi]^{-1}$ matrices. The unification of these two powerful analysis techniques is explicated in Reference [1].

10A.4 FORMULATION AND EVALUATION OF THE STRAIN ENERGY EXPRESSION

As mentioned in the previous section, the displacement interpolation functions provide the source of the strain representations in the finite element. The strain representations, in turn, are used to create the strain energy expression. In the next section, the finite element stiffness matrix is formed from the strain energy expression developed in this section by transforming it from strain gradient coordinates to nodal displacement coordinates with Eq. 10A.7.

The objective of this section is to form the strain energy expression with strain gradient quantities as the independent variables for the example finite element presented in this Appendix.

In order to prepare for forming the strain energy expression for the 10-node element, the three strain components given by Eq. 10.12 are reproduced here as:

$$\begin{Bmatrix} \varepsilon_x\ (x,y) \\ \varepsilon_y\ (x,y) \\ \gamma_{xy}\ (x,y) \end{Bmatrix} = \begin{bmatrix} [T_{Strain}]_{RB} & [T_{Strain}]_6 & [T_{Strain}]_{10} \end{bmatrix} \begin{Bmatrix} \{\varepsilon_{,}\}_{RB} \\ \{\varepsilon,\}_6 \\ \{\varepsilon,\}_{10} \end{Bmatrix} \qquad \text{(Eq. 10A.8)}$$

where

$$[T_{Strain}]_{RB} = \begin{bmatrix} 0 & 0 & 0 \\ 0 & 0 & 0 \\ 0 & 0 & 0 \end{bmatrix}$$

$$[T_{Strain}]_6 = \begin{bmatrix} 1 & 0 & 0 & x & 0 & 0 & y & 0 & 0 \\ 0 & 1 & 0 & 0 & x & 0 & 0 & y & 0 \\ 0 & 0 & 1 & 0 & 0 & x & 0 & 0 & y \end{bmatrix}$$

$$[T_{Strain}]_{10} = \begin{bmatrix} x^2/2 & xy & y^2/2 & 0 & 0 & 0 & 0 & 0 \\ 0 & 0 & 0 & x^2/2 & xy & y^2/2 & 0 & 0 \\ 0 & 0 & xy & xy & -y^2/2 & 0 & x^2/2 & y^2/2 \end{bmatrix}$$

The rigid body motions do not produce any strains by their very definitions. The rigid body motions are included in Eq. 10A.8 in preparation for forming the element stiffness matrix in the next section.

As we will see, any term in the strain energy expression that is related to a rigid body motion will produce no strain energy. This produces several terms that are equal to zero. These zero terms are identified as a result of the use the physically interpretable notation.

The strain energy for a planar region experiencing plane stress in the continuum is given by the following integral expression:

$$SE = 1/2 \int_\Omega \begin{Bmatrix} \varepsilon_x (x,y) \\ \varepsilon_y (x,y) \\ \gamma_{xy} (x,y) \end{Bmatrix}^T [E] \begin{Bmatrix} \varepsilon_x (x,y) \\ \varepsilon_y (x,y) \\ \gamma_{xy} (x,y) \end{Bmatrix} d\Omega$$

where (Eq. 10A.9)

$$[E] = \frac{E}{(1-v^2)} \begin{bmatrix} 1 & v & 0 \\ v & 1 & 0 \\ 0 & 0 & a \end{bmatrix}$$

$$a = \frac{1-v}{2}$$

The parameters E and v are Young's modulus and Poisson's ratio, respectively. They are given values of 30 and 0.3 in this example.

The strain energy expression for the 10-node element is formed by substituting Eq. 10A.8 into Eq. 10A.9. When this is done, the result is the following:

$$SE = 1/2 \begin{Bmatrix} \{\varepsilon,\}_{RB} \\ \{\varepsilon,\}_6 \\ \{\varepsilon,\}_{10} \end{Bmatrix}^T \int_\Omega \begin{Bmatrix} [T_{Strain}]_{RB} \\ [T_{Strain}]_6 \\ [T_{Strain}]_{10} \end{Bmatrix} [E] \begin{Bmatrix} [T_{Strain}]_{RB} \\ [T_{Strain}]_6 \\ [T_{Strain}]_{10} \end{Bmatrix}^T d\Omega \begin{Bmatrix} \{\varepsilon,\}_{RB} \\ \{\varepsilon,\}_6 \\ \{\varepsilon,\}_{10} \end{Bmatrix}$$
(Eq. 10A.10)

The strain gradient quantities are not included under the integral sign because they are not functions of x or y. They are the independent variables for the strain energy expression. It is in this step that the discrete approximation of the continuous problem is introduced.

When the integrand of Eq. 10A.10 is expanded, the integrals in the strain energy expression become:

$$\bar{U} = \int_\Omega \begin{bmatrix} [T_{Strain}]_{RB}^T [E][T_{Strain}]_{RB} & [T_{Strain}]_{RB}^T [E][T_{Strain}]_6 & [T_{Strain}]_{RB}^T [E][T_{Strain}]_{10} \\ [T_{Strain}]_6^T [E][T_{Strain}]_{RB} & [T_{Strain}]_6^T [E][T_{Strain}]_6 & [T_{Strain}]_6^T [E][T_{Strain}]_{10} \\ [T_{Strain}]_{10}^T [E][T_{Strain}]_{RB} & [T_{Strain}]_{10}^T [E][T_{Strain}]_6 & [T_{Strain}]_{10}^T [E][T_{Strain}]_{10} \end{bmatrix} d\Omega$$
(Eq. 10A.11)

This expression is not as imposing as it may appear because of the use of the physically interpretable strain gradient notation. As defined in Eq. 10A.8, the matrix $[T_{Strain}]_{RB}$ introduces the strains due to the rigid body motions. Since the rigid body motions, by definition, do not produce any deformations, the matrix $[T_{Strain}]_{RB}$ is the null matrix. Consequently, the first row and the first column of Eq. 10A.11 contains only zeros. Thus, Eq. 10A.11 reduces to the following:

$$\bar{U} = \int_\Omega \begin{bmatrix} [\bar{U}]_{11} & [\bar{U}]_{12} \\ [\bar{U}]_{12}^T & [\bar{U}]_{22} \end{bmatrix} d\Omega \qquad \text{(Eq. 10A.12)}$$

where

$$\bar{U}_{11} = \begin{bmatrix} 0 & 0 & 0 \\ 0 & 0 & 0 \\ 0 & 0 & 0 \end{bmatrix}$$

$$\bar{U}_{12} = \begin{bmatrix} 0 & 0 & 0 & 0 & 0 & 0 & 0 & 0 & 0 & 0 & 0 & 0 & 0 & 0 & 0 & 0 & 0 \\ 0 & 0 & 0 & 0 & 0 & 0 & 0 & 0 & 0 & 0 & 0 & 0 & 0 & 0 & 0 & 0 & 0 \\ 0 & 0 & 0 & 0 & 0 & 0 & 0 & 0 & 0 & 0 & 0 & 0 & 0 & 0 & 0 & 0 & 0 \end{bmatrix}$$

$$\bar{U}_{22} = \int_\Omega \begin{bmatrix} [T_{Strain}]_6^T [E][T_{Strain}]_6 & [T_{Strain}]_6^T [E][T_{Strain}]_{10} \\ [T_{Strain}]_{10}^T [E][T_{Strain}]_6 & [T_{Strain}]_{10}^T [E][T_{Strain}]_6 \end{bmatrix} d\Omega$$

Since the displacement interpolation functions are cubic polynomials and the strain expressions are derivatives of the displacement interpolation functions, the highest power that the strain polynomial representation will contain is a quadratic term. When the matrix in the fourth quadrant of Eq. 10A.12 is expanded it is found that the following 15 integrals must be evaluated:

The specific integrals that must be evaluated are easily identified using Pascal's triangle. The 15 relatively simple integrals that must be evaluated are the following:

$$I_1 = \int_\Omega dA$$

$$I_2 = \int_\Omega x \, dA \quad I_3 = \int_\Omega y \, dA$$

$$I_4 = \int_\Omega x^2 \, dA \quad I_5 = \int_\Omega xy \, dA \quad I_6 = \int_\Omega y^2 \, dA \qquad \text{(Eq. 10A.13)}$$

$$I_7 = \int_\Omega x^3 \, dA \quad I_8 = \int_\Omega x^2 y \, dA \quad I_9 = \int_\Omega xy^2 \, dA \quad I_{10} = \int_\Omega y^3 \, dA$$

$$I_{11} = \int_\Omega x^4 \, dA \quad I_{12} = \int_\Omega x^3 y \, dA \quad I_{13} = \int_\Omega x^2 y^2 \, dA \quad I_{14} = \int_\Omega xy^3 \, dA \quad I_{15} = \int_\Omega y^4 \, dA$$

The first six integrals are the area, the two first moments of the area, and the three second moments of area for the 10-node element relative to the local origin. These are the same integrals that must be evaluated during the formulation of a six-node linear strain element.

The additional nine integrals are added due to the presence of the three quadratic terms, x^2, xy, and y^2, in the representations of the three strain components. It should be noted that this relatively small number of 15 integrals contrasts to the 210 integrals that must be evaluated for a 10-node quadratic strain element formed using the isoparametric approach.[8]

When the integrals given by Eq. 10A.13 are used to replace the polynomial terms in the fourth quadrant of Eq. 10A.12, the result is presented as Table 10A.4.

The numerical values of the 15 integrals for the example problem are presented in Table 10A.5. In this case, the local origin is at the centroid of this isosceles triangle. As a result, several of the integral values are equal to zero. For example, the second and third integrals are the first moments of inertia. As can be seen, these values are equal to zero as should be the case.

When the values of these integrals are substituted into Eq. 10A.10 or into the matrix given in Table 10A.4, the result for the strain energy expression is contained in Table 10A.6.

This matrix contained in Table 10A.6 is presented in the form of its quadrants for the ease of presentation. Since the matrix is symmetric only following three quadrants are presented as Tables 10A.6a–c.

The independent variables for this matrix are the strain gradient quantities. In order to form a stiffness matrix with independent variables that are nodal displacements, the transformation presented as Eq. 10A.7 must be applied to the strain energy expression given by Eq. 10A.11. This operation is performed in the next Section.

10A.5 FORMULATION OF THE STIFFNESS MATRIX

The objective of this section is to generate the final form the finite element stiffness matrix for the 10-node element. Since the strain energy expression contained in Eq. 10.14 is expressed in terms of strain gradient quantities, it must be transformed so the independent variables are nodal displacements so the elements can be assembled using the usual procedures.

We are now in a position to form the stiffness matrix for the 10-node element. When Eq. 10A.7 is substituted into Eq. 10A.11, the strain energy for the 10-node element expressed in terms of strain gradient coordinates is the following:

$$SE = 1/2\{\varepsilon,\}^T [\bar{U}]\{\varepsilon,\}$$

(Eq. 10A.14)

---

[8] The number of integrals for an isoparametric element are identified in the following way. The 20×20 stiffness matrix is symmetric. So there are a total of $(400 - 20)/2 = 380/2 = 190$ off-diagonal terms. There are then 20 diagonal elements, so a total of $190 + 20 = 210$ integrals must be evaluated.

Table 10A.4. The fourth quadrant of eq. 10A.12 Designated as u_{22}

I_1	vI_1	0	I_2	vI_2	0	I_3	vI_3	0	$I_4/2$	I_5	$I_6/2$	$vI_4/2$	I_5	$vI_6/2$	0	0
vI_1	I_1	0	vI_2	I_2	0	vI_3	I_3	0	$vI_4/2$	I_5	$vI_6/2$	$I_4/2$	I_5	$I_6/2$	0	0
0	0	aI_1	0	0	aI_2	0	0	aI_3	0	0	I_5	I_5	0	0	$aI_4/2$	$aI_6/2$
I_2	vI_2	0	I_4	vI_4	0	I_5	I_5	0	$I_7/2$	I_8	$I_9/2$	$vI_7/2$	vI_8	$vI_9/2$	0	0
vI_2	I_2	0	vI_4	I_4	0	I_5	I_5	0	$vI_7/2$	vI_8	$vI_9/2$	$I_7/2$	I_8	$I_9/2$	0	0
0	0	aI_2	0	0	aI_4	0	0	I_5	0	0	aI_8	aI_8	0	0	$aI_7/2$	$aI_9/2$
I_3	I_3v	0	I_5	I_5	0	I_6	vI_6	0	$I_8/2$	I_9	$I_{10}/2$	$vI_8/2$	vI_9	$vI_{10}/2$	0	0
vI_3	I_3	0	I_5	I_5	0	vI_6	I_6	0	$vI_8/2$	vI_9	$vI_{10}/2$	$I_8/2$	I_9	$I_{10}/2$	0	0
0	0	aI_3	0	0	I_5	0	0	aI_6	0	0	aI_9	aI_9	0	0	$aI_8/2$	$aI_{10}/2$
$I_4/2$	$vI_4/2$	0	$I_7/2$	$vI_7/2$	0	$I_8/2$	$vI_8/2$	0	$I_{11}/4$	$I_{12}/2$	$I_{13}/4$	$vI_{11}/4$	$vI_{12}/2$	$vI_{13}/4$	0	0
I_5	I_5	0	I_8	vI_8	I_5	I_9	vI_9	0	$I_{12}/2$	I_{13}	$I_{14}/2$	$vI_{12}/2$	vI_{13}	$vI_{14}/2$	0	0
$I_6/2$	$vI_6/2$	I_5	$I_9/2$	$vI_9/2$	aI_8	$I_{10}/2$	$vI_{10}/2$	aI_9	$I_{13}/4$	$I_{14}/2$	$I_{15}/4+aI_{13}$	$vI_{13}/4+aI_{13}$	$vI_{14}/2$	$vI_{15}/4$	$aI_{12}/2$	$aI_{14}/2$
$vI_4/2$	$I_4/2$	I_5	$vI_7/2$	$I_7/2$	aI_8	$vI_8/2$	$I_8/2$	aI_9	$vI_{11}/4$	$vI_{12}/2$	$(v/4+a)I_{13}$	$I_{11}/4+aI_{13}$	$I_{12}/2$	$I_{13}/4$	$aI_{12}/2$	$aI_{14}/2$
I_5	I_5	0	vI_8	I_8	0	vI_9	I_9	0	$vI_{12}/2$	vI_{13}	$vI_{14}/2$	$I_{12}/2$	I_{13}	$I_{14}/2$	0	0
$vI_6/2$	$I_6/2$	0	$vI_9/2$	$I_9/2$	0	$vI_{10}/2$	$I_{10}/2$	0	$vI_{13}/4$	$vI_{14}/2$	$vI_{15}/4$	$I_{13}/4$	$I_{14}/2$	$I_{15}/4$	0	0
0	0	$aI_4/2$	0	0	$aI_7/2$	0	0	$aI_8/2$	0	0	$aI_{12}/2$	$aI_{12}/2$	0	0	$aI_{11}/4$	$aI_{13}/4$
0	0	$aI_6/2$	0	0	$aI_9/2$	0	0	$aI_{10}/2$	0	0	$aI_{14}/2$	$aI_{14}/2$	0	0	$aI_{13}/4$	$aI_{15}/4$

Table 10A.5. Integral values contained in the strain energy matrix

Integral no.	1	2	3	4	5	6	7	8	9	10	11	12	13	14	15
Integral value	4.5	0.0	0.0	1.6875	0.0	2.25	0.0	–0.675	0.0	0.9	1.5188	0.0	0.675	0.0	13.5

This strain energy expression is transformed to nodal coordinates with the application of Eq. 10A.7. When this transformation is applied, the result is the following:

$$SE = 1/2\{u\}^T [\Phi]^{-T} [\bar{U}] [\Phi]^{-1} \{u\} \tag{Eq. 10A.15}$$

When the principle of minimum potential energy is applied as done in Reference [1], the stiffness matrix for the 10-node finite element is found to be the following:

$$[K]_{10\,\text{Node}} = [\Phi]^{-T} [\bar{U}][\Phi]^{-1} \tag{Eq. 10A.16}$$

When the stiffness matrix for the 10-node element is formed according to Eq. 10A.15, the result is presented in Table 10A.7.

This Table is presented in Quadrant form on the page following Table 10A.7 to provide an alternate presentation.

10A.6   SUMMARY AND CONCLUSION

This Appendix has presented a step-by-step formulation of the stiffness matrix for a 10-node element. The procedure has been checked for this example by computing the stiffness matrix with three other locations for the local origin. The stiffness matrix was identical to the one presented when the local origins were in the following locations: (1) Node 1, on the lower edge between Nodes 6 and 7, and at a location outside of the domain of the 10-node element. As would be expected, all of the intermediate matrices are significantly different even though the final result is identical for each case.

One final check was performed. The stiffness matrix for a similar element with one-half the size of the example case was computed. It was, as expected, identical to the original stiffness matrix. This result validates the fact that planar elements are independent of size.

Table 10A.6. U matrix—strain energy in strain gradient coordinates given by Eqs. 10A.11 and 12

1	0.00	0.00	0.00	0.00	0.00	0.00	0.00	0.00	0.00	0.00	0.00	0.00	0.00	0.00	0.00	0.00	0.00	0.00	0.00	0.00
2	0.00	0.00	0.00	0.00	0.00	0.00	0.00	0.00	0.00	0.00	0.00	0.00	0.00	0.00	0.00	0.00	0.00	0.00	0.00	0.00
3	0.00	0.00	0.00	0.00	0.00	0.00	0.00	0.00	0.00	0.00	0.00	0.00	0.00	0.00	0.00	0.00	0.00	0.00	0.00	0.00
4	0.00	0.00	0.00	4.50	1.35	0.00	0.00	0.00	0.00	0.00	0.00	0.00	0.84	0.00	1.13	0.25	0.00	0.34	0.00	0.00
5	0.00	0.00	0.00	1.35	4.50	0.00	0.00	0.00	0.00	0.00	0.00	0.00	0.25	0.00	0.34	0.84	0.00	1.13	0.30	0.00
6	0.00	0.00	0.00	0.00	0.00	1.58	0.00	0.00	0.00	0.00	0.00	0.00	0.00	0.00	0.00	0.00	0.00	0.00	0.00	0.39
7	0.00	0.00	0.00	0.00	0.00	0.00	0.51	1.69	0.00	0.00	0.00	0.00	0.00	-0.68	0.00	0.00	0.00	0.00	0.00	0.00
8	0.00	0.00	0.00	0.00	0.00	0.00	1.69	0.51	0.00	0.00	0.00	0.00	0.00	-0.20	0.00	0.00	-0.20	0.00	0.00	0.00
9	0.00	0.00	0.00	0.00	0.00	0.00	0.00	0.00	0.59	0.00	0.00	0.00	0.00	0.00	-0.24	0.00	-0.68	0.00	0.00	0.00
10	0.00	0.00	0.00	0.00	0.00	0.00	0.00	0.00	0.00	2.25	0.68	0.00	-0.34	0.00	0.45	-0.10	0.00	0.14	0.00	0.00
11	0.00	0.00	0.00	0.00	0.00	0.00	0.00	0.00	0.00	0.68	2.25	0.00	-0.10	0.00	0.14	-0.34	0.00	0.45	0.00	0.00
12	0.00	0.00	0.00	0.00	0.00	0.00	0.00	0.00	0.00	0.00	0.00	0.79	0.00	0.00	0.00	0.00	0.00	0.00	-0.12	0.16
13	0.00	0.00	0.00	0.84	0.25	0.00	0.00	0.00	0.00	-0.34	-0.10	0.00	0.38	0.00	0.17	0.11	0.00	0.05	0.00	0.00
14	0.00	0.00	0.00	0.00	0.00	0.00	-0.68	-0.20	0.00	0.00	0.00	0.00	0.00	0.68	0.00	0.00	0.20	0.00	0.00	0.00
15	0.00	0.00	0.00	1.13	0.34	0.00	0.00	0.00	-0.24	0.45	0.14	0.00	0.17	0.00	0.91	0.29	0.00	0.20	0.00	0.00
16	0.00	0.00	0.00	0.25	0.84	0.00	0.00	0.00	0.00	-0.10	-0.34	0.00	0.11	0.00	0.29	0.62	0.00	0.17	0.00	0.00
17	0.00	0.00	0.00	0.00	0.00	0.00	0.00	-0.20	-0.68	0.00	0.00	0.00	0.00	0.20	0.00	0.00	0.68	0.00	0.00	0.00
18	0.00	0.00	0.00	0.34	1.13	0.00	0.00	0.00	0.00	0.14	0.45	0.00	0.05	0.00	0.20	0.17	0.00	0.68	0.00	0.00
19	0.00	0.00	0.00	0.00	0.30	0.00	0.00	0.00	0.00	0.00	0.00	-0.12	0.00	0.00	0.00	0.00	0.00	0.00	0.13	0.06
20	0.00	0.00	0.00	0.00	0.00	0.39	0.00	0.00	0.00	0.00	0.00	0.16	0.00	0.00	0.00	0.00	0.00	0.00	0.06	0.24

Table 10A.6a. Quadrant 1 of U matrix—strain energy in strain gradient coordinates given by Eqs. 10A.11 and 12

1	0.00	0.00	0.00	0.00	0.00	0.00	0.00	0.00	0.00	0.00
2	0.00	0.00	0.00	0.00	0.00	0.00	0.00	0.00	0.00	0.00
3	0.00	0.00	0.00	0.00	0.00	0.00	0.00	0.00	0.00	0.00
4	0.00	0.00	0.00	4.50	1.35	0.00	0.00	0.00	0.00	0.00
5	0.00	0.00	0.00	1.35	4.50	0.00	0.00	0.00	0.00	0.00
6	0.00	0.00	0.00	0.00	0.00	1.58	0.00	0.00	0.00	0.00
7	0.00	0.00	0.00	0.00	0.00	0.00	1.69	0.51	0.00	0.00
8	0.00	0.00	0.00	0.00	0.00	0.00	0.51	1.69	0.00	0.00
9	0.00	0.00	0.00	0.00	0.00	0.00	0.00	0.00	0.59	0.00
10	0.00	0.00	0.00	0.00	0.00	0.00	0.00	0.00	0.00	2.25

Table 10A.6b. Quadrant 2 of U matrix—strain energy in strain gradient coordinates given by Eqs. 10A.11 and 12

1	0.00	0.00	0.00	0.00	0.00	0.00	0.00	0.00	0.00	0.00
2	0.00	0.00	0.00	0.00	0.00	0.00	0.00	0.00	0.00	0.00
3	0.00	0.00	0.00	0.00	0.00	0.00	0.00	0.00	0.00	0.00
4	0.00	0.00	0.84	0.00	1.13	0.25	0.00	0.34	0.00	0.00
5	0.00	0.00	0.25	0.00	0.34	0.84	0.00	1.13	0.00	0.00
6	0.00	0.00	0.00	0.00	0.00	0.00	0.00	0.00	0.30	0.39
7	0.00	0.00	0.00	−0.68	0.00	0.00	−0.20	0.00	0.00	0.00
8	0.00	0.00	0.00	−0.20	0.00	0.00	−0.68	0.00	0.00	0.00
9	0.00	0.00	0.00	0.00	−0.24	−0.24	0.00	0.00	0.00	0.00
10	0.68	0.00	−0.34	0.00	0.45	−0.10	0.00	0.14	0.00	0.00

Table 10A.6c. Quadrant 4 of U matrix—strain energy in strain gradient coordinates given by Eqs. 10A.11 and 12

11	2.25	0.00	−0.10	0.00	0.14	−0.34	0.00	0.45	0.00	0.00
12	0.00	0.79	0.00	0.00	0.00	0.00	0.00	0.00	−0.12	0.16
13	−0.10	0.00	0.38	0.00	0.17	0.11	0.00	0.05	0.00	0.00
14	0.00	0.00	0.00	0.68	0.00	0.00	0.20	0.00	0.00	0.00
15	0.14	0.00	0.17	0.00	0.91	0.29	0.00	0.20	0.00	0.00
16	−0.34	0.00	0.11	0.00	0.29	0.62	0.00	0.17	0.00	0.00
17	0.00	0.00	0.00	0.20	0.00	0.00	0.68	0.00	0.00	0.00
18	0.45	0.00	0.05	0.00	0.20	0.17	0.00	0.68	0.00	0.00
19	0.00	−0.12	0.00	0.00	0.00	0.00	0.00	0.00	0.13	0.06
20	0.00	0.16	0.00	0.00	0.00	0.00	0.00	0.00	0.06	0.24

Table 10A.7. Ten-node element stiffness matrix

	1	2	3	4	5	6	7	8	9	10	11	12	13	14	15	16	17	18	19	20
1	5.08	1.52	-0.88	-0.02	-0.17	-0.29	1.03	1.12	-0.92	-2.21	-6.70	-0.06	3.46	0.22	-0.45	-0.13	-0.13	-0.45	0.00	0.00
2	1.52	2.80	0.02	-0.10	-0.34	-0.48	1.29	1.90	-2.61	-3.67	0.33	-0.54	0.05	0.58	-0.13	-0.25	-0.25	-0.13	0.00	0.00
3	-0.88	0.02	5.08	-1.52	-0.17	0.29	-0.45	0.13	-0.45	0.13	3.46	-0.22	-6.70	0.06	-0.92	2.21	1.03	-1.12	-0.00	-0.00
4	-0.02	-0.10	-1.52	2.80	0.34	-0.48	0.13	-0.25	0.13	-0.25	-0.05	0.58	-0.33	-0.54	2.61	-3.67	-1.29	1.90	0.00	-0.00
5	-0.17	-0.34	-0.17	0.34	1.63	0.00	-1.23	-2.74	0.72	1.15	-0.14	-0.00	-0.14	0.00	0.72	-1.15	-1.23	2.74	0.00	-0.00
6	-0.29	-0.48	0.29	-0.48	0.00	4.67	-2.35	-3.50	0.99	2.06	-0.00	-0.41	0.00	-0.41	-0.99	2.06	2.35	-3.50	0.00	-0.00
7	1.03	1.29	-0.45	0.13	-1.23	-2.35	23.41	-0.00	-6.63	-3.89	0.65	-1.21	0.65	-1.21	3.38	-0.00	-16.92	0.00	-3.89	7.23
8	1.12	1.90	0.13	-0.25	-2.74	-3.50	-0.00	20.40	-3.34	-9.64	-1.21	1.85	-1.21	1.85	0.00	0.37	0.00	-1.85	7.23	-11.13
9	-0.92	-2.61	-0.45	0.13	0.72	0.99	-6.63	-3.34	23.41	-0.00	-3.25	6.03	0.65	6.03	3.38	-0.00	-0.00	-0.00	-20.31	-0.00
10	-2.21	-3.67	0.13	-0.25	1.15	2.06	-3.89	-9.64	-0.00	20.40	6.03	-9.27	-1.21	-9.27	-0.00	0.37	0.37	-0.00	-0.00	-2.23
11	-6.70	0.33	3.46	-0.05	-0.14	-0.00	0.65	-1.21	-3.25	6.03	23.41	-0.00	-14.84	0.28	0.65	1.21	0.65	1.21	-3.89	-7.23
12	-0.06	-0.54	-0.22	0.58	-0.00	-0.41	-1.21	1.85	6.03	-9.27	-0.00	20.40	0.28	-5.19	1.21	1.85	1.21	1.85	7.23	-11.13
13	3.46	0.05	-6.70	-0.33	-0.14	0.00	0.65	-1.21	0.65	-1.21	-14.84	0.28	23.41	-0.00	-3.25	-6.03	-6.63	3.34	0.00	7.23
14	0.22	0.58	0.06	-0.54	0.00	-0.41	-1.21	1.85	6.03	-9.27	0.28	-5.19	-0.00	20.40	-6.03	-9.27	-0.00	-9.64	-20.31	-11.13
15	-0.45	-0.13	-0.92	2.61	0.72	-0.99	3.38	0.00	3.38	-0.00	0.65	1.21	-3.25	-6.03	23.41	-0.00	-6.63	3.89	-3.89	0.00
16	-0.13	-0.25	2.21	-3.67	-1.15	2.06	-0.00	0.37	-0.00	0.37	1.21	1.85	-6.03	-9.27	-0.00	20.40	3.89	-9.64	0.00	-2.23
17	-0.13	-0.25	1.03	-1.29	-1.23	2.35	-16.92	0.00	-0.00	0.37	0.65	1.21	-6.63	-0.00	-6.63	3.89	23.41	0.00	-3.89	-7.23
18	-0.45	-0.13	-1.12	1.90	2.74	-3.50	0.00	-1.85	-0.00	-0.00	1.21	1.85	3.34	-9.64	3.89	-9.64	0.00	20.40	-7.23	-11.13
19	0.00	0.00	-0.00	0.00	0.00	0.00	-3.89	7.23	-20.31	-0.00	-3.89	7.23	0.00	-20.31	-3.89	0.00	-3.89	-7.23	56.19	0.00
20	0.00	0.00	-0.00	-0.00	-0.00	-0.00	7.23	-11.13	-0.00	-2.23	-7.23	-11.13	7.23	-11.13	0.00	-2.23	-7.23	-11.13	0.00	48.96

Table 10A.7a. Quadrant 1 of the 10-node stiffness matrix

1	5.08	1.52	−0.88	−0.02	−0.17	−0.29	1.03	1.12	−0.92	−2.21
2	1.52	2.80	0.02	−0.10	−0.34	−0.48	1.29	1.90	−2.61	−3.67
3	−0.88	0.02	5.08	−1.52	−0.17	0.29	−0.45	0.13	−0.45	0.13
4	−0.02	−0.10	−1.52	2.80	0.34	−0.48	0.13	−0.25	0.13	−0.25
5	−0.17	−0.34	−0.17	0.34	1.63	0.00	−1.23	−2.74	0.72	1.15
6	−0.29	−0.48	0.29	−0.48	0.00	4.67	−2.35	−3.50	0.99	2.06
7	1.03	1.29	−0.45	0.13	−1.23	−2.35	23.41	−0.00	−6.63	−3.89
8	1.12	1.90	0.13	−0.25	−2.74	−3.50	0.00	20.40	−3.34	−9.64
9	−0.92	−2.61	−0.45	0.13	0.72	0.99	−6.63	−3.34	23.41	−0.00
10	−2.21	−3.67	0.13	−0.25	1.15	2.06	−3.89	−9.64	−0.00	20.40

Table 10A.7b. Quadrant 2 of the 10-node stiffness matrix

1	−6.70	−0.06	3.46	0.22	−0.45	−0.13	−0.45	−0.13	0.00	0.00
2	0.33	−0.54	0.05	0.58	−0.13	−0.25	−0.13	−0.25	0.00	0.00
3	3.46	−0.22	−6.70	0.06	−0.92	2.21	1.03	−1.12	−0.00	−0.00
4	−0.05	0.58	−0.33	−0.54	2.61	−3.67	−1.29	1.90	0.00	−0.00
5	−0.14	−0.00	−0.14	0.00	0.72	−1.15	−1.23	2.74	0.00	−0.00
6	−0.00	−0.41	0.00	−0.41	−0.99	2.06	2.35	−3.50	0.00	−0.00
7	0.65	−1.21	0.65	−1.21	3.38	−0.00	−16.92	0.00	−3.89	7.23
8	−1.21	1.85	−1.21	1.85	0.00	0.37	−0.00	−1.85	7.23	−11.13
9	−3.25	6.03	0.65	−1.21	3.38	−0.00	3.38	−0.00	−20.31	−0.00
10	6.03	−9.27	−1.21	1.85	−0.00	0.37	0.00	0.37	−0.00	−2.23

Table 10A.7c. Quadrant 4 of the 10-Node stiffness matrix

11	23.41	−0.00	−14.84	−0.28	0.65	1.21	0.65	1.21	−3.89	−7.23
12	−0.00	20.40	0.28	−5.19	1.21	1.85	1.21	1.85	−7.23	−11.13
13	−14.84	0.28	23.41	−0.00	−3.25	−6.03	0.65	1.21	−3.89	7.23
14	−0.28	−5.19	−0.00	20.40	−6.03	−9.27	1.21	1.85	7.23	−11.13
15	0.65	1.21	−3.25	−6.03	23.41	−0.00	−6.63	3.34	−20.31	0.00
16	1.21	1.85	−6.03	−9.27	−0.00	20.40	3.89	−9.64	0.00	−2.23
17	0.65	1.21	0.65	1.21	−6.63	3.89	23.41	0.00	−3.89	−7.23
18	1.21	1.85	1.21	1.85	3.34	−9.64	0.00	20.40	−7.23	−11.13
19	−3.89	−7.23	−3.89	7.23	−20.31	0.00	−3.89	−7.23	56.19	0.00
20	−7.23	−11.13	7.23	−11.13	0.00	−2.23	−7.23	−11.13	0.00	48.96

MATLAB FORMULATION OF THE 10-NODE ELEMENT STIFFNESS MATRIX

10B.1 INTRODUCTION

This Appendix contains annotated versions of the m-files used to form the stiffness matrix for a 10-node plain stress **triangle**. In general, a 10-node element can have one or more curved edges. However, the m-file that integrates the 15 integrals required for the formulation of the stiffness matrix handles only straight edges. If curved edges are to be represented, the m-file entitled Green-Integrals10 must be modified.

This presentation is closely keyed to the numerical example in the previous Appendix and, consequently, to the main text of the chapter.

It should be noted that no attempt has been made to present an elegant or efficient code for forming this stiffness matrix. This Appendix attempts to resolve any questions that might have arisen in the main text or in the first Appendix. This objective is approached by including comments and using semifriendly notation to clarify the code that forms the stiffness matrix for the 10-node element.

The program consists of a driver program, three m-files that perform computations, and two m-files that present data in user-friendly forms. The computational m-files are the following:

1. FormPhi: Forms the phi matrix and its inverse.
2. GreenIntegrals10: Evaluates the 15 integrals that occur in strain energy expression for the 10-node stiffness matrix using Green's theorem.
3. UKernel: Forms the strain energy expression for the element in strain gradient coordinates. This function places the 15 integrals in their proper locations and multiplies them by the correct parameters. This strain energy expression is transformed using phi inverse to form the stiffness matrix for the 10-node element in the last step in the driver program.

The two m-files that present data are the following:

1. ElementPlot: Plots the element geometry and the nodal points.
2. PrintTableReady: This m-file prints matrices in a form that can be converted in Word to tables. These tables are used in the main text and in Appendix 10A. The rows are numbered and the matrix elements are separated by commas in this function for use with the "convert" option in the Table function of Word.

Annotated versions of the driver program and the m-files of the functions called by the driver program are presented next.

10B.2   DRIVER PROGRAM FOR FORMING THE STIFFNESS MATRIX FOR A 10-NODE ELEMENT

```
% Form K for 10 node triangle with straight edges.
% A line preceded by % is a comment. It is not operational.
clc % Clear screen
clear all % Clear memory
close all % Close figures
format compact % Produces output that is single spaced
%
% Input Variables
%  xn = x locations of nodal points in the 10 node element.
%  yn = y locations of nodal points in the 10 node element.
%       Element geometry and nodal locations are identified in the figure shown below.
% pr = The numerical value for Poisson's ratio.
% E = The numerical value for Young's modulus.
%
% Internal Variables
% AxRB & AyRB = Polynomial terms for rigid body motions in the x and y directions.
% Ax6 & Ay6 = Polynomial terms for strain states in a 6 node element.
% Ax10 & Ay10 = Polynomial terms for additional strain states in a 10 node element.
%   See Eq. 10.5 in the main text for these equations and the associated discussion.
%
% Output Variables
% Figure 1 is a plot of the element geometry and nodal locations.
%  FigNo = incrementing figure number, if needed.
% Phi = Transformation from nodal to strain gradient (SG) coords.
% PhiInv = Transformation from SG to nodal coords.
%   Transforms strain energy matrix U10x10 to K, the 10-node stiffness matrix.
% Integrals = The numerical values of the 15 integrals that appear in the strain
```

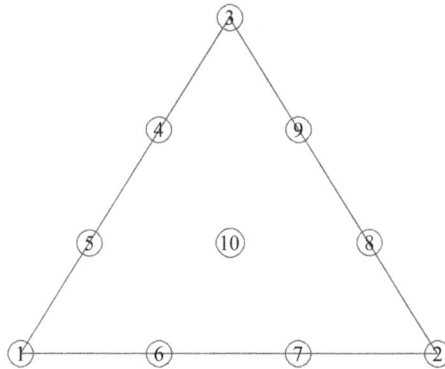

Figure 10B.1. Element nodal numbering.

%   energy expression. Green's theorem is used in this calculation.
%   Other approaches could be used.
% U10x10 = The 20x20 strain energy expression in SG coordinates.
%   This matrix can be viewed as the stiffness matrix in SG coords.
%   It will be transformed using PhiInv to form the element stiffness matrix.
% Kfinal = The 20x20 finite element stiffness matrix being sought.
%
% Introduce the input nodal coordinates for the ten node element being formed.
display('Nodal Coordinates') % The display operation puts text in output.
% The following are the locations of Node 1 2 3 4 5 6 7 8 9 10
xn=[-1.5 1.5 0.0 -0.5 -1.0 -0.5 0.5 1.0 0.5 0.0] % xn are the x locations.
yn=[-1.0 -1.0 2.0 1.0 0.0 -1.0 -1.0 0.0 1.0 0.0] % yn are the y locations.
pr=0.3 % Poisson's ratio
E=10 % Young's Modulus
%
% Plot the element geometry and the nodal locations.
FigNo=1; % A counter.
[FigNo]=**ElementPlot**(xn,yn,FigNo)
%
syms x y % Identify x and y and any function
%   formed with x and y as symbolic variables.
%
% Put in displacement interpolation functions for strain gradient coordinates.
% Polynomial terms for rigid body motions in planar elements.
% Order of rigid body motions.
% [urb, vrb, rrb]; In 3-D, the rotations are labeled p, q and r. rrb designates a rotation
%   about the z-axis in the x-y plane.
AxRB=[1 0 -y]; % Rigid body contributions to displacements in the x direction.

AyRB=[0 1 x]; % Rigid body contributions to displacements in the y direction.
%
% Polynomial terms for six-node element displacements, less rigid body terms.
% Order of strain gradient coordinates (g for gamma, shear strain).
% [ex ey g ex,x ey,x g,x ex,y ey,y g,y]; e.g., ex,x = $\varepsilon_{x,x}$ and g,x y = $\gamma_{xy,y}$.
Ax6=[x 0 y/2 x^2/2 -y^2/2 0 x*y 0 y^2/2];
Ay6=[0 y x/2 0 x*y x^2/2 -x^2/2 y^2/2 0];
%
% Additional polynomial terms for 10-node element deformations.
% Order of added strain gradient coordinates.
% [ex,xx ex,xy ex,yy ey,xx ey,xy ey,yy g,xx g,yy]
% The term g,xy is not there because of compatibility, i.e., g,xy = ex,yy + ey,xx
% See Eq. 10.11 and the associated discussion in the main text.
%
Ax10=[x^3/6 x^2*y/2 x*y^2/2 0 -y^3/6 0 0 y^3/6];
Ay10=[0 -x^3/6 0 x^2*y/2 x*y^2/2 y^3/6 x^3/6 0];
% End of displacement interpolation polynomial formulation.
%
% Now form the phi matrix and its inverse.
% See Eq. 10.20 and the associated discussion in the main text.
[Phi,PhiInv]=**FormPhi**(xn,yn,AxRB,AyRB,Ax6,Ay6,Ax10,Ay10)
%
% Prepare output for converting to a table in Word.
%[check]=PrintTableReady(Phi,RowStart,ColumnStart,RowEnd,ColumnEnd,TableNo)
[check]=PrintTableReady(Phi,1,1,20,20,0)
[check]=PrintTableReady(PhiInv,1,1,20,20,0)
%
% Form the numerical values for the 15 integrals needed for element formulation.
% See Section 10.5 in the main text.
[Integrals]=**GreenIntegrals10**(xn,yn)
%
% Form strain energy matrix in strain gradient coordinates.
%
% See Section 10.4 in the main text.
[U10x10]=**UKernel**(Ax6,Ay6,Ax10,Ay10,Integrals,pr)
% The effect of E, Young's modulus, is included later.
%
% Prepare output for converting U10x10 to a table in Word.
[check]=PrintTableReady(U10x10,1,1,20,20,0) % Full matrix
[check]=PrintTableReady(U10x10,1,1,10,10,1) % First quadrant
[check]=PrintTableReady(U10x10,1,11,10,20,2) % Second quadrant

[check]=PrintTableReady(U10x10,11,11,20,20,3) % Fourth quadrant
%
% Now form the final stiffness matrix is local nodal coordinates.
display('This is the elemental stiffness matrix for a 10 node element')
% See Eq. 10.21 in the main text.
Kfinal=PhiInv'*U10x10*PhiInv;
Kfinal=(E/(1-pr^2))*Kfinal
%
% Prepare output for converting to a table in Word.
[check]=PrintTableReady(Kfinal,1,1,10,10,1)
[check]=PrintTableReady(Kfinal,1,11,10,20,2)
[check]=PrintTableReady(Kfinal,11,11,20,20,3)
%
display('This is the elemental stiffness matrix for a 10 node element')
End of Driver for forming Ten Node Stiffness Matrix

10B.3   FORM PHI AND PHI INVERSE FOR 10-NODE ELEMENT

function [Phi,PhiInv]=FormPhi(xn,yn,AxRB,AyRB,Ax6,Ay6,Ax10,Ay10)
% Form the phi matrix and its inverse.
%
% Input Variables
% xn = x locations of nodal points in ten node element.
% yn = y locations of nodal points in ten node element.
% AxRB & AyRB = Polynomial terms for rigid body motions in the x and y directions.
% Ax6 & Ay6 = Polynomial terms for strain states in 6 node element.
% Ax10 & Ay10 = Polynomial terms for additional strain states in 10 node element.
%
% Output Variables
% Phi = Transformation from nodal displacement to strain gradient coordinates.
% PhiInv = Transformation from strain gradient coordinates to nodal displacements.
%   This matrix transforms U10x10 to K, the desired stiffness matrix.
%
syms x y % Identify x and y and any derived functions as a symbolic variable.
% Form the displacement interpolation polynomials in terms of strain gradient
% coordinates: Ax10Total = rigid body motions + disp 6 node + disp 10 node.
%   Ay10Total = rigid body motions + disp 6 node + disp 10 node.
%
Ax10Total=[AxRB,Ax6,Ax10]; %Displacement interpolation function in x direction.
Ay10Total=[AyRB,Ay6,Ay10]; %Displacement interpolation function in x direction.

```
%
% Form Phi matrix
% Substitute the numerical locations of the 10 nodes, one at a time into Phi.
%
for i=1:10
    Ax10sub=subs(Ax10Total,x,xn(i)); % x numerical values in u interpolation function.
    Ax10sub=subs(Ax10sub,y,yn(i)); % y numerical values in u interpolation function.
Phi(2*i-1,:)=Ax10sub; % Place in alternate rows of the Phi matrix, 1, 3, 5, …
    Ay10sub=subs(Ay10Total,x,xn(i)); % x numerical values in v interpolation function.
    Ay10sub=subs(Ay10sub,y,yn(i)); % y numerical values in v interpolation function.
Phi(2*i,:)=Ay10sub; % Place in alternate rows of the Phi matrix, 2, 4, 6, …
end
%
% Phi is now formed, form Phi inverse.
PhiInv=inv(Phi);
%
display('Phi matrix and phi inverse are formed.')
```

End Form Phi and Phi Inverse for Ten Node Element

10B.4   FORM INTEGRALS IN STIFFNESS MATRIX USING GREEN'S THEOREM

```
function [Integrals]=GreenIntegrals10(xn,yn)
% This program finds the 15 integrals for a 10 node triangle with
%   straight edges. It handles a vertical edge.
%   Curved edges would take some logic to get every case.
%
% If used in a FE program, perform the symbolic formulation outside of the program.
%   Then, embed the results in the element formulation process.
%
display('Entering GreenIntegrals10')
%
% Nota Bene: Note Well: These integrals are formed using local coordinates.
%   Phi and integrals must be formed using the same coordinate system.
%
% These symbolic variables are used in the integration process.
syms  x y x1 x2 x3 y1 y2 y3 b1 b2 b3
%
% Input Variables
% xn = x locations of the element nodes.
% yn = y locations of the element nodes.
%
```

% Output Variables
% Integrals = 15 integrals in stiffness kernel.
%
% Internal Variables
% Use slope, y-intercept form of straight line in the integration process.
% m = slopes of three lines, m(1,1) ,m(1,2), m(1,3)
% M = Vector of slopes, [m(1,1), m(1,2), m(1,3)]
% b = y intercepts, y(1,1), y(1,2), y(1,3)
% B = vector of y intercepts, [y(1,1), y(1,2), y(1,3)]
% Y1,Y2,Y3 = Symbolic equations of element edges.
% Y = Vector of [Y1, Y2, Y3]
% IntegralT = Storage for summing the 3 integral components found for the
%   three edges as they are formed for each of the 15 integrals.
% xp=[xn(1,1),xn(1,2),xn(1,3),xn(1,1)]; % Description follows.
% Xp = symbolic equivalent of xp,(1,4). % Description follows.
% xp is a vector of the numerical x locations of the corner nodes for use as integration
% limits for the line integrals in Green's theorem. The first node is added to the end
%   because these vectors are used for the limits of integration, i.e., Node 1 to Node 2,
%   Node 2 to Node 3, and Node 3 to Node 1.
% yp is a vector of the numerical y locations of the corner nodes for use in finding
%   the slope-y intercept form of the equation for the edges of the triangle.
%
xp=[xn(1,1),xn(1,2),xn(1,3),xn(1,1)];
Xp=[x1 x2 x3 x1];% Symbolic nodal locations.
yp = [yn(1,1),yn(1,2),yn(1,3)]
% Will substitute numerical values after symbolic integration.
%
% Form the equation of the straight line for each of the 3 edges.
% The slope y intercept form of the equation for a line is y= m*x+b.
%   Now compute the m and b parameters from the nodal coordinates.
%
% The equations for the lines are used to replace y in the functions being
% integrated using Green's theorem. As a result, the functions
%   being integrated contain only x terms.
%
% Compute the slopes for the three lines.
%
% Assume slope is zero to cover the infinite slope case in integral.
%   It will be changed if it is not zero. The integration is equal to zero
%   for a vertical line since the integration limits are the x values at the
%   two ends of the line which are equal to each other for a vertical line.

```
%
m1=0;
m2=0;
m3=0;
%
if((xp(1,2)-xp(1,1))~=0)% Check for vertical line. If vertical, m1 = 0.
m1=(yp(1,2)-yp(1,1))/(xp(1,2)-xp(1,1)); % Compute slope.
end
%
if((xp(1,3)-xp(1,2))~=0)% Check for vertical line. If vertical, m2 = 0.
m2=(yp(1,3)-yp(1,2))/(xp(1,3)-xp(1,2)); % Compute slope.
end
%
if((xp(1,1)-xp(1,3))~=0)% Check for vertical line. If vertical, m3 = 0.
m3=(yp(1,1)-yp(1,3))/(xp(1,1)-xp(1,3)); % Compute slope.
end
%
% Put slopes for the boundary edges in vector form for looping.
M=[m1,m2,m3];
%
% Calculate the y intercepts for the three boundary lines
%  in numerical form.
b1=yp(1,1)-m1*xp(1,1);
b2=yp(1,2)-m2*xp(1,2);
b3=yp(1,3)-m3*xp(1,3);
%
%  Put y intercepts in vector form for looping.
B=[b1,b2,b3];
%
% Form symbolic equations for the lines that make up the edges of the triangle.
%
Y1=m1*x+b1;
Y2=m2*x+b2;
Y3=m3*x+b3;
%
Y=[Y1,Y2,Y3]; % Vector of equations of boundary lines.
% Put in vector form for looping.
%
% Variables used in integration.
X=[x1,x2,x3,x1]; % Symbolic x nodal locations.
XX=[x1,x2,x3]; % Symbolic x nodal locations.
xnodes=[xp(1,1),xp(1,2),xp(1,3)];% For use in evaluating symbolic integrals numerically.
YY=[y1,y2,y3]; % Symbolic y nodal locations.
```

ynodes=[yp(1,1),yp(1,2),yp(1,3)];% For use in evaluating symbolic integrals numerically.
%
% Compute the 15 integrals using Green's theorem.
% See Ref. 1 for derivation and explanation of Green's theorem.
%
% Form the integrals for the 3 edges of the triangle.
%   Integrand M in area integral (dy dx) becomes $-\partial M/\partial y$ in line integral
%   with integration over dx in Green's theorem. The area integrands are
%   identified in the main text in Section 10.5.
%
% The integrand matrix has 15 rows and 3 columns.
% The integrands for the 3 lines are formed separately in the three columns.
% The 15 integrands needed for the strain energy expression are in the 15 rows.
%
% This operation substitutes the equation for Y into the y's in the integrands.
% Now the integrands are in terms of x only.
%
% The integrands in the matrix are for Green's theorem. The comment following each
%   line identifies the integrand in an area integral and its equivalent in the line
%   integrals for Green's theorem.
Integrand=[-Y(1),-Y(2),-Y(3);% Row 1; 1 becomes –y.
        -Y(1)*x,-Y(2)*x,-Y(3)*x;% Row 2; x becomes –x y.
        (-Y(1)^2)/2,(-Y(2)^2)/2,(-Y(3)^2)/2;% Row 3; y becomes $-y^2/2$
        -x^2*Y(1),-x^2*Y(2),-x^2*Y(3);% Row 4; x^2 becomes $-x^2 y$
        (-x*Y(1)^2)/2,(-x*Y(2)^2)/2,(-x*Y(3)^2)/2;% Row 5; xy becomes $-x\,y^2/2$
        (-Y(1)^3)/3,(-Y(2)^3)/3,(-Y(3)^3)/3;% Row 6; y^2 becomes $-y^3/3$
        -x^3*Y(1), -x^3*Y(2), -x^3*Y(3);% Row 7; x^3 becomes $-x^3 y$
        (-x^2*Y(1)^2)/2,(-x^2*Y(2)^2)/2,(-x^2*Y(3)^2)/2;% Row 8; $x^2 y$ becomes $-x^2 y^2/2$
        (-x*Y(1)^3)/3,(-x*Y(2)^3)/3,(-x*Y(3)^3)/3;% Row 9; $x y^2$ becomes $-x y^3/3$
        -Y(1)^4,-Y(2)^4,-Y(3)^4;% Row 10; y^3 becomes $-y^4/4$
        -x^4*Y(1),-x^4*Y(2),-x^4*Y(3);% Row 11; x^4 becomes $-x^4 y$
        (-x^3*Y(1)^2)/2,(-x^3*Y(2)^2)/2,(-x^3*Y(3)^2)/2;% Row 12; $x^3 y$ becomes $-x^3 y^2/2$
        (-x^2*Y(1)^3)/3,(-x^2*Y(2)^3)/3,(-x^2*Y(3)^3)/3;% Row 13; $x^2 y^2$ becomes $-x^2y^3/3$
        (-x*Y(1)^4)/4,(-x*Y(2)^4)/4,(-x*Y(3)^4)/4;% Row 14; xy^3 becomes $-xy^4/4$
        -Y(1)^5,-Y(2)^5,-Y(3)^5];% Row 15; y^4 becomes $-y^5/5$
%
% Now compute the 15 integrals. This is done by summing the integrals for the 3 lines % that make
up the triangle.
%
for ij=1:15 % One loop for each integral.
sammtot1=0; % Total Integral, Sum of the three components.
samjj=zeros(1,3); % Components of integral for the three edges.

```
for i=1:3 % Loop on number of edges on triangle, i.e., 3.
    Integral=int(Integrand(ij,i),x,X(i),X(i+1)); % Integral on each line.
        if(xp(i+1)~=xp(i))% Check for vertical line.
            IntTT=subs(Integral,XX,xnodes);% Substitute x coordinates locations.
            samjj(1,i)=subs(IntTT,YY,ynodes);
            sammtot1=sammtot1+samjj(1,i);
        end
end
Integrals(ij)=sammtot1;
end
display('Exiting GreenIntegrals10')
```

End - Form Integrals in Stiffness Matrix using Green's Theorem

10B.5   FORM STRAIN ENERGY KERNEL FOR 10-NODE ELEMENT

```
function [U10x10]=UKernel(Ax6,Ay6,Ax10,Ay10,Ints,pr)
% Form the U matrix for the 10 node triangle
% in strain gradient coordinates.
%
```
% This function first forms the U_{22} matrix identified in Eq. 10.16. Then, it adds
% the zero matrices identified in Eq. 10.16 to form the strain energy matrix that
% is transformed to the finite element stiffness matrix being sought.
```
%
```
% Input Variables
% Ax6 & Ay6 = Polynomial terms for strain states in 6 node element.
% Ax10 & Ay10 = Polynomial terms for additional strain states in 10 node element.
% Ints = The numerical value of the 15 integrals that exist for a 10 node element.
% pr = the numerical value of Poisson's ratio. It is used in a parameter in the
% constitutive matrix, anum = (1-pr)/2.
```
%
```
% **Note:** AxRB and AyRB are not used because rigid body motions produce
% no   strain energy.
```
%
```
% Output Variables
% U10x10 = The strain energy matrix in strain gradient coordinates.
%   It will be transformed to nodal displacement cords using PhiInv.
```
%
anum=(1-pr)/2; % a numerical value in the constitutive matrix.
%
format compact
syms x y n a % Parameters in matrix formulation.
```

syms I1 I2 I3 I4 I5 I6 I7 I8 I9 I10 I11 I12 I13 I14 I15
% These quantities are the symbolic representations of the numerical values
%   of the integrals, I_1, I_2, ..., I_{15}.
%
syms A B C D E F G H K L M N P Q
% These are the **14** intermediate identifiers for the integrals, I_2, I_3, ..., I_{15}.
%   Their function will be discussed later when they are used.
%
% Form the vector of polynomial terms in the displacement interpolation due
%   to deformations. The rigid body motions produce no strain energy.
%
AxTot=[Ax6,Ax10]
AyTot=[Ay6,Ay10]
%
% Take derivatives of displacements for normal strains.
Axx=diff(AxTot,x);
Ayy=diff(AyTot,y);
% Take derivatives for shear strains.
Axy=diff(AxTot,y);
Ayx=diff(AyTot,x);
% Form shear strain expression.
gammaxy=Axy+Ayx;
% Assemble TepsTot, the 3 rows of **T**ransformation **eps**ilon (strains) **Tot**al.
TepsTot=[Axx;Ayy;gammaxy];
%
% Form constitutive matrix.
% This is the full constitutive matrix.
%   EY=(E/(1-n^2))*[1 n 0;n 1 0;0 0 a]% The first term is a multiplier.
% The multiplier (E/(1-n^2)) will be introduced numerically in the final step
%   of the main program.
% Until then, only the matrix portion of the constitutive matrix will be used.
EY=[1 n 0;n 1 0;0 0 a];
%
% Now form U10, the strain energy expression in strain gradient notation.
% Note that this expression does not contain the multiplier for the
%   constitutive relation, (E/(1-n^2)). It is added in the last stage
%   of the main program. It would complicate the symbolic expressions
%   if it were included here.
%
% Form strain energy **integrand**.
TTT=TepsTot.'; % Form transpose of the strain representation.

```
%
U10=TTT*EY*TepsTot;
% This is a symbolic expression with polynomial terms from constants
%    through x and y up to x^3 and y^3. The various terms are repeated throughout
%      the matrix. However, there are only 15 different polynomial terms. They
%      correspond to the 15 different integrals computed in GreenIntegrals10.
%        The polynomial terms are being replaced by symbolic expressions for the
%          integrals. In a later step, the symbolic integrals will be replaced by numerical
%            values in preparation for forming the stiffness matrix for the element.
%
% This code for replacing the polynomial expressions by symbolic representations of the
%    integrals was formed "iteratively" by visual inspection. That is why it is done in
%      two steps. The caps (A, B, ...) which initially replaced the polynomial terms
%        were easier to identify than the similar expressions for the integrals, e.g., I15.
%
% The first step in this process identifies the fourth power of y terms as Q.
%    Then, Q is replaced by the integral it designates, I15. This symbolic value
%    is later replaced by the numerical value of I15 for this element.
%
U10Q=subs(U10,y^4,Q);% Q=y^4
U10Q=subs(U10Q,Q,I15);

U10P=subs(U10Q,x^4,P);% P=x^4
U10P=subs(U10P,P,I11);

% This composite term, a term with both x and y, contains more steps.
%    This is the case because the x and y terms could be present in a variety of ways.
%      This variety was identified by visual inspection.
U10N=subs(U10P,x^3*y,N);% N=y*x^3
U10N=subs(U10N,x^3*a*y,a*N);
U10N=subs(U10N,x^3*n*y,n*N);
U10N=subs(U10N,N,I12);

U10M=subs(U10N,x*y^3,M);% M=x*y^3
U10M=subs(U10M,x*a*y^3,a*M);
U10M=subs(U10M,x*n*y^3,n*M);
U10M=subs(U10M,M,I14);

U10H=subs(U10M,x^2*a*y^2,a*H);% H=x^2*y^2
U10H=subs(U10H,x^2*n*y^2,n*H);
U10H=subs(U10H,x^2*y^2,H);
U10H=subs(U10H,H,I13);
```

```
U10K=subs(U10H,x*y^2,K);% K=x*y^2
U10K=subs(U10K,x*a*y^2,a*K);
U10K=subs(U10K,x*n*y^2,n*K);
U10K=subs(U10K,K,I9);

U10L=subs(U10K,x^2*y,L);% L=x^2*y
U10L=subs(U10L,x^2*a*y,a*L);
U10L=subs(U10L,x^2*n*y,n*L);
U10L=subs(U10L,L,I8);

U10A=subs(U10L,x^2,A);% A=x^2
U10A=subs(U10A,A,I4);

U10B=subs(U10A,y^2,B);% B=y^2
U10B=subs(U10B,B,I6);

U10C=subs(U10B,x*y,C);% C=x*y
U10C=subs(U10C,x*a*y,a*C);%C=x*y
U10C=subs(U10C,x*n*y,n*C);%C=x*y
U10C=subs(U10C,C,I5);

U10D=subs(U10C,x^3,D); % D=x^3
U10D=subs(U10D,D,I7);

U10E=subs(U10D,y^3,E); % E=y^3
U10E=subs(U10E,E,I10);
U10E=subs(U10E,x,I2);
U10E=subs(U10E,y,I3);

% The following terms were constants, i.e., either 1, n or a. Their locations were
%   identified by the visual inspection of U10E, the matrix formed in the previous
%     operations. Then, the following Matlab operations were added to replace the
%     constants with the expression containing I1.
%
U10E(1,1)=I1;
U10E(2,2)=I1;
U10E(1,2)=n*I1;
U10E(2,1)=n*I1;
U10E(3,3)=a*I1;
%
% The numerical values of the integrals (Ints) are substituted for the symbolic values
%   of the integrals in the next two steps.
%
```

% Form a vector of symbolic integrals for one step replacement
Isym=[I1,I2,I3,I4,I5,I6,I7,I8,I9,I10,I11,I12,I13,I14,I15];
%
% Substitute numerical values for parameters.
U10num=subs(U10E,Isym,Ints); % insert numerical values for integrals.
%
U10num=subs(U10num,n,pr); % Insert numerical value for Poisson's ratio.%
%
U10num=subs(U10num,a,anum); Insert numerical value for anum=(1-pr)/2.
%
% Change from symbolic variable to a double precision numerical value.
U10num=double(U10num);
%
% Now put in the zero rows and columns due to the rigid body motions.
% Form zero matrices.
Z11=zeros(3,3);
Z12=zeros(3,17);
Z21=zeros(17,3);
% Form final 20x20 strain energy expression is strain gradient coordinates.
U10x10=[Z11,Z12;Z21,U10num];
End Form Strain Energy Kernel for Ten Node Element

10B.6   PLOT GEOMETRY AND NODES FOR 10-NODE ELEMENT

function [FigNo]=ElementPlot(xn,yn,FigNo)
% This function is used in the m-file Tri10plot
display('Entering Element Plot')
figure(FigNo);
sam1=[xn(1),xn(2),xn(3) xn(1)];
sam2=[yn(1),yn(2),yn(3) yn(1)];
plot(sam1,sam2)
hold on
plot(xn,yn,'o','markersize',8,'markerfacecolor','r')
xlabel('X Coordinate')
ylabel('Y Coordinate')
title('Ten Node Element')
a=[1.1*min(sam1),1.1*max(sam1)];
b=[1.1*min(sam2),1.1*max(sam2)];
axis([a b])
%fgh=num2string(FigNo) % Could use this to get figure number in printout.
display(['Figure ',num2str(FigNo),(' is a plot of the ten node element')])

%fprintf('Plot of ten node triangle as Figure%2i george',FigNo)
% I cannot get an end to this code.
FigNo=FigNo+1;
End Plot Geometry and Nodes for Ten Node Element

10B.7 FUNCTION TO TRANSFORM MATLAB MATRICES TO FORM FOR USE IN WORD

```
function[check]=PrintTableReady(Dinput,RowStart,ColumnStart,RowEnd,...
ColumnEnd,TableNo)
% This function prints arrays that are convertible to tables in word.
%
% Dinput is the input array to be put in a table (with commas as separators).
D=Dinput;
%
% The number of displayed digits can be modified by fixing "sam."
%
% DeltaInteger is the element row numbering.
% e.g., if start at 1, DeltaInteger =1, etc.
%
% RowStart is the row in the matrix to start the table.
% ColumnStart is the column in the matrix to start table.
%
% RowEnd is the row in the matrix to stop the table.
% ColumnEnd is the collumn in the matrix to stop the table.
%
display('Start of PrintTableReady')
display('This is Table Number')
check=TableNo
%
display('Input size of matrix D')
[nr,nc]=size(D) % Find size of matrix being input.
%
%sam=('%7.4f,'); % 4 decimal places, floating point.
%sam2=('%7.4f\n'); % Starts a new line.
sam=('%7.2f,'); % 2 decimal places, floating point.
sam2=('%7.2f\n'); % Starts a new line.
%
%for i=1:nr
for i=RowStart:RowEnd
    fprintf('%2.0f,',i);
```

```
%for j=1:nc-1
 for j=ColumnStart:ColumnEnd-1
   fprintf(sam,D(i,j));
end
     fprintf(sam2,D(i,j+1));
   end
% From Intro to Matlab, Palm, p.95.
```

End Function to transform Matlab matrices to form for use in Word

CHAPTER 11

PERFORMANCE-BASED REFINEMENT GUIDES

11.1   INTRODUCTION

A new approach for forming refinement guides is developed in this chapter. This is accomplished by comparing an approximation of the exact solution that is emerging from the finite element result to the modeling capabilities of the individual finite elements. This physically based approach contrasts with refinement guides currently in use that correlate the level of refinement to the magnitude of the estimated error. The new approach is outlined with the aid of Fig. 11.1.

The key step in this process is to create a strain distribution that is closer to the exact solution than the finite element result. In Fig. 11.1a, the improved solution is labeled as the finite difference approximation because the strains at the nodes are extracted using finite difference templates. The resulting strain distribution differs from the finite element result in two ways. It is continuous and it is more complex on the domain of a single element than the element can represent unless the finite element result is exact.

In the case shown in Fig. 11.1 for a model formed with three-node elements, the approximation of the exact solution contains curvature, a complexity that a three-node linear strain element cannot represent. Another approach for forming an improved solution is based on the interelement jumps in the finite element strain representations. This alternate approach is presented and demonstrated in later sections.

The differences between the improved and the finite element strain representations are used to identify the modeling deficiencies in the individual elements. These differences are quantified by decomposing the two strain representations using the physically interpretable notation presented in Chapters 3, 8, and 10. Then, the level of refinement needed to produce a solution that satisfies the termination criterion is computed. The number of subdivisions given to an element is found by comparing the modeling capabilities of the element to the complexities of the approximation of the exact solution.

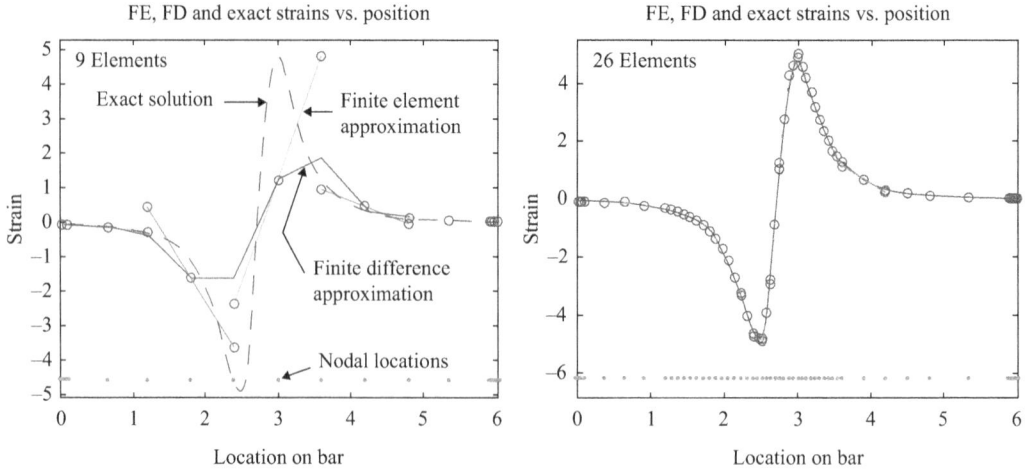

Figure 11.1. Finite element, exact, and smoothed strain representations.

The result of adaptively refining the initial model with one application of this refinement guide is shown in Fig. 11.1b. As can be seen, the three strain representations contained in the refined model are nearly identical. That is to say, the higher complexity of the exact solution is closely represented by the finite element solution and the improved strain distribution formed from the finite element result. The refined model satisfied the termination criterion with one iteration of the adaptive refinement procedure.

The dots along the bottom of the strain distributions identify the nodal locations of the finite elements that comprise the model. In Fig. 11.1b, the concentration of the nodes in the center of the model indicates that the elements added to the model by the refinement process were placed in the locations where they were needed. As will be seen in the examples that follow, the modifications to the finite element model identified by this refinement guide lead to the rapid satisfaction of the termination criterion.

The objective of this chapter is to develop and demonstrate refinement guides based on the *in situ* element evaluation process described above.

The refinement guides developed here have two significant advantages over existing approaches. On one hand, the level of refinement is directly related to the cause of the modeling errors. That is to say, the refinement guide attempts to define a level of refinement so that the polynomial bases of the subdivisions can represent the complexity of the exact solution to within the specified level of accuracy. On the other hand, the new refinement guide is computed from point-wise quantities so no integrals are required for its computation [1–3].

An overview of the development of the finite difference approach to smoothing is presented in the next section. The theoretical background that provides the basis of this development is contained in the following two Sections. The development of the procedure for quantifying the discretization errors and forming the refinement guides follows. Finally, a series of illustrative examples are used to demonstrate the efficacy of the refinement guides developed here.

11.2 THEORETICAL OVERVIEW FOR FINITE DIFFERENCE SMOOTHING

Modeling errors occur in finite element solutions when the polynomial basis for an individual finite element cannot capture the complexity of the exact result that exists on its domain. The refinement guide developed here compares the modeling capabilities of a single element to an improved strain representation that exists on the domain of the element being evaluated. The improved solution is a smoothed, higher order strain approximation that is derived from the current finite element results. This smoothed solution is taken to be an approximation of the exact solution that is emerging from the finite element model. The comparison between the smoothed strain representation and the modeling capabilities of the individual elements identifies the number of subdivisions an element needs to capture the higher order strain approximation to within the desired level of accuracy.

The most direct approach for extracting the higher order strain representation using finite difference operators is described as follows. First, **a higher order finite difference template** is placed at the center of the element being evaluated [4]. Then, the displacements for the element being evaluated and the displacements for selected interior nodes of the adjacent elements are introduced into the template in order to extract the higher order strain terms.

However, in this presentation, a two-step process is used to identify the higher order strain quantities contained in the improved solution rather than the one-step process just described. The advantage of the two-step process is that a lower order finite difference template is used instead of a higher order one. In the first step of this process, the interelement strains are computed. Then, the interelement strains found in the first step and the nodal strains on the interior of the element being evaluated are inserted into a central difference template in order to estimate the higher order strain terms that exist in the smoothed solution.[1]

The estimates of the modeling deficiencies in the individual finite elements are identified as the differences between the smoothed and the finite element strain distributions. These differences are computed by subtracting the strain gradient form of the Taylor series representations of the two approximate strain representations from each other. In the three-node bar elements used in the examples presented here, the first three terms of the smoothed strain representation are subtracted from the two-term finite element strain representation over the domain of an individual finite element.

---

[1] The process of using two applications of lower order finite difference templates is similar to the procedure followed in most Runge–Kutta numerical integration schemes for solving differential equations. Higher order differential equations are reduced to a set of first-order equations so the solution process is applicable to any order of differential equation.

The strain representation contained in a three-node bar element consists of the following two physically interpretable Taylor series coefficients, namely, ε_x and $d\varepsilon_x/dx$. The finite difference representation of the smoothed strain used in these examples also contains the next term in the Taylor series expansion. The coefficient of this higher order term, $d^2\varepsilon_x/dx^2$, is the curvature that exists in the smoothed strain representation on the domain of the element being evaluated.

When this subtraction is performed, the estimated error in the strain representation for the finite element being evaluated is given as:

$$\Delta\varepsilon_x = \left((\varepsilon_x)_0 + (\varepsilon_{x,x})_0 x + (\varepsilon_{x,xx})_0 x^2/2 \right)_{\substack{\text{Smoothed} \\ \text{Solution}}} - \left((\varepsilon_x)_0 + (\varepsilon_{x,x})_0 x \right)_{\substack{\text{Finite} \\ \text{Element}}} \quad \text{(Eq. 11.1)}$$

The Taylor series coefficients in Eq. 11.1 are expressed in strain gradient notation where: $(\varepsilon_{x,x})_0 = (d\varepsilon_x/dx)_0$ and $(\varepsilon_{x,xx})_0 = (d^2\varepsilon_x/dx^2)_0$.

The constant strain terms have the same value for both approximate representations because both expansions use the interior node of the three-node element as the local origin. The equality of the strains at the local origin in the center of an element for the two approximate strain representations can be seen in Fig. 11.1a.

Since the constant strain coefficients are identical for the two strain representations, they cancel each other when the two strain representations are subtracted. When Eq. 11.1 is simplified, the estimated modeling deficiency in the element being evaluated reduces to the following:

$$\Delta\varepsilon_x = \left[\left((\varepsilon_{x,x})_0 \right)_{\substack{\text{Smoothed} \\ \text{Solution}}} - \left((\varepsilon_{x,x})_0 \right)_{\substack{\text{Finite} \\ \text{Element}}} \right] x + \left[\left((\varepsilon_{x,xx})_0 \right)_{\substack{\text{Smoothed} \\ \text{Solution}}} \right] x^2/2 \quad \text{(Eq. 11.2)}$$

The estimated strain modeling error identified by Eq. 11.2 consists of both a first- and a second-order term. The first-order term contains the difference between the smoothed and the finite element slope representations. The second-order term is the most important component of this quantification of the inability of the finite element being analyzed to represent the smoothed solution. This term is important because it represents the magnitude of the curvature that exists in the smoothed solution.

Since a linear strain element cannot represent curvature, several lower order elements must be used to approximate this higher order term with straight-line segments. The refinement guide presented here estimates the number of the lower order elements that are needed to represent the higher order strain state to within the desired level of accuracy defined by the termination criterion. The characteristics of this representation of higher order strain states are discussed in detail in Chapter 9.

The three strain gradient components contained in Eq. 11.2 are labeled in Fig. 11.2 for the fourth element from the left in the finite element model. As can be seen in this figure, there is a significant difference between the finite element slope representation and the finite difference slope representation. This difference is captured in the first term of Eq. 11.2.

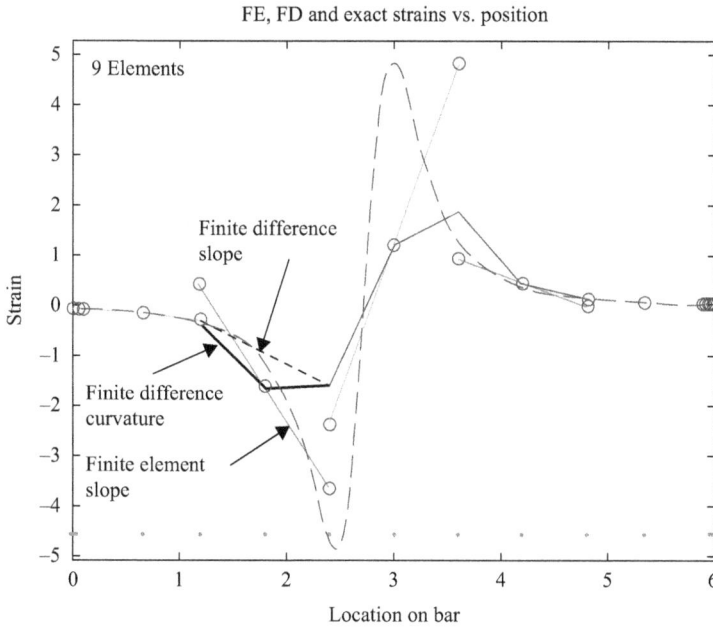

Figure 11.2. Finite difference parameters.

An approximation of the curvature in the smoothed strain representation is also shown for this element in Fig. 11.2. This approximation consists of the two heavy lines labeled as the "finite difference curvature." Since curvature is related to the rate of change of the slope, the finite difference approximation of the curvature for the finite element strain representation is a function of the difference between the slopes of the two segments of the heavy line shown in Fig. 11.2.

The level of refinement identified for an individual element is found by relating the maximum difference between the two approximate solutions given by Eq. 11.2 to the modeling capability of the element being evaluated via the termination criterion. This maximum difference for an element of length $2h$ is given by the following relationship:

$$|\Delta \varepsilon_x| = \text{abs}\left[\left((\varepsilon_{x,x})_0\right)_{\substack{\text{Smoothed} \\ \text{Strain}}} - \left((\varepsilon_{x,x})_0\right)_{\substack{\text{Finite} \\ \text{Element}}} \right] h + \text{abs}\left[\left((\varepsilon_{x,xx})_0\right)_{\substack{\text{Smoothed} \\ \text{Strain}}} \right] h^2/2 \quad \text{(Eq. 11.3)}$$

The level of refinement is found by identifying the maximum length for h that gives $|\Delta \varepsilon_x|$ a value that is less than or equal to the termination criterion. The variable h is used with Eq. 11.3 instead of the overall length because h is the distance from the local origin of this Taylor series expression. The process of identifying the number of elements to include in the refinement is be formalized in the next section.

11.3 DEVELOPMENT OF THE REFINEMENT GUIDE

The refinement guide developed here uses an *a posteriori* approach to estimate the level of refinement needed to rapidly improve the finite element model. This approach estimates the number of subdivisions an individual element needs to be given in order to satisfy the termination criterion. This is accomplished by comparing the modeling capability of the element to an improved strain representation that is extracted from the finite element result. The level of refinement identified is designed to satisfy the termination criterion in a small number of iterations.

Equation 11.3 estimates the inability of the modeling capabilities of an individual finite element to represent the exact solution being sought in the analysis as a function of the length of an element. This relationship, in conjunction with the termination criterion, is used to estimate the number of subdivisions to be given to an element so that the model produces acceptable results.

That is to say, the refinement guide is formed by satisfying the following condition:

$$\Delta\varepsilon_x(h) < \text{Termination Criterion} \qquad \text{(Eq. 11.4)}$$

The length of the element subdivisions is equal to $2h$. The maximum difference between the two approximate solutions will exist at one end of the element or the other, which is a distance h from the local origin of the element.

The number of elements that replaces the element being subdivided is found by dividing its length by the length of an element that will possess an acceptable error. The element being evaluated is designated as Element n and it has a length given as L_n. The number of subdivisions to be given to Element n is computed as follows:

$$N = \frac{L_n}{2h} \qquad \text{(Eq. 11.5)}$$

For example, let us say that Eq. 11.3 is either less than or equal to the termination criterion when h is equal to one-fifth of the original length of the element. This means that the element being evaluated must be divided into $N = L_n/2\,(L_n/5)$ elements. Since N for this case is equal to 2.5, the element being evaluated will be divided into three equal-length elements in the refined model.

An example of the results produced by this refinement guide is presented in Table 11.1. This table contains the refinements applied to the initial model shown in Figs. 11.1a and 11.2, the latter figure being an augmented version of Fig. 11.1a. The numbers in the second row of this table are found by following the process associated with Eq. 11.5.

Table 11.1. Element subdivisions for the initial nine-element model

El. no.	1	2	3	4	5	6	7	8	9
No. divisions	1	1	2	7	10	2	1	1	1

The termination criterion for this case is 4% of the maximum strain in the current finite element model. The results for this example are given in more detail in the next section. This discussion is designed to clarify the way in which the level of refinement is identified.

The contents of this table show that five of the nine elements in the initial model are not subdivided and that four are modified. The elements designated as Element 3 and Element 6 are each divided into two elements. Element 4, the element shown in Fig. 11.2 to define the strain gradient components of Eq. 11.3, is divided into seven elements. In the next Section, this element is shown to span a region that contains one inflection point and the minimum point for the strain distribution.

Element 5 is subdivided into 10 elements. In the next section, this element is shown to cover a region that contains two inflection points and the maximum point for the strain distribution. In other words, the level of refinement for a given element identified by this refinement guide can be seen to be related to the complexity of the exact solution that an element is being asked to represent.

Figure 11.1b presents the results produced by the model formed from the refinements contained in Table 11.1. As can be seen in this figure, these refinements produce a new model with 26 elements that represent the exact solution very well. This model satisfies the termination criterion of 4% of the maximum absolute strain in the finite element model. That is to say, the performance of this refinement guide exceeds expectations. The initial model is sufficiently refined in regions with excessive discretization error so that a second iteration is not needed to obtain a result with an acceptable level of error.

11.4 PROBLEM DESCRIPTION

This section describes the longitudinal bar problem with fixed ends that is being approximated in this Chapter. As can be seen in Fig. 11.1, the loading is such that the strains are nearly zero at both ends of the bar. The loading condition used in this problem was chosen because it produces a complex strain distribution on the interior of the bar and nearly zero strains on the boundaries. As a result, the boundary conditions have little if any effect on this problem [5–7].

In order to further reduce any boundary effects on the complex strain distribution on the interior of the bar, two very small elements are included in the model at both ends to absorb any small strain modeling errors. As can be seen in Table 11.1, the small elements on the boundaries are performing as expected.

None of the small elements on the boundary of the bar have been subdivided. This means that the errors in these elements satisfy the termination criterion, so the errors are insignificant. The presence of the elements designed to reduce any boundary effect make this model different from the control problem used in earlier chapters. These models are formed with three-node linear strain elements.

The longitudinal bar problem is loaded with a distributed load that produces a displacement in the exact solution that is a Runge function. This loading condition is chosen because

Runge functions are difficult for polynomial representations to capture [8]. The difficulty arises from the fact that Runge functions are rational polynomials, that is, one polynomial is divided by another polynomial. As such, these functions contain an infinite number of nonzero Taylor series terms. As a result, a finite number of polynomial terms cannot represent a Runge function exactly.

The displacement in the exact solution is based on the Runge function shown in Fig. 11.3: $f(x) = 300/(x + 30.0/2.0)^2$. This function is symmetric about the center of the domain. In order to make the problem more complex, the symmetry in the displacement function is eliminated by moving the displacement slightly away from the center of the bar. The off-center displacement function is the following: $f(x) = 300/(x + 30.0/2.2)^2$.

The loading condition that produces the displacement given by this Runge function is found by integrating this displacement function twice. When these two integrations are performed, the resulting "high-demand" loading condition is shown in Fig. 11.4a. The strain distribution produced by this loading is presented in Fig. 11.4b.

As can be seen in Fig. 11.4b, this complex strain distribution presents a difficult task for linear strain elements. It contains three inflection points, a maximum point, and a minimum point. The inflection points are indicated by the x's in Fig. 11.4b. These modeling complexities and the near-zero value for the strain on the boundaries are the reasons for choosing this problem to demonstrate this approach to adaptive refinement.

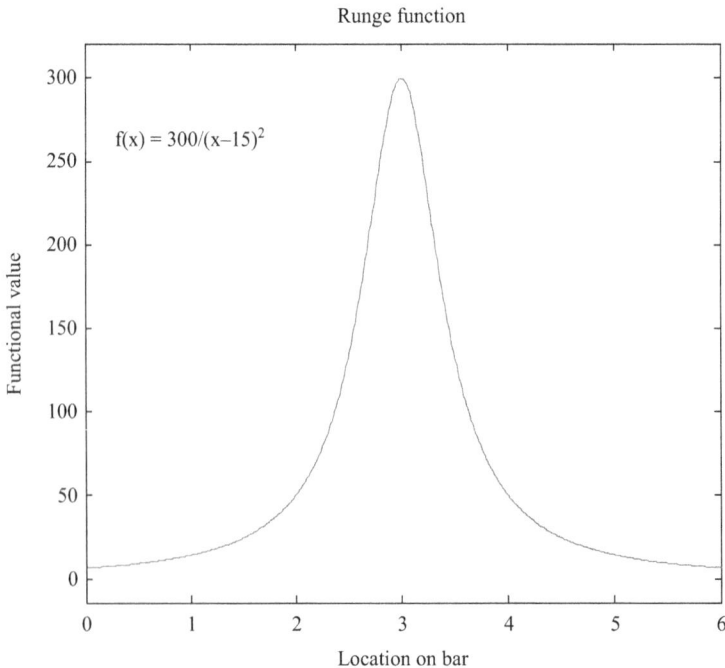

Figure 11.3. A Runge function.

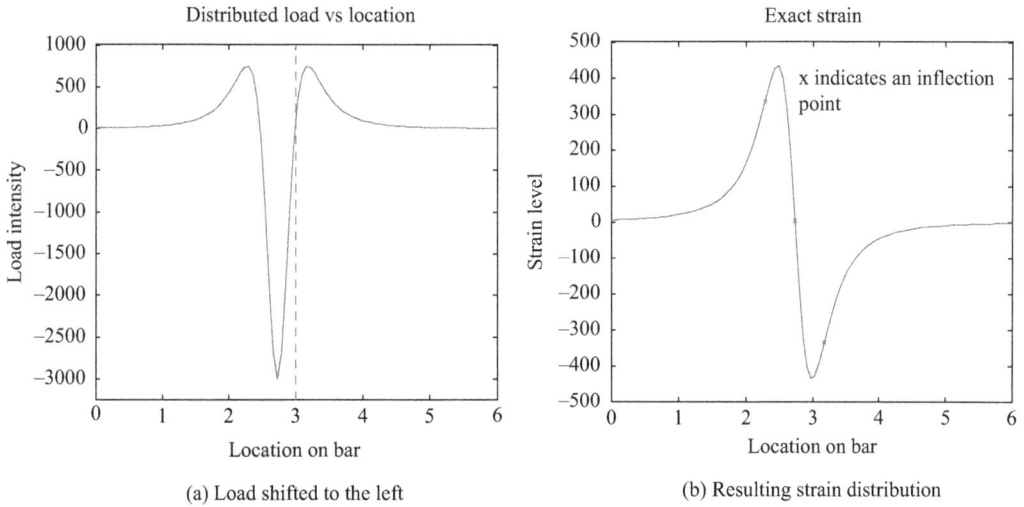

Figure 11.4. A "high-demand" loading condition and resulting strain distribution.

11.5 EXAMPLES OF ADAPTIVE REFINEMENT

This section contains three examples of problems that have been refined under the guidance of the approach developed here. The first objective of these examples is to demonstrate the effectiveness of the refinement guide. The second is to demonstrate the sensitivity of the refinement guide to the size of the discretization errors in a model and to the restriction imposed by the termination criterion.

The first example is the problem described in the previous Section and which is used in the Introduction to give an overview of this development. For the convenience of the reader, Fig. 11.1 is reproduced here as Fig. 11.5. Figure 11.5a presents the finite element and the smoothed solution. The smoothed solution is formed with finite difference strain approximations extracted from the finite element displacements for the initial nine-element model. These approximate solutions are superimposed on the exact strain distribution in this figure. As can be seen, there is a significant difference between these two approximate strain representations where they both vary greatly from the exact solution.

The differences between the two approximate solutions are computed with Eq. 11.3 and a termination criterion of 4% of the maximum strain in the finite element model. These differences are used in this example to form the refinement guides for the individual finite elements. The refinements for the individual elements contained in this problem are shown in Table 11.2. This table is identical to Table 11.1. It is reproduced here for the convenience of the reader.

Table 11.2 indicates that five of the elements in the initial model satisfy the termination criterion and, therefore, will not be subdivided. Two of the original elements will be subdivided into two elements. However, the two elements in the center of the model that need to represent the most complex region are subdivided into seven and 10 elements, respectively.

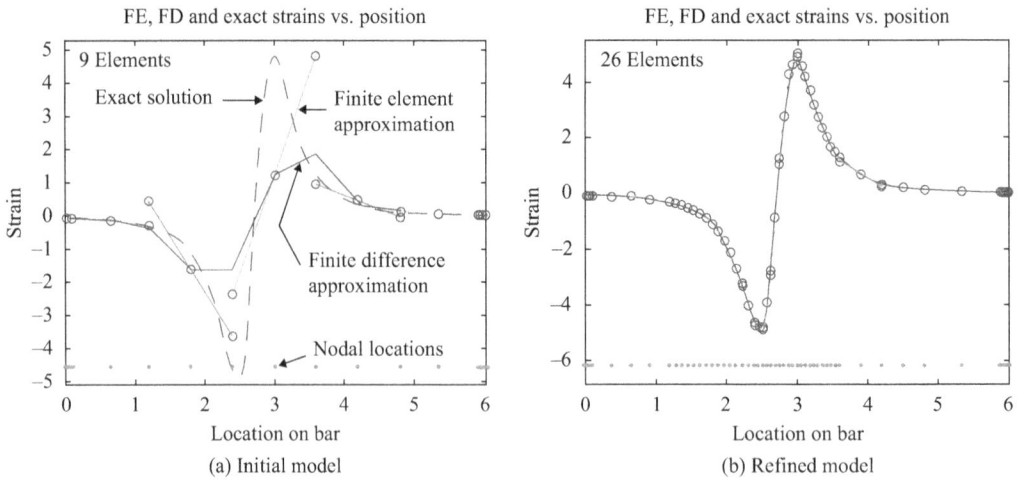

Figure 11.5. Initial and refined finite element models.

Table 11.2. Element subdivisions for initial nine-element model

El. no.	1	2	3	4	5	6	7	8	9
No. divisions	1	1	2	7	10	2	1	1	1

Figure 11.5b contains the same three strain representations as Fig. 11.2b for the refinement model containing 26 elements. As can be seen, the three strain representations are close to each other. There are no significant differences between the two approximate solutions given by the finite element and the smoothed approximations. Since the refined model satisfies the specified termination criterion of 4%, little difference exists between the two approximations and the exact result. The similarity of the three strain distributions in the refined model is shown in detail with a close-up of the three strain representations of the minimum and maximum strains in Fig. 11.6. As can be seen, the interelement jumps in the strain, which form the basis of the termination criterion, are small.

The refinement guide will now be applied to an initial model that better represents the exact solution than the initial model for the first example. The initial model for this case consists of 15 three-node elements instead of the nine elements contained in the initial model for the previous example. As was the case in the previous example, the elements are of equal length.

This example is designed to show the sensitivity of the refinement guides to the level of error in the finite element strain representations being evaluated. The strain representations for the initial model are shown in Fig. 11.7a. By comparing Fig. 11.7a with Fig. 11.5a, it can be seen that the interelement jumps in strain are significantly smaller for this 15-element model than for the nine-element model.

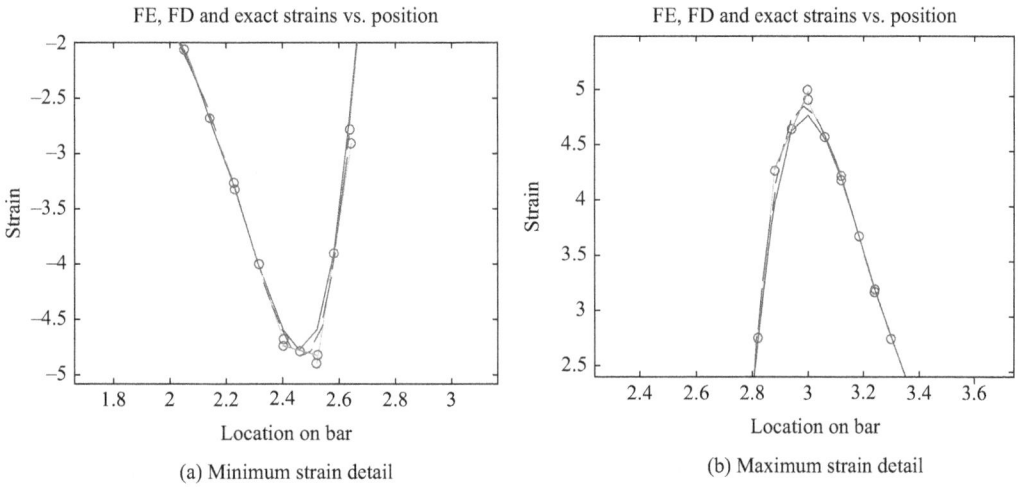

(a) Minimum strain detail

(b) Maximum strain detail

Figure 11.6. Strain representations at the extreme points—26-element model.

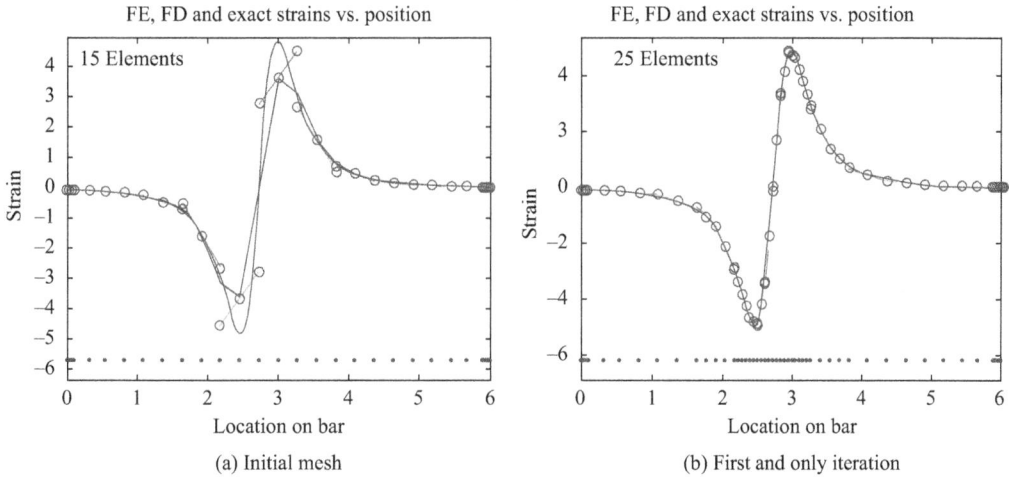

(a) Initial mesh

(b) First and only iteration

Figure 11.7. Initial and refined finite element models.

The refinements identified by the refinement procedure for this model are shown in Table 11.3. In this case, the refinement guide added 10 new elements to the initial model, which contrasts with the 17 elements that were added to the previous example.

Table 11.3. Refinements for 15-element model, termination criterion of 4%

El. no.	1	2	3	4	5	6	7	8	9	10	11	12	13	14	15
No. divisions	1	1	1	1	1	2	5	5	2	1	1	1	1	1	1

The strain results for the final model are shown in Fig. 11.7b. As can be seen when Fig. 11.7a is compared to Fig.11.7b, the elements were added in the regions that needed refinement. This model satisfies the termination criterion. As was the case in the previous example, it took only one cycle of the adaptive refinement procedure for the model to satisfy the termination criterion.

In the final example, the previous model with 15 elements in the initial model is evaluated with a more restrictive termination criterion. The termination criterion is reduced from 4% to 2% of the maximum absolute strain in the finite element model. The objective of this example is to demonstrate the sensitivity of the refinement guides to the termination criterion. The refinements are identified in Table 11.4.

The initial strain distribution is, of course, the same as that shown in Fig. 11.7a. The final strain distribution for the model with 33 elements is presented in Fig. 11.8a. As can be seen, elements are added only in the critical regions of the maximum and minimum strains. This final result is contrasted with the final result for the same initial model with the 4% termination criterion that is presented in Fig. 11.8b. Again, the termination criterion was satisfied with one application of the adaptive refinement procedure.

The strain models for the critical points are shown with close-ups in Fig. 11.9. When these close-ups are compared to the magnified strains in the critical regions for the 25-element model shown in Fig. 11.6, the strain distributions in these regions more closely match the exact result than do the distributions for the coarser model shown in Fig. 11.6. This example has demonstrated that the refinement guides are sensitive to the termination criterion.

Table 11.4. Refinements for 15-element model, termination criterion of 2%

El. no.	1	2	3	4	5	6	7	8	9	10	11	12	13	14	15
No. divisions	1	1	1	1	1	2	9	9	2	1	1	1	1	1	1

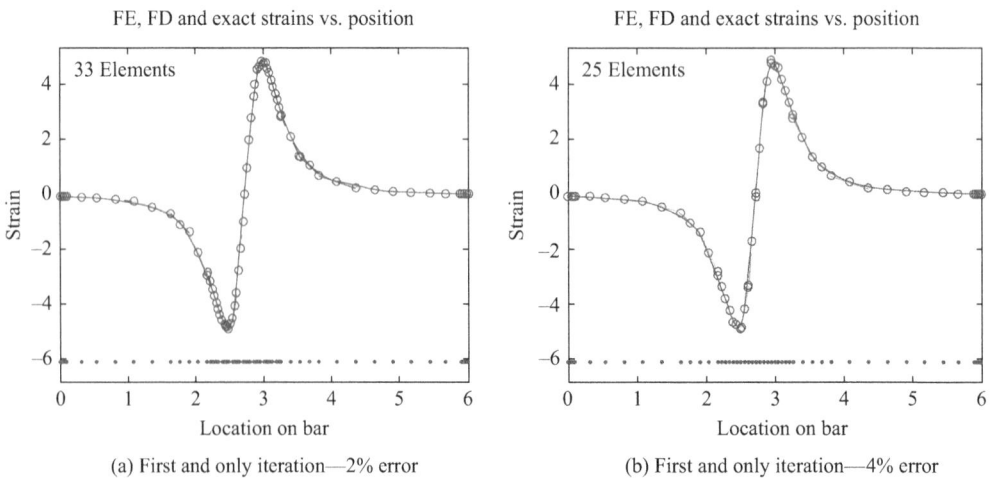

FE, FD and exact strains vs. position — 33 Elements

(a) First and only iteration—2% error

FE, FD and exact strains vs. position — 25 Elements

(b) First and only iteration—4% error

Figure 11.8. Refined finite element models—initial 15 elements.

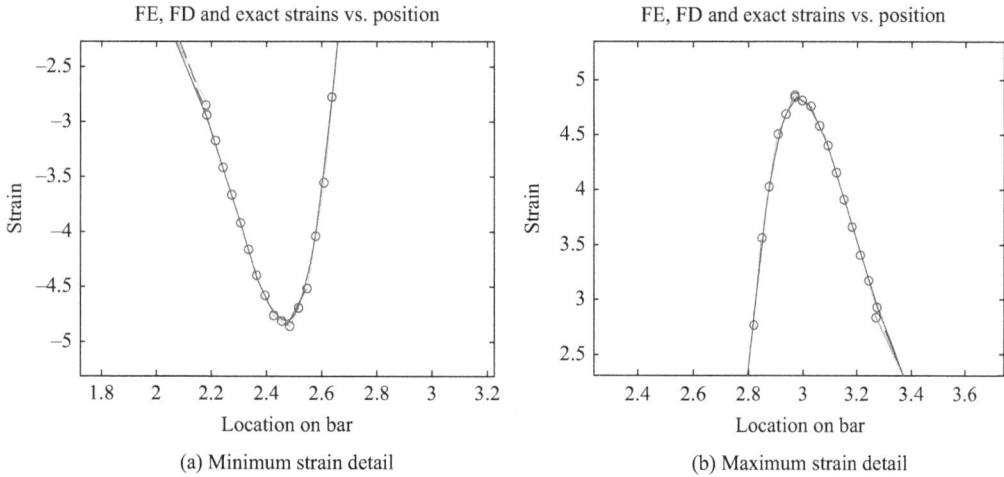

Figure 11.9. Strain representations at the extreme points in the 33-element model.

11.6 AN EFFICIENT REFINEMENT GUIDE BASED ON NODAL AVERAGING

This Section presents and demonstrates a refinement guide that forms the smoothed solution using a simpler approach than that used in the refinement guide developed earlier. The smoothed strain representation is formed by averaging the strains at the interelement nodes. This approach is identical to the one used in the Zienkiewicz and Zhu (Z/Z) error estimator discussed in Chapter 7. Once the smoothed solution is formed, the procedure for generating this refinement guide is identical to the procedures presented in Sections 11.2 and 11.3 for the refinement guide just demonstrated.

The approach presented in this Section has significant practical advantages, particularly in the case of multidimensional problems. When the nodal strains are found by applying a finite difference template to the nodal displacements, the locations of the nodes surrounding the point of interest must be identified in order to form the necessary finite difference template. In the multidimensional case, this is **not** a minor bookkeeping problem.

In contrast, when the higher order strain representations are formed for the individual elements from the average nodal strains, no such topological information is required. The nodal strain quantities for the individual elements are readily available. Similarly, the nodal coordinates of the individual elements needed to form the finite difference template at the local origin of the individual elements are part of the finite element model. As a result, the higher order strain representations for the individual elements are readily available.

The efficacy of this approach is demonstrated by applying the refinement guide developed in this section to the examples used earlier in the chapter.

The strain distribution formed by connecting the averages of the nodal strain for the nine-element model that was evaluated earlier with straight lines is shown in Fig. 11.10a. The strain distribution

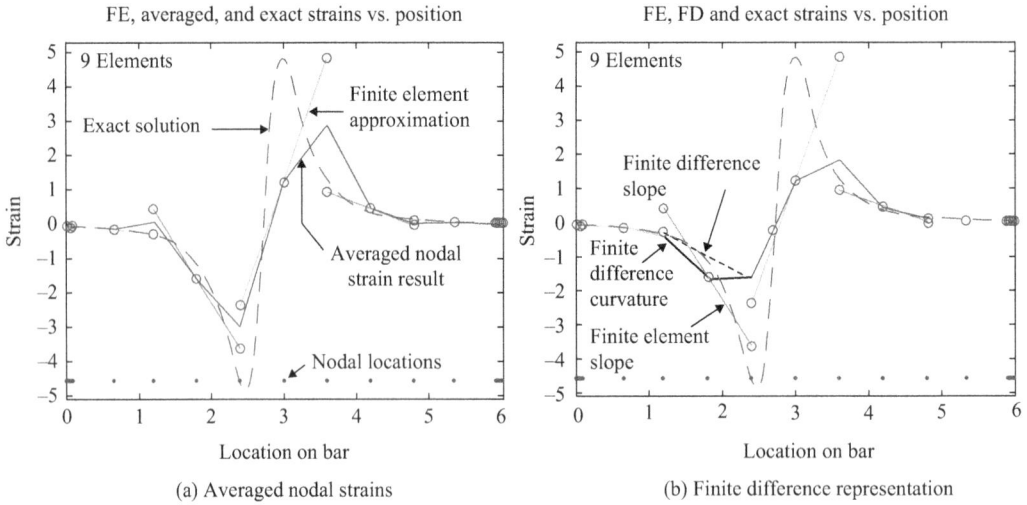

Figure 11.10. Smoothed strain comparison.

formed using the finite difference templates is shown in Fig. 11.10b. As can be seen when Figs. 11.10a and 11.10b are compared, the two approaches have both similarities and differences.

The two strain distributions are nearly identical when the finite element model accurately represents the exact solution. This is to be expected since both approximate solutions will converge to the exact solution in the limit. However, the averaged and finite difference strain representations can be very different when the finite element solution is highly inaccurate. The effect of this difference is seen in the examples that follow.

When the refinement guide for the nine-element model is formed using the averaged nodal strain approach, the refinements produced by this approach are identified in Table 11.5. When the subdivisions identified in row 2 of this table are summed, we see that the refined model will contain 18 elements. As we will see, the refinements identified here did not produce a model that satisfies the termination criterion.

When the results contained in Table 11.5 are compared to the refinements identified by the finite difference approach to smoothing contained in Table 11.2, which is reproduced here for the convenience of the reader, the similarities and differences just discussed can be seen.

When the two tables are compared, we see the following. The three elements that are closely representing the exact solution at the two ends of the model are either not refined or are only subdivided

Table 11.5. Element subdivisions for initial nine-element model

El. no.	1	2	3	4	5	6	7	8	9
No. divisions	1	1	2	2	4	5	1	1	1

Table 11.2. Element subdivisions for initial nine-element model

El. no.	1	2	3	4	5	6	7	8	9
No. divisions	1	1	2	7	10	2	1	1	1

into two elements by both refinement guides. This is to be expected since both smoothed solutions are nearly identical when the approximate result closely represents the exact solution.

However, the two refinement guides subdivide the elements containing high levels of error in the center of the model by significantly different amounts. The refinement guide based on two applications of finite difference smoothing give these elements a larger number of subdivisions. In fact, the adaptive refinement process guided by this refinement procedure needs only one application of the adaptive refinement process to satisfy the termination criterion. The final model for this case contains 26 elements.

When the 18-element model identified by the refinements contained in Table 11.5 is solved, the strain distribution produced by this model is shown in Fig. 11.11a. As can be seen, there are significant interelement jumps in the strain distribution. This model does not satisfy the termination criterion. This means that the adaptive refinement process must be applied again.

When the 18-element model is evaluated, the refinements to the model are identified in Table 11.6. This produces a model with 22 elements. The strain distribution for this refined model is shown in Fig. 11.11b. This model satisfies the convergence criterion. As can be seen, this refinement guide introduces new elements where they are needed.

When Fig. 11.11b is compared to Fig. 11.5.b, we see that, although the smoothing formed using nodal averaging requires two iterations of adaptive refinement, it produces an acceptable model with fewer elements.

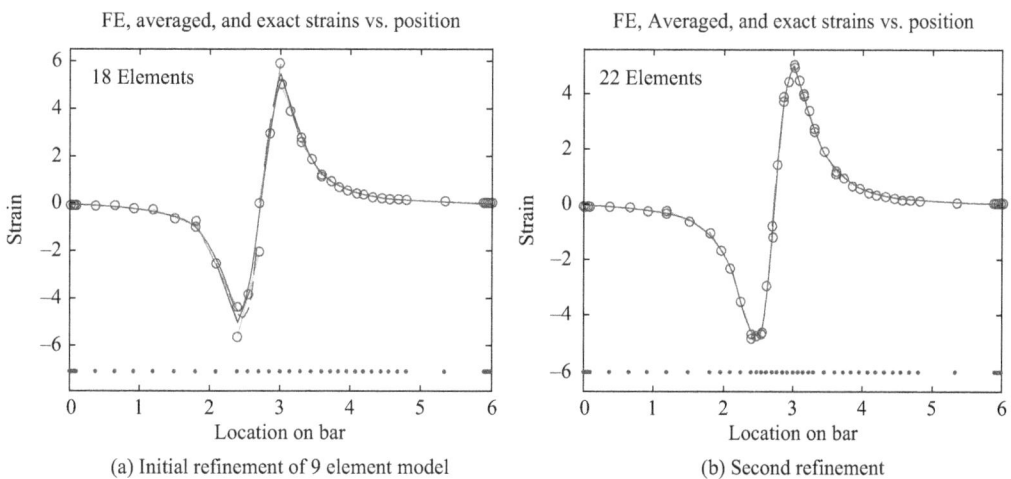

(a) Initial refinement of 9 element model   (b) Second refinement

Figure 11.11. Refinements of the nine-element model—4% termination criterion.

Table 11.6. Element subdivisions for initial 18-element model

El. no.	1	2	3	4	5	6	7	8	9	10	11	12	13	14	15	16	17	18
No. divisions	1	1	1	1	1	2	2	2	2	1	1	1	1	1	1	1	1	1

11.7   FURTHER COMPARISONS OF THE REFINEMENT GUIDES

The performance of the two versions of the refinement guide will be compared in this Section with the problem that has 15 elements in the initial model. This set of examples compares the behavior of the refinement guides when the errors in the initial model are smaller than in the case where the initial model has nine elements.

The strain distribution formed using the averaging approach and the finite difference approach for the initial model of this problem are presented in Figs. 11.12a and 11.12b, respectively. The performance of the two refinement guides is compared for two different termination criteria, namely 4% and 2%.

When this initial model is refined with the nodal averaging approach and a termination criterion of 4%, the refinements identified for the individual elements are presented in Table 11.7.

Table 11.7. Element Subdivisions for Initial 15-Element Model—4%

El. no.	1	2	3	4	5	6	7	8	9	10	11	12	13	14	15
No. divisions	1	1	1	1	1	3	6	6	3	1	1	1	1	1	1

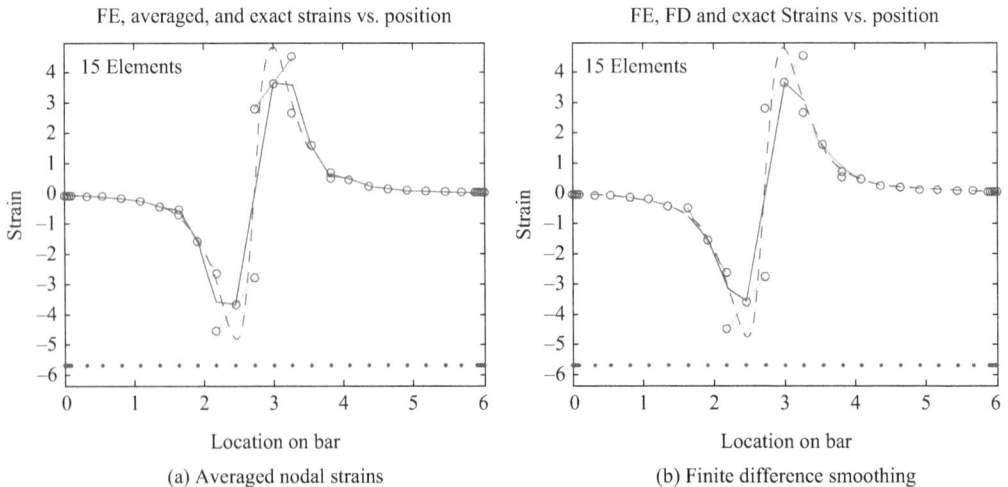

Figure 11.12. Smoothed strain comparison.

This produces a final model containing 29 elements. The strain distribution for this case is shown in Fig. 11.13a. This model satisfies the termination criterion.

The number of subdivisions found for the individual elements when the finite difference approach provides the smoothing is presented in Table 11.3. This table is reproduced here for the convenience of the reader. This refinement produced a model with 25 elements. The strain distribution for this refined model is shown in Fig. 11.13b.

When Tables 11.7 and 11.3 are compared, we see that they are very similar. Both refinement procedures identify the same elements as satisfying the termination criterion, that is, the elements that are not subdivided. The refinement guide that uses nodal averaging in the first step subdivides each of the remaining elements to a higher degree. One more element is contained in the refinement of each element for the averaging approach than for the finite difference approach. Both of the refinement guides produce models that satisfied the termination criterion in a single application of the adaptive refinement process.

The difference in the number of elements in the center for the averaging approach can be seen in Fig. 11.13 by the difference in the concentration of the nodes shown at the bottom of these figures. When the distributions of nodes for the two final models are compared, it appears that the finite difference approach concentrates the elements where they are needed better than the averaging approach.

Table 11.3. Refinements for 15-element model, termination criterion of 4%

El. no.	1	2	3	4	5	6	7	8	9	10	11	12	13	14	15
No. divisions	1	1	1	1	1	2	5	5	2	1	1	1	1	1	1

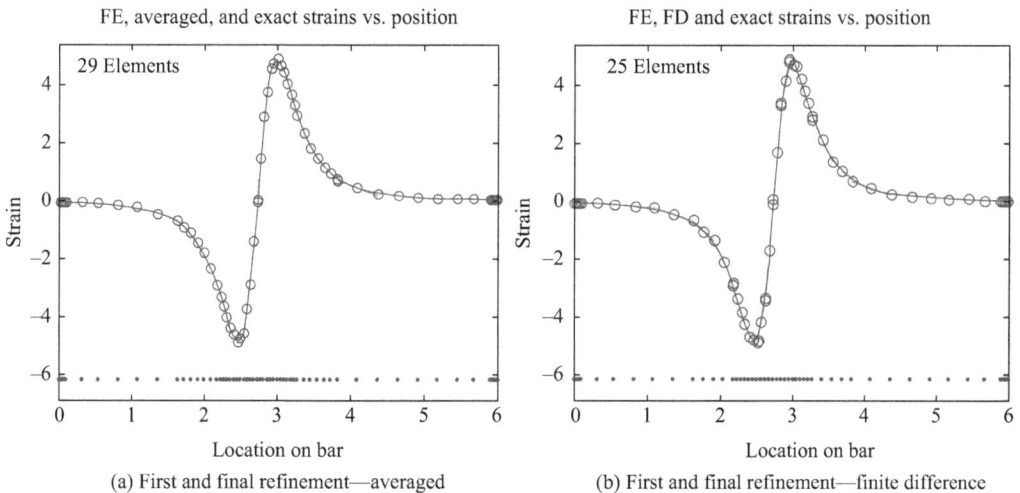

FE, averaged, and exact strains vs. position

FE, FD and exact strains vs. position

(a) First and final refinement—averaged

(b) First and final refinement—finite difference

Figure 11.13. Refinement of the 15-element model—4% termination criterion.

When the initial model with 15 elements is refined with the nodal averaging approach and a termination criterion of 2%, the refinements identified for the individual elements are presented in Table 11.8. This produces a final model containing 43 elements. The strain distribution for this case is shown in Fig. 11.14a. This model satisfies the termination criterion.

The number of subdivisions found for the individual elements when the finite difference approach provides the smoothing is presented in Table 11.4. This table is reproduced here for the convenience of the reader. This refinement produced a model with 33 elements. The strain distribution for this refined model is shown in Fig. 11.14b.

When Tables 11.8 and 11.4 are compared, they are seen to have a similar structure. Both refinement procedures identify the same elements as satisfying the termination criterion. In this case, the first four elements and the last five elements are not subdivided by either refinement guide. Again, the refinement guide that uses nodal averaging in the first step subdivides each of the remaining elements to a higher degree. In this case, either two or three additional elements are introduced for each of the four elements representing the most complex portion of the exact solution. Both of the refinement guides produce models that satisfy the termination criterion in one application of the adaptive refinement process.

The difference in the number of elements in the center for the two refinement procedures can be seen in Fig. 11.14 by the difference in the concentration of the nodes shown at the bottom of these figures. Again, the finite difference approach concentrates the added elements in the final model where they are needed better than the averaging approach.

Table 11.8. Element subdivisions for initial 15. element model—2%

El. no.	1	2	3	4	5	6	7	8	9	10	11	12	13	14	15
No. divisions	1	1	1	1	1	5	11	11	5	1	1	1	1	1	1

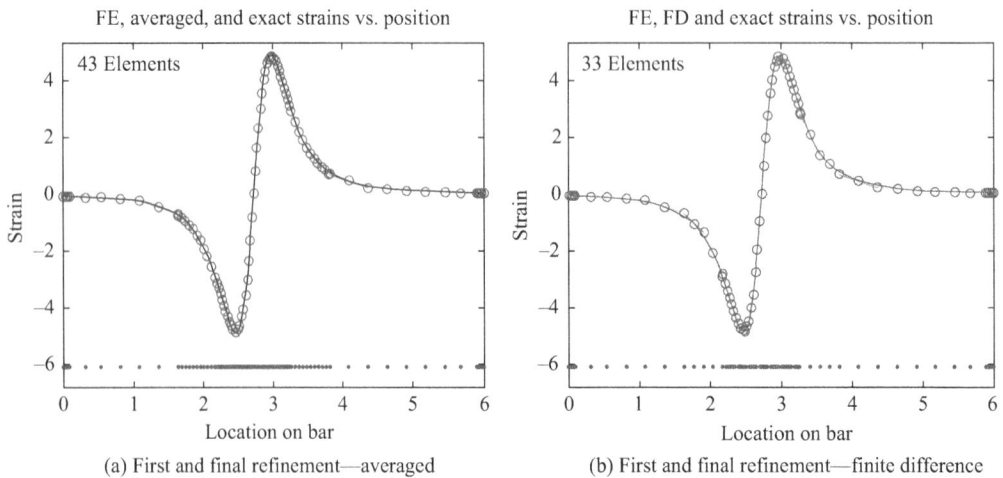

(a) First and final refinement—averaged

(b) First and final refinement—finite difference

Figure 11.14. Refinement of the 15-element model—2% termination criterion.

Table 11.4. Refinements for 15-element model, termination criterion of 2%

El. no.	1	2	3	4	5	6	7	8	9	10	11	12	13	14	15
No. divisions	1	1	1	1	1	2	9	9	2	1	1	1	1	1	1

11.8 SUMMARY AND CONCLUSION

The development presented here forms refinement guides in a totally new way. This approach compares the modeling capability of an individual element to an approximation of a higher order strain representation on the domain of the individual elements. The level of refinement depends on the differences between the finite element and the higher order strain representation in the element and the termination criterion. The refinement guide attempts to identify the level of refinement needed to produce a model that satisfies the termination criterion in very few iterations. In several of the examples presented here, only one iteration of the adaptive refinement process was needed to produce an acceptable result.

This refinement guide is designed to provide a rational alternative to refinement guides that are correlated to the magnitude of the error estimator. In other words, the refinement guides developed here are based on the causes of the modeling errors and not on the effect of the errors. This, of course, means that the new approach opens the door to further developments, as is usually the case when the basic understanding of a problem emerges.

A higher order strain representation is formed that is closer to the exact strain distribution than is the discontinuous finite result. The improved strain distribution is formed in two different ways. In one approach, the improved solution is formed by two applications of finite difference templates. In the first application, nodal strains are found by applying finite difference templates to a set of nodal displacements. In the second application of the finite difference templates, the nodal strains found in the first application are used to find the improved, higher order strain distribution. In the second approach, the first application of the finite difference templates to find nodal strains is replaced by averaging the nodal strains of the finite element result.

Both approaches produce rapidly converging results. The use of the averaging approach is computationally simpler, particularly for higher dimension models.

The development of this refinement guide utilizes many of the improvements to computational mechanics that result from the use of physically interpretable notation:

1. The notation eliminates the separation between the solution technique and the problem being solved because the problem is expressed in terms of the quantities being sought in the analysis.
2. The modeling capabilities of individual elements can be identified during the formulation process.
3. The procedures used in Item 2 are extended to form an improved, higher order strain representation that is taken to be an approximation of the exact solution that is emerging from the finite element model.

4. The finite element and the finite difference methods can be formulated from the same basis, so the finite difference method can be used to evaluate any finite element model. This follows from the fact that the finite difference method can be used to solve practically any problem that the finite element method can solve.

11.9 REFERENCES

1. Akin, J. E. *Finite Element Analysis with Error Estimators* (Chapter 5–Error Estimates for Elliptic Problems), Oxford: Elsevier, 2005.
2. Zienkiewicz, O. C. and Zhu, J. Z. "A Simple Error Estimator and Adaptive Process for Practical Engineering Analysis," *International Journal for Numerical Methods in Engineering* 24 (1987): 337–57. DOI: http://dx.doi.org/10.1002/nme.1620240206.
3. Kelly, D. W. "The Self-Equilibration of Residuals and Complementary *A-Posteriori* Error Estimates in the Finite Element Method," *International Journal for Numerical Methods in Engineering* 20 (1984): 1491–506. DOI: http://dx.doi.org/10.1002/nme.1620200811.
4. Dow, J. O. *A Unified Approach to the Finite Element Method and Error Analysis Procedures*, New York: Academic Press, 1999.
5. Dow, J. O., Ho, T. H. and Cabiness, H. D. "A Generalized Finite Element Evaluation Procedure," *Journal of Structural Engineering* 111(ST2) 1985: 435–52. DOI: http://dx.doi.org/10.1061/(ASCE)0733-9445(1985)111:2(435).
6. Dow, J. O., Jones, M. S. and Harwood, S. A. "A Generalized Finite Difference Method for Solid Mechanics," *International Journal for Partial Differential Equations* 6(2) (1990): 137–152. DOI: http://dx.doi.org/10.1002/num.1690060204.
7. Dow, J. O., Jones, M. S. and Harwood, S. A. "A New Approach to Boundary Modeling for Finite Difference Applications in solid Mechanics Problems," *International Journal for Numerical Methods in Engineering* 30(1) (1990): 99–113. DOI: http://dx.doi.org/10.1002/nme.1620300107.
8. Van Loan, C. F. *Introduction to Scientific Computing*, 2nd Ed., (Chapter 2–Polynomial Interpolation), Upper Saddle River, NJ: Prentice-Hall, 2000.

CHAPTER 12

SUMMARY AND RESEARCH RECOMMENDATIONS

12.1   INTRODUCTION

This concluding chapter puts the developments presented in this book in context and identifies research opportunities in computational mechanics that derive from these advances. These objectives are accomplished in the following five Sections.

The **first** section summarizes the primary objective of this book, which is to provide an overview of adaptive refinement and develop improvements to its components. The **second** section reviews the significance of the reinterpretation of the displacement interpolation functions in terms of the physically interpretable strain gradient quantities that have been presented in this book.

The **third** and **fourth** sections outline the simplifications and extensions to the finite element and finite difference methods made possible by using the physically interpretable notation. These advances, in turn, allow the improvements to the adaptive refinement process to be based on physical concepts.

Finally, the **fifth** section offers recommendations for future work and research topics that have been made available because of the advances contained in this book.

12.2   AN OVERVIEW OF ADVANCES IN ADAPTIVE REFINEMENT

The advances in adaptive refinement represent extensions of the two principal approaches to the analysis of errors in finite element results, namely, the recovery and residual approaches. The basics of these approaches to error analysis have been presented in Chapters 4 and 7, respectively. In this section, the advances in adaptive refinement are outlined in the context of extensions of the two approaches to error analysis.

The recovery technique forms, that is, recovers an improved strain distribution from the finite element solution. The improved strain distribution is formed by imposing continuity, in one form or another, on the discontinuous finite element strain representations. Then, the error contained in an element is estimated by computing a metric based on the difference between the improved strain result and the discontinuous finite element strain distribution. In this work, an extension of the recovery approach has been used to form a new type of refinement guide that directly identifies the strain modeling deficiencies in an individual element.

This new refinement guide extends the recovery approach by first decomposing the improved solution into the strain gradient components of the physically interpretable notation. Then, the modeling deficiencies in the element being evaluated are identified by comparing the modeling capabilities of the element expressed in terms of strain gradient quantities to the decomposition of the improved solution. As a result of this comparison, an estimate is made of the number of sub-divisions that needs to be given to the element being evaluated so that the estimated error in the subsequent model ends up below the termination criterion.

The new refinement guide discussed above also provides an implicit improvement to the error estimation process. Until the development of this refinement guide, the degree of refinement was based on correlations with the magnitudes of the error estimates. With the advent of refinement guides based on physical principles, the error estimators can be returned to their primary role of quantifying the accuracy of the finite element results. The error estimators need not be directly involved in iden-tifying the level of refinement needed to rapidly improve the finite element model.

As a consequence of this change in the role of the error estimator, the accuracy of a solution can best be evaluated if the estimated errors are expressed in terms of quantities of direct interest in solid mechanics. Then, the decision of whether or not to refine an element can be made for reasons that directly relate to meaningful engineering criteria.

The two new error estimators that have been developed in this work satisfy the requirements described above. The new error estimators identify the error in terms of quantities of direct impor-tance to solid mechanics by quantifying the failure of the finite element result to satisfy the governing differential equations being solved. As such, these error estimators can be classified as point-wise extensions of the residual approach to error analysis.

The error estimator developed in Chapter 4 estimates the discretization errors in the finite ele-ment model in terms of the interelement jumps that exist in the discontinuous strain representations of a finite element solution. In this case, the point-wise errors are expressed as a percentage of a speci-fied strain value. Thus, the termination criterion can be related directly to a failure criterion.

In the second form of residual error measure developed in Chapter 8, the failure of the finite ele-ment result to satisfy the finite difference approximation to the differential equation being solved is computed at selected nodal points. This point-wise quantity is expressed in units that are equivalent to an applied distributed load. This means that the level of error can be directly related to the level of loading on the problem being analyzed by comparing this fictitious load to the level of strains produced by the actual load. This error estimate has the advantage that the error is directly related to the element being evaluated. This contrasts with the interelement jumps in strain that are aggregates of the errors existing in all of the intersecting elements at a given node.

12.3 DISPLACEMENT INTERPOLATION FUNCTIONS REVISITED: A REINTERPRETATION

All of the developments presented in this book result from a reinterpretation of the standard form of the finite element displacement interpolation functions. In this reinterpretation, the arbitrary coefficients of the displacement interpolation polynomials are replaced with coefficients expressed in terms of quantities that possess physical meaning, namely, rigid body motions and strain quantities. Two insights into the nature of the interpolation polynomials resulted in this reinterpretation and subsequent improvement of the interpolation functions.

In the first insight, the finite element displacement interpolation functions are recognized as truncated Taylor series expansions. Since the finite difference derivative approximations are also formed from truncated Taylor series expansions, finite difference templates can be formed at every node of an irregular mesh in a finite element model. This means that the developments presented here that utilize a finite difference computation are generally applicable. Furthermore, this means that the two methods have been unified.[1] As a result, the finite element and the finite difference methods can represent the same problems, which, in turn, means that the finite difference method has been infused with new life.

The fact that the two methods share a common basis is exploited in this work in order to improve the adaptive refinement process in two ways. In Chapter 8, this relationship is used to create a residual-based point-wise error estimator. Later in Chapter 11, the smoothed or improved solution that forms the basis for one of the refinement guides developed there uses finite difference approximations formed with irregular templates.

The second insight into the nature of the displacement interpolation functions recognizes that the coefficients of the Taylor series expansions can be expressed in terms of the quantities that cause displacements in the continuum, namely, rigid body motions and strain quantities. The coefficients can be interpreted in terms of these physical quantities because the coefficients of the Taylor series expansions are derivatives of the displacements with respect to spatial quantities.

For example, the coefficient of the x term for the displacement in the x direction, du/dx, can be interpreted as a normal strain term, that is, $\varepsilon_x = du/dx$. It should be noted that the rigid body displacements are considered as zero-th order strain quantities. The coefficients of the Taylor series expansions are related to strain gradient notation in Chapter 3.

---

[1] Readers familiar with the finite difference method will recognize that the use of irregular templates instead of templates with evenly spaced nodes means that there is a reduction in the order of the error in the finite difference approximation of the derivatives. This is of no great significance because given the availability of irregular meshes, finite difference models can be adaptively refined using modifications of the techniques presented here. The primary advantage of this recognition is the ability to handle irregular boundaries and fictitious nodes with a rational, easily apprehensible process. These capabilities, which are detailed in Reference [1], may inject new life into the application of the finite difference method in solid mechanics.

12.4   ADVANCES IN THE FINITE ELEMENT METHOD

The improvements to adaptive refinement developed here are based on advances made in the finite element and finite difference methods. These advances are made possible because of the insights gained by using physically interpretable notation. The objective of this Section is to outline the improvements and advances in the finite element method that have resulted from this notation.

When the strain approximations needed for computing the stiffness matrix are created using the new notation, the strain modeling characteristics of the element become available for visual inspection. In addition to identifying the modeling capabilities of the element, the visual inspection allows modeling errors to be identified, and sometimes, corrected. The errors consist of violations of the principles of mechanics explicitly seen due to the physical nature of the notation. The ability to identify modeling errors in this way was demonstrated for the case of a four-node element in Chapter 3.

The insight into the strain modeling characteristics of an element provided by this notation is like examining an x-ray or an MRI of the element's inner structure. These insights allow the procedure for forming the element stiffness matrices to be both simplified and improved. The major significance of the new formulation techniques is that it renders the isoparametric formulation procedure obsolete as was discussed in detail in Chapter 5.[2] The new approach improves on the isoparametric approach in four ways.

On the computational level, the number of integrals that must be evaluated is significantly reduced. For example, the number of integrals that must be evaluated for a six-node linear strain element decreases from 76 to 6 simple integrals. The use of strain gradient notation in element formulation is outlined in Chapter 5 and presented in detail in Chapter 10 and in Reference [1].

On the pedagogical level, the reduction in the number of integrals that must be evaluated simplifies the element formulation process by dispensing with two detours that exist in the isoparametric approach. The remaining integrals can be evaluated without resorting to Gauss quadrature.[3] This, in turn, eliminates the need to map the actual geometry of the element onto a "regular shape." Anyone familiar with the formulation of isoparametric elements is acquainted with the disruption of the theoretical development that occurs when detours must be made to introduce Gauss quadrature and the isoparametric mapping.

---

[2] The isoparametric approach was developed soon after the inception of the finite element method. The large number of integrals and the limited computer capacity in the era warranted the use of the isoparametric approach. This approach has been taken as the standard approach ever since. Element formulation has not been seriously revisited until the advent of the physically interpretable notation that has been presented here.

[3] In the implementation of the new formulation process shown in Reference [1], the necessary integrals are evaluated using Green's theorem, a topic studied in undergraduate calculus. However, other evaluation techniques can be used.

The third benefit of the new formulation procedure is more a matter of elegance and modeling efficiency than anything else. When an isoparametric element possesses a "nonstandard" shape, for example, a quadrilateral that is not a rectangle, the isoparametric mapping introduces errors into the integrals contained in the strain energy expression as shown in Chapter 5. These errors may be small and can be ameliorated by the adaptive refinement process. However, now that a better formulation procedure is available, the very fact that the isoparametric process can introduce errors into the finite element model disqualifies it as the standard approach.

The fourth advantage of the new element formulation procedure is the ability to identify the strain modeling capabilities and deficiencies of the individual elements by visual inspection. These improvements to the finite element method, which were demonstrated for the four-node element in Chapter 5, are summarized in the following paragraphs.

When the representations of the three strain components for the four-node quadrilateral are formed from the displacement polynomials, they contain modeling errors of both commission and omission. The errors due to commission are known as parasitic shear. This error is caused by the existence of normal strain terms in the shear strain expression. The presence of these errone-ous terms in the shear strain expression can be identified by visual inspection as they represent an obvious violation of the constitutive relationship. Normal strains are not coupled to the shear strains in the constitutive relationship, so normal strains are "parasitically" attached to the shear strain term.[4]

Unfortunately, the errors of omission in a four-node element cannot be corrected. They result from the absence of terms that add the effects of Poisson's ratio to the element, thereby making the element overly stiff. As a result of this modeling error, a four-node element ends up being no more effective than a three-node triangle. A four-node element can only correctly represent rigid body motions and constant strain states, which is also the modeling capability of a three-node linear strain triangle. These observations are made possible due to the clear identi-fication of the modeling characteristics of an element produced by the physically interpretable notation.

As a consequence of the new formulation procedure, higher order elements become more attractive. The new formulation procedure reduces the computational effort required to form higher order elements. This is significant because finite element models formed with higher order elements are more efficient than models formed with lower order elements when the strain distribution is complex. This difference in modeling efficiency is demonstrated in Chapter 9.

---

[4] An approach known as reduced-order Gauss quadrature is sometimes applied in an attempt to eliminate the parasitic shear terms. However, it should be noted that unless extreme care is taken the use of reduced-order Gauss quadrature introduces other modeling errors into higher order elements. For example, the rote applica-tion of reduced-order Gauss quadrature can introduce "spurious zero energy modes" into eight-and nine-node elements. It is not clear whether the introduction of other errors can be avoided in higher order elements if the physically interpretable notation is not used. Furthermore, using this trick can introduce erroneous behavior in some plate elements that is not eliminated by model refinement. These ideas are discussed in detail in Reference [1].

12.5   ADVANCES IN THE FINITE DIFFERENCE METHOD

Some of the improvements to adaptive refinement that have been presented here are based on simplifications and improvements made to the finite difference method. As mentioned earlier, these advances are also largely due to insights resulting from the use of the physically interpretable notation that have been outlined earlier in this chapter. The improvements to the finite difference method are summarized in this section.

In the case of the finite difference method, the reinterpretation of the displacement interpolation functions as truncated Taylor series expansions with coefficients expressed in terms of strain quantities has resulted in two major advancements. The first is the recognition that the finite difference method can be applied using irregular meshes. The second rationalizes the treatment of boundary condition models in the finite difference method. On a larger scale, these improvements and those discussed in Reference [1] mean that finite difference models are able to represent any problem that can be modeled with the finite element method. Also, adaptive refinement procedures similar to those developed here for the finite element method can be applied to finite difference models.

The ability to form finite difference models with irregular meshes follows directly from the recognition that finite element interpolation functions are actually truncated Taylor series expansions. The ability to form finite difference approximations for irregular meshes is used in Chapter 9 to create a residual-based point-wise error estimator and to form the smoothed solution that is used in one version of the refinement guides developed in Chapter 11.

In the finite difference method, when a central difference template is applied on the boundary, nodes appear outside the boundary of the problem being analyzed. These nodes are called "fictitious nodes." They are given this name because they lie outside the physical boundary of the problem and are not part of the problem's domain. In most presentations of the finite difference method, these auxiliary nodes are included in the analysis by a process that is not clearly defined.

The idea of fictitious nodes is not addressed in this book because their inclusion in the one-dimensional problems is not instructive for higher dimensions. However, with the use of the physically interpretable notation, the fictitious node can be directly related to the boundary condition that exists in a problem. As a result, the fictitious nodes can be incorporated into the model with straightforward physical arguments. The details for their role in one- and two-dimensional problems are covered in Reference [1].

12.6   RECOMMENDATIONS FOR FUTURE WORK AND RESEARCH OPPORTUNITIES

The primary function of this work has been to present an overview of error analysis and to introduce improvements to adaptive refinement. The background that allowed this function to be fulfilled contains results that simplify and extend the capabilities of the finite element and the finite difference methods.

The objective of this Section is to identify research opportunities that derive from the developments presented here. Many of the research opportunities address the extensions of the developments presented here to multidimensions. This is because the work presented here focuses on one-dimensional applications in order to make this presentation easily accessible and compact. Even though only a few two-dimensional applications are discussed here, the simplicity of the one-dimensional case does not detract from the exposition of the theoretical bases of the concepts. With the exception of the new approach to refinement guides, many of the developments contained in this book are covered in a multidimensional setting in Reference [1].

These research opportunities will be presented in two categories: (1) Finite Element Research Opportunities and (2) Finite Difference Research Opportunities.

Research Opportunities in the Finite Element Method:

1. Extend the refinement guides developed in Chapter 11 to two and three dimensions.
2. Develop other strategies for defining the level of refinement in high-error elements.
3. Identify failure criteria for different materials for use in forming the error estimators and termination criteria.
4. Identify the efficacy for using higher order elements in finite element models as a way to speed convergence to acceptable solutions.
5. Use strain gradient notation to evaluate plate elements for inherent errors (see Reference [1] for details).

Research Opportunities in the Finite Difference Method:

1. Include nonlinear material properties in finite difference models of continuum mechanics problems. This eliminates the need for the stepwise solution of a series of linear problems and exports the difficulties with the nonlinearities to the equation solver.
2. Formulate and solve curved beam problems using the finite difference method as a precursor for solving shell problems.
3. Formulate and solve shell problems using the finite difference method as a way to eliminate problems with rigid body motions that exist in finite element models.
4. Develop an error estimator that compares the interelement jumps in the strains found at the nodes using forward or backward finite difference operators with the smoothed strains produced by the finite difference model.
5. Determine the feasibility of using forward or backward finite difference templates to introduce boundary conditions into finite difference models.
6. Implement adaptive refinement procedures in finite difference models using the transition elements introduced in Reference [1].

12.7   REFERENCE

1. Dow, J. O. *A Unified Approach to the Finite Element Method and Error Analysis Procedures*, New York: Academic Press, 1999.
This book puts the finite element method, the finite difference method, and error analysis on a common Taylor series basis. It develops the alternate element formulation procedure mentioned and applied here.

INDEX

www.ingramcontent.com/pod-product-compliance
Lightning Source LLC
Chambersburg PA
CBHW082004190326
41458CB00010B/3067